Flora Indica

OR DESCRIPTIONS OF

Indian Plants

BY WILLIAM ROXBURGH

WITH ADDITIONS AND OBSERVATIONS
BY NATHANIEL WALLICH

Flora Indica

OR DESCRIPTIONS OF

Indian Plants

BY WILLIAM ROXBURGH

WITH ADDITIONS AND OBSERVATIONS

BY NATHANIEL WALLICH

Superintendent of the Botanic Garden, Calcutta

EDITED BY

WILLIAM CAREY

VOLUME ONE

FACSIMILE REPRINT OF THE FIRST EDITION

WITH AN INTRODUCTION BY

D. H. NICOLSON

Curator, Department of Botany, Smithsonian Institution

New York

ORIOLE EDITIONS

1975

FACSIMILE OF THE FIRST EDITION OF 1820, 1824

PUBLISHED IN 1975 BY ORIOLE EDITIONS, INC.

WITH AN INTRODUCTION BY D. H. NICOLSON

DEPARTMENT OF BOTANY, SMITHSONIAN INSTITUTION

LIBRARY OF CONGRESS CATALOG CARD NUMBER: 74-31524

ISBN 0-88211-077-2

PRINTED IN THE UNITED STATES OF AMERICA

by SENTRY PRESS, NEW YORK, N. Y. 10013

FLORA INDICA;

OR

DESCRIPTIONS

OF

Indian Plants,

BY THE LATE

WILLIAM ROXBURGH, M. D. F. R. S. E. &c. &c.

EDITED

BY WILLIAM CAREY, D. D.

TO WHICH ARE ADDED

DESCRIPTIONS OF PLANTS

MORE RECENTLY DISCOVERED

BY NATHANIEL WALLICH, M. D. F. L. S. &c.

Superintendent of the Botanic Garden, Calcutta.

VOL I.

"ALL THY WORKS PRAISE THEE O LORD." *David.*

SERAMPORE:

PRINTED AT THE MISSION PRESS.

1820.

INTRODUCTION

The People Involved

William Roxburgh (1751-1815) came to Madras in 1776 and met Johann Gerhard Koenig, an avid botanist. Roxburgh developed experimental gardens of possible crops, studied wild plants and accumulated a large collection of drawings of plants. In 1793 he became Director of the Botanic Garden in Calcutta which had been founded by his predecessor, Robert Kyd. In 1814 ill health forced him to leave India and he died the next year.

William Carey (1761-1834) was a superb linguist who was the first to translate the New Testament into Bengali. He came to India in 1783 as missionary. He was a friend of Roxburgh and became Roxburgh's editor.

Nathaniel Wallich (1786-1854) became Director of the Calcutta Botanic Garden soon after Roxburgh's death. Wallich was a dedicated collector and spent much time in the field.

Publication of the FLORA INDICA

In 1813 Roxburgh turned over the completed manuscript of his *Flora Indica* to William Carey. In 1816, after Roxburgh's death and the arrival of Wallich, Carey went to Wallich to discuss publication of Roxburgh's manuscript. Wallich agreed, saying "Publish and I will edit it and bring it up to date."

The first volume was published in 1820 and the second volume in 1824. The scope was enlarged by comments by Wallich on Roxburgh's taxa and by the inclusion of many new species being described by Wallich from his collections in Nepal and elsewhere. Wallich was unable to continue the pace and publication ceased.

Carey and Roxburgh's children, James and Bruce, were disappointed that publication had stopped and were somewhat disturbed by the substantial additions being made by Wallich. Consequently, in 1830 they decided to publish the entire Roxburghian manuscript on flowering plants without Wallich's additions and completed it in 1832. This became the second edition and includes all Roxburgh's taxa described in the first edition and those not published in the first edition. This second edition also omits all the taxa published by Wallich in the first edition.

The second edition has been reprinted several times, the best known being the one by C. B. Clarke in 1874. However, the first edition with its earlier publications of Roxburgh's and Wallich's taxa has never been reprinted, is quite rare and found only in older botanical libraries.

Significance of the First Edition

One difficulty with Joseph Hooker's superb *Flora of British India* is that it does not cite dates of publication, although botanists must use the oldest available epithets. When citing Roxburgh's *Flora Indica,* Hooker and his collaborators always cited the second edition but frequently did not cite the first edition. The omission of the earlier publication can cause difficulties as, for example, *Mariscus microcephalus* Presl (1828) which would have been correctly named by C. B. Clarke (in J. Hooker, *Fl. Brit. Ind.* 6: 624. 1893) if the citation in synonymy of "*Cyperus spinulosus Roxb. Fl. Ind.* 1: 203 [1832]" were the earliest publication of the latter binomial. However, Clarke omitted citation of the earlier publication in the first edition, obscuring the fact that the epithet "*spinulosus*" (1820) is clearly priorable to "*microcephalus*" (1828).

Also, the index to most published plant binomials, *Index Kewensis,* commonly only cites the second (1832) edition of Roxburgh's *Flora Indica* when the first edition (1820 or 1824) should be cited, e.g., *Ixora stricta* Roxb. *Fl. Ind.* 1: 379 [1832] cited instead of 1:383 [1820], *Commelina salicifolia* Roxb. *Fl. Ind.* 1 172 [1832] cited instead of 1: 176: [1820], and *Sonerila Maculata* Roxb. *Fl. Ind.* 1: 177 [1832] cited instead of 1:180 [1820].

Thus, this first reprint of the first edition of Roxburgh's *Flora*

Indica makes a needed, rare, major, frequently and easily overlooked work available to botanists.

November, 1974

D. H. NICOLSON, *Curator*
Department of Botany
Smithsonian Institution
Washington, D. C.

PREFACE

ON the occasion of publishing a Botanical Work which owes its origin to the labors of others, the Editor deems it proper briefly to state the circumstances which have thus led him to publish the manuscript of his much valued friend the late Dr. Roxburgh.

With this eminent botanist and friend of science, the Editor was for many years on terms of the closest intimacy; and this work having been frequently the subject of conversation, he is enabled to say, that Dr. Roxburgh repeatedly expressed to him his earnest wish that it might be published in some future time; and his apprehensions that otherwise, imperfect and garbled copies of the work would be published by others not acquainted with him; or that if delayed, a part of his labors might be lost by others publishing descriptions of the same plants which he had previously named and described.

Desirous as he was to publish this work however, he felt unwilling to attempt it, till his return to Europe should give him an opportunity of consulting those eminent botanists in Britain who do so much honor to the science, and enable him fully to avail himself of their remarks; a step which he deemed necessary to the perfection of the work, although no immediate opportunity appeared likely to offer.

In a little time the illness which unhappily terminated in depriving the world of his labors, compelled him to undertake a voyage for the sake of his health; which he at first intended should have been only to the Cape. His declining state of health however, induced him to proceed as far as St. Helena. During his stay in that island, he, ever attentive to the interests of science, improved the opportunity by describing most of the plants he found growing there. After his arrival in England, whither he was at length obliged to proceed, he entertained considerable hopes of being able to put this work to the press; and once wrote to the Editor, that in preparing it for this purpose he hoped to procure the assistance of one of the first botanists of the present day, Mr. Robert Brown. Death however soon prevented his accomplishing this object which he had so much at heart.

The Editor had been permitted by Dr. Roxburgh to take a copy of the manuscript after it had received the latest alterations and additions which he made to it while he remained in India, the last part of it being returned to him after he had gone on board the ship with the view of leaving the country. On his arrival at Ceylon he met with and described a few plants there; copies of which descriptions, he kindly sent thence to the Editor at Serampore.

After waiting more than a year after the news of Dr. Roxburgh's death had reached India, the Editor, being unable to learn that any one thought of publishing this work, which his deceased friend felt so anxious to put to press in an unmutilated state, before the labors of others should render it useless and obsolete, he formed the resolution of taking the risk of the publication on himself; and having secured the co-operation of his various friends in India, he immediately put it to press.

The principal motives which urged him to this step, were these. He in the first instance wished to secure to an old and highly valued Friend the well-earned reward of upwards of thirty years labor, in the description of more than Two Thousand Plants, natives of Eastern Asia; a work in which none had anticipated him to an equal extent, and which will form a monument to his memory more lasting than the most pompous inscription engraved on tablets of brass; but of which he would have been completely and unjustly deprived, had the work been suffered to remain unpublished. This, was likewise, strengthened by his desire to promote the interests of a science to which he has been attached from the earliest period of his life, and which must have suffered an almost irreparable loss had the labors of Dr. Roxburgh been suffered to die with him.

To this measure the Editor felt encouraged by another circumstance which he esteems it his duty to mention particularly. On his communicating his wish to publish the Manuscript of his deceased friend, to Dr. Nathaniel Wallich, the present Superintendent of the Honorable Company's Botanic Garden, whom he feels honored in numbering among his most intimate friends, and whose ability and indefatigable zeal in conducting that Institution are so well known, he not only consented to assist in superintending this work of his esteemed predecessor and friend, as it passed through the press; but in the most generous manner promised to insert therein, those numerous plants which have since been communicated to him from all parts of Asia, and particularly from Nepal*a*, a country the productions of which are as yet scarcely known to the European world. This will prove such an addition to the original work of the late Dr. Roxburgh, while it is fully congenial with its nature, that it cannot but form a most desirable acquisition to every lover of this pleasing science, both in Asia and in Europe.

In thus publishing the work of his deceased friend, the Editor feels assured that no one will impute to him motives of a mercenary nature, who considers the expense and labor of printing a work, of this nature, the limited sale it must necessarily have, and the little probability there is of a *second* edition of it being called for either in India or Europe: as in every new systematic work on general botany, or new editions of preceding ones, all the plants described in particular Floras published in the intermediate time, are invariably inserted. Hence no work of this kind can be published, at least in India, with any well-founded expectation of its being a profitable speculation. Should any of the relatives of his late friend have an idea of its being made such in this instance, however, the Editor will cheerfully give up to them the whole edition, on their defraying the expenses attending its printing and publication.

To those acquainted with Botany, it is well known that many changes have taken place in the nomenclature of plants, since the late Dr. Roxburgh added the final touch to his manuscripts. In many instances whole orders have undergone a complete revision; genera have been frequently divided into two or three; specific names have been changed; and plants have been removed from one genus to another. The Editor however thought it best to publish his late friend's manuscript in the state in which he left it; and to add in notes such alterations as have been made by botanists with whose improvements the author was unacquainted, or who have written since his death. The whole work therefore, corresponds exactly with Dr. Roxburgh's manuscript after it had received his last correction, with the exceptions just mentioned, and of such corrections of the language as were necessary, which in no instance have been suffered to make any alteration in the sense.

The Editor feels gratified in bearing testimony to the disinterested and highly liberal manner in which Dr. Wallich has contributed his assistance throughout the whole of this work; and to his generosity in enriching it with a number of plants described by himself; and easily to be recognised by the initials N. W. in the first volume. The amount of these, especially among scitamineæ and grasses would have been much more considerable had there been time to insert the recent vast accessions of plants, which the Garden has received from all parts of India, and which from his earnest solicitude to promote the interests of this pleasing and important science, may be reasonably expected to bear a far greater proportion to the whole in each succeeding volume. It is intended to terminate the work with an additional volume, which shall include all the plants which may be discovered during the progress of it through the press.

The Editor now submits the work to the relatives of his deceased Friend, as the best tribute of respect he is able to show to his memory;—and to the candid consideration of all who are attached to the interests of the science both in India and Europe, remaining,

RESPECTFULLY THEIRS,

W. CAREY.

Class 1*st*.

MONANDRIA MONOGYNIA.

CANNA Schreb. gen. No. 1.

Anther single, attached to the edge of the petal-like filament. *Style* spatulate, growing to the tube of the corolla. *Stigma* linear. *Capsule* 3-celled: *seeds* several, naked.

C. *indica. Linn. sp. pl. ed. Willd.* 1. 3.

Leaves ovate-lanceolate. Segments of the inner border of the corol lanceolate, and bidentate.

Sans. वनकेली, V*una*-kelee, सर्व्वजया, S*urva*-j*u*ya.

Béng. Lal S*urva*-j*u*ya, the red variety, and Shwet*a* S*urva*-j*u*ya, the yellow.

Teling. Kristma-tamara.

Katu-bala. *Rheed. mal.* 11. *p.* 85. *t.* 43.

Cannacorus. *Rumph. amb.* 5. *p.* 177, *t.* 71. *fig* 2.

Common over *India*, and in flower and seed most part of the year.

Observation. The parts of the corolla are exactly alike in both varieties. *C. lutea* of Roscoe, must therefore be different, as in that, the inner limb of the corolla is bifid.

PHRYNIUM. Willd.

Anther single, terminal on a short erect filament. *Style* growing to the tube of the corolla. *Stigma* infundibuliform. *Capsule* 3-celled, 3-valved. *Seeds* solitary, arilled at the base. *Embryo* uncinate and furnished with a perisperm.

1. P. *dichotomum. R.*

Shrubby, dichotomous. Leaves cordate.

Thalia *cannaformis. Linn. sp. pl. ed. Willd.* 1. 16.

Arundastrum. Tonchat seytam. *Rumph. amb.* 4. *p.* 22. *t.* 7.

Donax *Arundastrum. Lourier. Cochin-ch.* 15.

Sans. वला, V*u*la, गोदनिका God*u*nik*a*, भद्रा, Bh*u*dr*a*, भद्रौदनी, Bh*u*droud*u*nee, शरकाष्ठिका, Sh*u*r*a*kasth*i*ka, कल्याणिनी, K*u*lyaninee, भद्रवला, Bh*u*dr*a*-bul*a*, घाटा, Ghat*a*, पाटी, Patee, वलह्वा, Vula-hwa.

Beng. Moocta-patee, Pattee-patee,* or Madar-patee.

This elegant shrub, is a native of *Bengal,* as well as of various other parts of *India,* &c. Flowering time the hot season; seeds ripen in the rains.

Root ramous, woody, perennial.—*Stems* straight, tapering and simple to the branches, about as thick as a man's thumb, and from three to five, or six feet high, of a beautiful, highly polished green colour. *Branches* numerous, dichotomous spreading in every direction, smooth like the stem, but jointed at every division.—*Leaves* alternate, petioled, ovate-cordate, smooth, entire, acute, from 4 to 8 inches long; veins numerous, very fine and parallel.—*Petioles* sheathing, except a small portion, one tenth, or less, at the apex, which is round, and may be called the proper petiole.—*Racemes* terminal, generally solitary, jointed, a little flexuous.—*Bractes* glumaceous, in pairs on the joints of the raceme, each pair embracing between them a two-flowered pedicel.—*Flowers* pair'd, on a common pedicel, from the alternate joints of the rachis, large, pure white.—*Pedicels* clavate, thick, white, and fleshy, near the apex two-parted; *divisions* unequal, with two glands at their ends like the rudiments of flowers.—*Calyx* above, three-leaved; *leaflets* oblong-conic,

* "Mats made of the split stems of this plant, being smooth, and particularly cool, "and refreshing, are termed in Hindi *Sital-pati,* which signifies a *cool mat,* whence the "plant itself is said to bear the same name. Suspecting, however, this to be a misap-"propriation of the term, I have enquired of natives of the eastern parts of *Bengal,* "who assure me, that the plant is named *Mucta-pata,* or Patti-pata, and the mat only "is called Sital-pati." *Colebrooke.*

tapering to an acute point, and much shorter than the tube of the corol.—*Corol* one-petalled, irregular. *Tube* funnel-shaped. *Border* double; *exterior* of three, equal, linear-oblong, recurved segments. *Inner* elevated far above the exterior on a continuation of the tube, and consists of five very unequal portions: the *exterior* two are larger, obcordate, with waved, expanding margins; the *third* or middle one stands opposite to the first two, is broad, obcordate, deeply emarginate, and of a middling size between the two exterior, and two interior segments; on the inside near the base a semilunar body which projects inward between the anther and stigma; the *inner* two small, and crown the mouth of the tube, closely embracing the stigma, and anther.—*Filament* short, attached to the margin of the smallest segment of the interior border of the corol. *Anther* erect, simple, ovate-oblong. In this species, as well as in *P. capitatum,* the pollen consists of remarkably large round grains.—*Germ* inferior, villous, 3-celled, with one *ovulum* in each, attached to the bottom of its cell. *Style* growing to the inside of the tube of the corol, and does not appear until above its mouth, when it enlarges, into an infundibuliform, hooked *stigma,* surrounded with a lobate, callous margin, and directed in an oblique manner toward the face of the anther, as if in search of it, though separated by a callous projection from the base of the middle division of the inner border of the corol.—*Capsule* broad-turbinate, size of a cherry, rugose, somewhat 3-lobed, 3-celled, 3-valved, opening from the apex.—*Seed* solitary, round, resting on a small, partial, scaly aril. *Integuments* two: *exterior* soft and spongy; *inner* nuciform.—*Perisperm* conform to the seed, horny, tinged blue; a fissure penetrates from the base pretty far up, separating the two extremities of the embryo. *Vitellus;* none has been discovered in any of the simple-anthered scitamineæ which have been examined by me. —*Embryo* pure white, hooked, with the convexity upward, which corresponds with the periphery of the seed.

2. P. *virgatum. R.*

Stems simple. Leaves bifarious, lanceolate. Spikes terminal filiform compound.

A native of *Malabar,* from thence sent to the late Dr. Anderson of *Madras* for the *Cardamom* plant, and by him presented to the Botanic Garden at *Calcutta,* where it blossoms during the rainy season.

Root a stout, crooked, tough, fibrous, horizontal main body, with numerous, thick, strong fibers therefrom; every part void of odour. —*Stems* numerous, erect, 6-8 feet high, slender, jointed, lower joint or two much swelled; invested in the sheaths of the leaves, fistulous; the largest are about as thick as a slender walking cane near the base, and at the top as thick as a common quill.—*Leaves* bifarious, short-petioled on their sheaths, lanceolate, polished; length from 6 to 18 inches.—*Sheaths* and *petioles* smooth and even, except the inside of the petioles, where there are a few long, soft hairs.—*Spikes* terminal, two or three together, and they are very generally two, or three cleft, very slender, and often near a foot long, jointed, with a two flowered pedicel at each joint.—*Flowers* pure white, inodorus.—*Bractes* two or three at each division; and also embracing the two-flowered pedicels, size very various, but the shape ensiform, and channelled.—*Calyx* of three, small, distinct, acute, conic, leaflets.—*Corol* with a very short tube; *exterior* border of three, nearly equal, oblong, reflexed segments; within are two which are unequal, and much larger; they form the conspicuous portion of the flower, one is obcordate, the other is obovate; inner segments 3, and smaller, the two longest of them form yellow margined hoods for the anther and stigma. —*Filament* inserted between two of the inner segments of the corol, erect. *Anther* one-celled, ovate.—*Germ* inferior, hairy, 3-celled, with one ovulum in each, attached to the base of the axis. *Style* growing to the short tube of the corol, above free, and uncinate. *Stigma* funnel-form.—*Capsule* oblong-obovate, size of a field-bean, rugose, when ripe pale yellow, 3-celled, 3-valved. *Partitions* opposed.—*Seed* solitary, oblong; a small lobate aril round the umbilicus at the base of the seed. *Integuments* two; *exterior* brown, and glossy, *interior* membranaceous.—*Perisperm* conform to the seed, and as in *P. capitatum*, &c. a brown substance runs up from the base, into the arch of the embryo.—*Embryo* simple, hooked, with the convexity upwards.

Obs. It is almost needless to observe that this plant has no relation to that which furnishes the *Cardamom* of *Malabar*, viz. my *Alpinia Cardamomum*.

3. P. *spicatum*. *R.*

Stemless. Leaves radical, bifarious, oblong. Spikes radical, shorter than the petioles, subcylindric, bifariously imbricated.

Plants of this small species, were sent by Mr. Felix Carey from *Pegu*, to the Botanic Garden at *Calcutta*, where they blossom in April and May.

Root perennial, flexuose, creeping horizontally deep in the earth, thick as a ratan; *fibres* numerous from the main body, and spread in all directions.—*Stem* none, the sheathing petioles unite from 6 to 12 inches in extent, and resemble one, as in the *Curcumas*.—*Leaves* in little radical tufts, of three or four each, sub-bifarious, petioled, linear-oblong; base broad and rounded; apex tapering, and sharp, both surfaces quite smooth; position nearly horizontal; length 6-10 inches, by 2-3 broad.—*Petioles*, the most exterior of the tuft sheathing, with their upper part round and smooth, those in the centre, (one or two) have no sheath, and are from apex to base round and smooth; general length of the whole from 6 to 12 inches, which is nearly the height of the whole plant, the position of the leaves being nearly horizontal.—*Spikes* solitary in the centre of the petioles, or lateral from their base, short-peduncled, sub-cylindric, acute, apex elevated but little above the soil, imbricated on opposite sides, with 4-8, alternate, bifarious bractes, or sheaths; each embraces a small spikelet of four or five pairs of sessile, small, nearly white flowers, on the back, or interior side of the spikelet, which expand in succession, beginning with the lower pair. Rachis flexuose, smooth.—*Bractes; exterior*, or *common*, one. at each joint of the spike, sub-cylindric, sheathing, smooth, and rather obtuse; *interior*, or *proper* one to each pair of flowers.—*Calyx* 3-leaved; *leaflets* lanceolate, smooth. —*Corol* with a long, very slender, smooth tube. *Border; exterior* of three reflexed, oblong, obtuse segments; within these are two larger, round-obovate, erect, curled portions. *Interior,* or *nectarium* small, and like the other species unequally 3-lobed.—*Anther* one-celled, attached to the inner border of the corol.—*Germ* inferior, 3-celled, with one *ovulum* in each, attached to the bottom of the cell.—*Style* blended with the tube of the corol, as in the other species described by me.—*Stigma* uncinate, thick, and fleshy.—*Seed vessel* (not seen.)

4. P. *imbricatum. R.*

Stemless. Leaves radical. Spikes simple, petiolar, solitary, short-peduncled; bractes imbricated, truncate, many-toothed.

Pittuli-pata the vernacular name in *Chittagong*, where it is common in its wild state. Flowers in March and April and the seeds ripen in August.

Root tuberous, &c. as in other *scitamineous* plants.—*Stem* none. —*Leaves* radical, long-petioled, from oblong, to broad lanceolar, tapering nearly as much at the base, as at the apex, smooth, entire, finely acuminate; 12-24 inches long, by 4-8 broad.—*Petioles* longer than the leaves, cylindric, smooth, the flower-bearing ones have a hairy joint near the middle, and from the joint up, a slit of between 2 and 3 inches in length opens on one side for the spike, its margins thin, and hairy.—*Spike* solitary, short-peduncled, simple, linear-oblong, 3-4 inches long, imbricated.—*Bractes; exterior* oval, truncate, with the transverse apex many-toothed, smooth, each embracing several small, pale, pink flowers, which expand in succession. *Inner bractes* small, and membranaceous.—*Calyx, corol,* and *stamen,* as in *P. parviflorum.*—*Germ* 3-celled, with one *ovulum* in each, attached to the bottom of its cell. *Style,* as in the other species, it grows to, and is blended with the tube of the corol, its apex free, thick, and hooked. *Stigma* gaping obliquely toward the anther.—*Capsule* oblong, size of a large field bean, rugose, 3-celled, 3-valved, opening from the apex. —*Seeds* solitary, sitting on a partial aril, and attached to the bottom of the cell; external tunic hard, brown, and polished.—*Perisperm* conform to the seed.—*Embryo* erect, simple, the apex hooked. *Radicle* inferior, close to the umbilicus of the seed.

Obs. The simple, short-peduncled, imbricated, linear-oblong spike, bursting from the middle of the petiole, readily points it out.

5. *P. parviflorum. R.*

Stemless. Leaves radical. Heads of flowers petiolary; bractes acute.

A native of the eastern parts of *Bengal.* With *P. capitatum,* it blossoms about the beginning of the rains.

Root tuberous, horizontal, perennial, hairy, in other respects as in *P. capitatum,* &c.—*Leaves* long-petioled, elliptically-oval, polished, entire, short, acute-pointed.—*Heads* of flowers sub-globular, sessile, always bursting from the inner side of the petioles, near, or above their middle, composed of many, imbricated sessile, distichous, compact, simple, or compound spikes.—*Bractes;* the *exterior* two are broad-ovate, and common to the whole head; the rest within smaller; all are smooth, and end in a small, rigid, acute point.—*Flowers* in pairs, small, nearly inconspicuous, yellow.—*Calyx, coral, and stamen,* as in *capitatum.*—*Germ* short pedicelled, smooth, but hairy round the insertion of the calyx, 3-celled, and in this species I have only found one of the cells fertile, in it is one *ovulum,* attached to the bottom of the cell. In *capitatum* all the three cells are fertile. Style as in the other species, it grows to the tube of the corol, its apex free and hooked. *Stigma* infundibuliform.

6. P. *capitatum. Linn. sp. pl. ed. Willd.* 1. 17.

Stemless. Leaves radical, long-petioled, oblong. Heads of flowers petiolary and terminal, glomerate. Bractes truncate, incurvate.

Naru-killa. *Rheed. mal.* 11. *p.* 67. *t.* 34.

Phyllodes *Placentaria. Lour. Cochin-ch.* 16. 17.

Kudali. Beng. which also signifies a *Plantain.*

A native of *Chittagong,* and of various other parts of India. From the former place it was sent to the Botanic Garden at *Calcutta,* by Dr. Buchanan in 1797. Flowering time the rainy season, the seeds ripen in the cool season.

Root perennial, tuberous like ginger, with long, fleshy fibres from the crown, and various other parts.—*Stem* none.—*Leaves* radical, long-petioled, oblong, entire, smooth on both sides; length from 6 to 18 inches, and broad in proportion. *Veins* numerous, fine and parallel.—*Petioles* longer than the leaves, slender, round, smooth, taper a little from the base, and are there expanded into a sheath for those immediately within; such as are destined to bear the flowers have a joint a little above the middle; immediately above this joint there is a swelling, which in due time is forced open on the inside by the growing flowers exactly as in our *Indian* species of *Pontederia*. It however sometimes appears, and even in the same plant, that some of the petioles, now scapes, extend no further than the flowers.—*Flowers* numerous, collected into a pretty large, sessile head, which bursts from the anterior margin of the jointed petioles, small, and nearly hid among the large bractes; when they first expand in the morning rose-colour'd, gradually becoming purple by the evening.—*Bractes*, calyx of Louriero's *Phyllodes*, several, collecting the small sessile flowers into several fasciculi of two, three, or four pair, each of which expand in succession. The exterior one of each fascicle is large, subcylindric, of a firm texture, and reddish colour; apices truncate, with scariose, incurved margin; the rest are smaller, and more or less pointed.—*Calyx* superior, three-leaved; *leaflets* ensiform, length of the tube of the corol, hairy about the points.—*Corol* one-petalled. *Tube* funnel-shaped, scarcely perforated. *Border* double; *exterior* of three, equal, purple, spatulate, recurvate segments; *interior* more elevated on a continuation of the tube, and consists of five, very unequal segments; the exterior two larger, round, curled, and of a pale rose colour; the innermost one small, and running down the side of the filament like a wing.—*Filament* solitary, short, inserted on the mouth of the tube close by the free, curved end of the style. *Anther* simple; grains of the pollen white, large, and spherical.—*Germ* inferior, oblong, very short pedicelled, hairy, 3-celled, with one *ovulum* in each, attached to the bottom of its cell. *Style* growing to, or blended with the tube of the corol. *Stigma* large, funnel-shaped, obliquely-incurved or hooked towards the anther.—*Capsule* short, turbinate, three-sided, pretty smooth, size of a gooseberry, three-celled, three-valved, opening from the apex.—*Seed* solitary, nearly round, smooth, dark grey, resting on a small, squamiform aril, which is attached to and surrounds the short umbilical cord which unites the seed to the bottom of the partition. *Integument* double; *exterior* spongy; *inner* thin and rather hard, adhering to the exterior, and to the perisperm.—*Perisperm* conform to the seed, horny, tinged with pale blue; a fissure replete with a brown substance penetrates from the base to a considerable distance upwards.—*Embryo* pure white, hooked, with the convexity upwards, which corresponds with the periphery of the seed.

HEDYCHIUM. Kön.

Corolla with a long, slender tube; both borders three-parted; inner resupinate. *Anther* double, naked. *Capsule* three-celled, three-valved. *Seeds* numerous, arilled. Embryo simple, and furnished with both perisperm and vitellus.

1. H. *coronarium. Linn. sp. pl. ed. Willd.* i. *p.* 10. *König in Retz. obs.* 3. 73.

Leaves lanceolate. Spikes compactly imbricated; segments of the cleft lip of the corol semi-lunar.

Gandsulium. *Rumph. amb.* 5. *t.* 69. *f.* 3.

Goruk-nathh*a* of the Bengalees, also by them called Doolal*a*-champ*a*.

A native of various parts of *Bengal* and the neighbouring provinces. Flowers during the rains; and the seeds ripen in the early part of the cool season.

Root horizontal, perennial, round, fleshy, marked with annular cicatrices, about as thick as a man's thumb, fleshy, with fibrous ramifications issuing from the cicatrices; taste simply herbaceous.—*Stems* erect, 3-4 feet high, round, every where covered with the sheaths of the leaves.—*Leaves* sessile, alternate, bifarious, lanceolate, tapering to an evanescent point; above, a deep smooth green; below, striated, pale, and slightly covered with depressed, soft, white hairs; 9-12 inches long, and about two broad.—*Sheaths* smooth on both sides, striated, terminating in a ligulated process, (as in many of the grasses) which is often two-parted.—*Spike* terminal, solitary, erect, linear-oblong, about the size of the thumb, compactly imbricated with many large, oval, concave, green, common, permanent bractes.—*Flowers* large, pure white,* very fragrant, 2-4 to each common bracte, but seldom more than one or two expand at the same time.—*Bractes*; besides the common exterior ones, already mentioned, there are as many interior as there are flowers in the fascicle; the largest of which is about half the length of the common bracte, and envelopes not only the most foreward (exterior) flower, but all those of the same fascicle, gradually diminishing in size; all have a sheathing base, are membranous, and striated.—*Calyx* superior, one-leafed, a little inflated, contracted at the mouth, a little hairy, striated, half the length of the tube of the corol. —*Corol* one-petalled. *Tube* long and slender, from the apex of the exterior bractes recurved. *Border* flat, double. *Exterior* three-parted; *divisions* equal, lanceolate, acute, membranaceous, striated. *Interior* resupinate, three-parted; *lower* two divisions obliquely oval, short-clawed; *upper* division broad, obcordate, banner-like, with the fissure deep, and the lobes roundish; margins waved; towards the base tinged yellow.—*Nectary* (of König and Retzius) two short, fleshy, subulate bodies embracing the base of the style.—*Filament* from the lower margin of the mouth of the tube, between the insertions of the inner border of the corol, ascending; upper side channelled for the style, and inserted into the anther a little above its base. *Anther* linear, sub-erect, two-lobed.—*Germ* obsoletely three-sided, three-celled, with many *ovula* in each, attached to the axis. *Style* slender, in length sufficient to elevate the large, hairy, perforated, glandular, green *stigma* just about the apex of the anther. —*Capsule* oblong, three-celled, three-valved, opening from the apex; inside orange coloured.—*Seeds* many in each cell, invested in their proper, multifid, crimson aril. *Integuments* two.—*Perisperm* and

* A variety deeply tinged with yellow, is found in some of the interior provinces; and another of a pale yellow colour: all are equally fragrant.

vitellus together conform to the seed; the former occupies the lower and exterior portion ; and the latter the upper and interior portion, and is in immediate contact with the embryo.—*Embryo* simple, sub-cylindric, nearly as long as the seed.

Obs. This, to me, is the most charming of all the plants of this natural order that I have yet met with; the great length of time it continues to throw out a profusion of large, beautiful, fragrant blossoms, makes it particularly desirable.

2. H. *angustifolium. R.*

Leaves linear-lanceolate. Spikes open; fascicles of flowers subtern, segments of the cleft lip oblong; the other five segments of the corol linear.

Beng. Bhoo-ada.

Native of *Chittagong, Silhet,* and the eastern parts of *Bengal,* where it flowers about the beginning of the rains, in June.

Root tuberous, horizontal, jointed, &c. as in the other scitamineous plants.—*Stems* erect, 3-6 feet high, entirely invested in the sheaths of the leaves.—*Leaves* bifarious, sessile on their smooth sheaths, narrow-lanceolate, very fine, I may say filiform, pointed, smooth on both sides ; length 10-14 inches, by 1-2 broad.—*Spike* terminal, erect, rigid, 6-18 inches long, smooth. Fascicles of flowers generally threefold, and spreading out considerably from the rachis.—*Bractes*, exterior of the fascicle cylindrical, linear ; interior, one to each flower, and smaller.—*Flowers* rather small, dull red *(lateritius,)* generally about four in the fascicle, and expanding in succession.—*Calyx* superior, cylindric, length of the túbe of the corol; mouth equally three-toothed.—*Corol;* tube slender, cylindric, about an inch long; 5 segments of the border linear (sub-filiform,) variously revolute. *Lip* lateral, with linear, channelled and two-parted *lamina ;* the *segments* thereof semi-oval.—*Filament* opposite to the lip, and twice its length, nearly straight, filiform, grooved for the style. *Anther* two-lobed, naked.—*Germ* oval, a little hairy, three-celled, with many *ovula* in each, attached to the axis. *Style* filiform, its base embraced by the usual two subulate bodies. *Stigma* funnel-form, projecting from the apex of the anther.

3. H. *gracile. R.*

Leaves lanceolar. Spikes terminal, open; flowers solitary, scattered; segments of the bifid, sessile lip, semi-lanceolar, the other five segments of the corol linear.

Khasee name, *Kattia Laphya.*

Obs. This is a slender species, about three feet in height, a native of the mountains on the North-East border of Bengal. Like the preceding two species it flowers in the rainy season. Its solitary-flowered spike of white* corols with their scarlet filament, and its lanceolar leaves, distinguish it from *H. angustifolium.*

Note. The following two species, discovered since the Author of this work left the Botanic Garden, are added by Mr. N. Wallich, the present superintendent of that institution.

4. H. *villosum. Wall.*

Spike elongated, villous, open. *Fascicles* approximate and copious, scattered or paired, three-flowered; *lip* short-clawed, bifid, of equal length with the five linear segments of the corol.

A native of the mountains North-East of Bengal, from whence our indefatigable collector of plants, *Mr. Matthew Robert Smith,* sent specimens to me in 1815. Flowering time the rainy season.

Khasee name, *Kattia Ram Rait.*

Stems upright, slender, smooth, as well as the leaves, from two to three feet.—*Leaves* flat, lanceolar-oblong, elevated from the back of the sheaths by a very short petiole; glaucous underneath. *Sheaths* terminated by an oblong, obtuse, closely adpressed, long ligula.—*Spike* terminal, erect, cylindric, obtuse, from 10-12 inches long; all its parts covered with a short, sericeous down.—*Bractes* oval, much shorter than the tube, of a reddish tint; *exterior* or *common,* three, seldom two-flowered, flat; *interior,* one to each flower, the base of which it embraces.—*Flowers* delightfully fragrant even when dry, less succulent than those of *H. coronarium* and *angusti-*

* There is a variety of this species with cream-coloured flowers. *Ed.*

folium, of a pale yellow colour.—*Calyx* very slender, somewhat shorter than the filiform, inch and half-long *tube* ; *mouth* obliquely bifid.—*Border* smooth ; *segments* linear, sub-clavate, spreading, half the length of the tube ; divisions of the *claw*, or superior segment of the interior border, linear-oblong, obtuse.—*Filament* as long as the tube, filiform, smooth, of a scarlet colour ; *Anther* very short, ovate, two-lobed.—*Germ* densely villous ; *stigma* slightly ciliated.

Obs. This beautiful species differs from *H. gracile*, *R.* (of which the Rev. Dr. W. Carey has favored me with a living specimen in blossom), in the length and villosity of its spikes, its copious fascicled flowers, and the equal length of all the segments of the border. It seems altogether to be a larger plant.

5. H. *speciosum*. *Wall.*

Spike elongated, open, smooth ; fascicles approximate scattered, two-flowered ; lip entire sessile, acute. A native of the same country with the preceding, and like all the species flowering in the rainy season.

Khasee name, *Kattia Lung Mutti.*

Leaves very large, oblong-lanceolate, attenuate at both ends, long-pointed ; margins slightly waved.—*Spike* terminal, large.—*Fascicles* spreading and numerous.—*Bractes; exterior* very large, obtuse, as well as the others, with a few white hairs at the end.—*Interior*, two, opposite ; that close to the rachis obtuse, with a few loothlets at the end ; the other much smaller, linear-lanceolate, acute.—*Flowers* pale yellow, fragrant.—*Calyx* cylindric, half the length of the tube ; *mouth* obliquely bifid, with two or three ovate, bearded teeth.—*Tube* upwards of two inches long, filiform.—*Exterior segments* linear, very long, of a greenish colour ; *interior* clavate. *Lip* of equal length with the interior segments, ovate, acute, tapering downwards.—*Filament* filiform, scarlet, somewhat longer than the tube ; *Anther* elongated, arched.

Obs. My examination of this stately plant has hitherto been confined to a well preserved spike and a few leaves only, which however point it out as the largest of the genus. The sessile, undivided, and acute lips distinguish it from the other species.

KÆMPFERIA. *Linn. gen. pl. ed. Schreb. No.* 9.

Corolla with a long, slender tube, and both borders three-parted. *Anther* double, with a two-lobed crest. *Capsule* three-celled, many-seeded.

1. K. *Galanga. Linn. sp. pl. ed. Willd.* 1. 15.

Leaves sessile, round-ovate-cordate. Spikes central. Upper segments of the inner border of the corol oval.

Katsjula Kelengu. *Rheed. mal.* 11. *p.* 81. *t.* 41.

Alpinia sessilis. *Kön. in Retz. obs.* 3. *p.* 62.

Soncorus. *Rumph. amb.* 5. *p.* 173. *t.* 69. *f.* 2.

Sans. चन्द्र मूलिका, Ch*u*ndr*a*-moo*l*ika.

Beng. Ch*u*ndr*a*-moola, or K*u*m*u*la.

A native of various parts of *India,** and common in gardens also. Flowering time the rainy season.

Root biennial, tuberous, with fleshy fibres from the tubers.—*Stem* none.—*Leaves* radical, petioled, spreading flat on the surface of the earth, round-ovate-cordate, between acute and obtuse; margins membranaceous, and waved; upper surface smooth, deep green; below taper and somewhat woolly, streaked length ways with ten or twelve slender lines.—*Petioles* hid in the earth; in short they are only cylindric sheaths embracing those within, and the fascicle of flowers, their mouths being crowned with a membranaceous process.—*Flowers* collected in small fascicles, of 6-12, within the sheaths of the leaves, expanding in succession; pure, pellucid white, except a purple spot on the centre of each of the divisions of the inner border, or lip.—*Bractes* three to each flower; a larger exterior; and two within at the sides; all are linear, acute, membranaceous, and half the length of the tube of the

* This plant, which is said to be very common on the mountainous districts beyond *Chittagong*, is called *Kumula*. I find on further inquiry that it is cultivated by the Joomma (mountain) Mugs, and by them brought down and sold in the markets to the people of *Bengal* under the above name, who use it as an ingredient in their be-tle. *Ed.*

corol.—*Calyx* length of the bractes.—*Corol. Tube* long, filiform. *Border* double, both three-parted. *Exterior* divisions linear, acute. *Interior*, upper two divisions ovate, erect; under one expanding, two-parted, with lobes bifid.—*Filament* short, apex, two-parted, and recurved. *Anther* replete with white pollen, crowned with a bidentate crest, or continuation of the filament above the anther. *Stigma* funnel-shaped.—*Nectarial* filaments of König two, filiform, erect, pretty long, embracing the lower part of the style.—*Pericarp;* I never saw it ripen.

Obs. The roots possess an agreeable, fragrant smell, and somewhat warm, bitterish, aromatic taste; the Hindoos use them not only as a perfume, but also medicinally. They are unknown to our best judges in London.

2. K. *rotunda. Linn. sp. pl. ed. Willd.* 1. 15.

Leaves oblong, coloured. Spike radical, appearing before the leaves. Upper segment of the inner border of the corol lanceolar, and acute.

Malan-Kua. *Rheed. mal.* 11. *p.* 17. *t.* 9.

Sans. भूचम्पका, Bhoo-chumpuca. *See Asiat. Res.* iv. 242.

Beng. and *Hind.* Bhooi-champa or Bhoo-champa.

This elegant plant is very generally found cultivated in gardens on account of the beauty and fragrance of its flowers, and though no doubt a native of various parts of India, yet I cannot say where it is indigenous. Flowering time in our gardens March and April, at which period the plant is totally destitute of leaves.

Root biennial, tuberous.—*Stem* none.—*Leaves* radical, petioled, oblong, waved, smooth, generally coloured underneath; about a foot long in a good soil, and 4-6 inches broad.—*Petioles* sheathing, united into a short stem, as in *Curcuma.*—*Scapes* just sufficient to elevate the flowers above the earth, embraced by a few common sheaths, of a greenish purple colour.—*Flowers* sessile, from 4 to 6 to the scape, or spike, very large, fragrant; colour of various shades of purple, and white.—*Bractes* two to each flower, surrounding the base of the germ; the inner one has its apex two-parted; the exterior, or longest is here only about half the length of the calyx.—*Calyx* above, one-leaved, as long as the tube of the corol, somewhat gibbous; apex generally two-toothed, and of a dotted, purplish colour.—*Corol. Tube* long, slender, cylindric, nearly erect, obliquely funnel-shaped towards the mouth. *Border* double, both three-parted. *Exterior* divisions drooping, linear, with margins involute, pure white. *Interior* border, the two uppermost divisions erect, lanceolar, acute; colour also white, with the margins slightly tinged with purple; the remaining inferior one is divided into two broad, obcordate, deflected lobes, of a deep purple colour, particularly towards the centre, and base.—*Nectarial filaments* of König as in *K. Galanga.*—*Filament* short, erect, broad, inserted on the base of the uppermost two interior divisions of the corol. *Anther* linear, and enlarged with an ovate, two-forked, coloured, somewhat recurved crest.—*Germ* ovate. *Style* filiform. *Stigma* funnel-shaped.

Obs. Woodville observes, at page 362 of his *Medical Botany*, that the roots of *Zedoaria longa*, and *rotunda* are both produced by the same species of plant, and are indiscriminately used in the shops: the former, he says, are brought to us in oblong pieces, about the thickness of the little finger; two or three inches in length, bent, rough, and angular; the latter are roundish, about an inch in diameter, of an ash colour on the outside, and white within. I have never met with any of the roots of this plant that agree with his description of the long sort, and they are too small for the round; so that I must conclude they are the produce of some other plant: at the same time Dr. Woodville's observations on the same page, give me reason to think my *Curcuma Zedoaria* may be the plant, which yields this round sort of *Zedoaria*.

3. K. *angustifolia. R.*

Leaves radical, linear-lanceolate, waved. Spikes hid in the centre of the leaves. Upper two segments of the inner border of the corol linear, oblong, obtuse.

Hind. Kunjan-boora, also Madu-nirbisha.

A native of *Bengal.* Flowering time the rainy season.

Root biennial, tuberous, with numerous ramous fibres from every part.—*Stems* none.—*Leaves* radical, numerous, sessile, linear-lanceolate, keeled; margins entire, and much waved; striated lengthways with darker and lighter coloured green, smooth in every part; 6-8 inches long, and about one broad.—*Spikes* hid in the centre of the leaves, bearing from 4 to 6 large, beautiful flowers in succession.—*Bractes* two or three to each flower; one or two interior embrace the germ laterally, the exterior one is larger, all much pointed, and of nearly the same length.—*Calyx* superior, one-leaved, half the length of the tube of the corol, opening obliquely, (spathe like,) near the apex, which is a single acute point.—*Corol; Tube* very long, and slender, elevating the border about two inches above the earth; border double, &c. as in *K. rotunda. Nectarial filaments* of König, *stamens,* and *pistil* as in the last quoted species.

Obs. The root is used as a medicine for cattle, by the people of *Bengal.*

4. K. *pandurata. R.*

Leaves petioled, broad-lanceolate, smooth. Spike central. Corol; with the upper segments of the inner border obovate, the inferior one much larger, and panduriform.

Zerumbet *claviculatum. Rumph. amb.* 5. *p.* 172. *t.* 69. *f.* 1; and I could almost wish to quote Manja-kua, *Rheed. mal.* 11. *p.* 19. *t.* ii. although referred to already by Linneus for *Curcuma rotunda,* a plant I have never met with, if this be not it.

From *Sumatra,* this beautiful plant was sent by Dr. Charles Campbell to the Botanic Garden at *Calcutta,* where it blossoms during the rains; chiefly in August.

Root perennial, tuberous like ginger, but of a tolerably bright yellow colour; from this, straight, slender, fusiform, paler coloured portions descend deep into the soil.—*Stems* scarcely any; height of the plant about a foot and a half.—*Leaves* bifarious, petioled, oblong, acute, smooth; deeper green on the upper surface; veins parallel, and pretty conspicuous; length from 6 to 12 inches.—*Petioles* above the bifid ligula, deeply channelled, and from 2 to 6 or 8 inches long; the lower portion sheathing, and forming something like a stem.—*Spike* hid in the centre of the sheaths of the leaves, unilateral, being imbricated on one side, (say the anterior,) with many, sessile, slender, lanceolate, colourless, one-flowered pairs of (*exterior* and *interior*) *scales* or *bractes*.—*Flowers* pretty large, expanding in succession, and rarely more than one at a time.—*Calyx* superior, membranaceous, irregularly three-toothed.—*Corol*; *Tube* long, and slender. *Border* double; *Exterior* of three, slender, lanceolate, nearly equal, straight, pale pink-coloured segments. *Inner* of three unequal portions; the two superior, or lateral, obovate, the colour and length of the exterior three; the lower, or lip, somewhat panduriform, much broader and longer than the former, and of a much deeper colour.—*Filament* very slender. *Anther* two-lobed, and crowned with an emarginate, recurved helmet.—*Germ* three-celled, each cell containing many *ovula*, attached to the axis. *Stigma* large, with a semilunate opening, which has the straight portion of its margin three-toothed, and the semicircular one even.—*Nectarial* scales long, and slender.

5. K. *ovalifolia*. *Roxb.*

Leaves oval. Spikes central, crest of the anthers laciniate.

A native of the peninsula of *Malacca*, and from thence sent, by Major Farquahar, to the Botanic Garden at *Calcutta*, where it blossoms in July. It is an intermediate species, both with respect to size and habit, between *K. Galanga* and *rotunda*, and ought to be placed between them in the system.

Root tuberous, as in the other species, and about biennial.—*Stems* very short. I may say scarcely any.—*Leaves* few, bifarious, patent, oval, with a short acute point, smooth on both sides, about 6 inches long, by 4 broad.—*Spikes* central, hid by the short sheaths of the leaves, few flowered; the flowers expand in succession, all pure white except the purple lip.—*Corol* nearly as in *rotunda* and *angustifolia*, the upper two segments of the inner border are obtusely lanceolate, as in the latter; the lip or nectarium of a deep variegated purple colour, the shape cuneate-obcordate, and divided from the apex nearly half the length into two long emarginate lobes.—*Filament* short. *Anther* two-lobed, and crowned with a large, laciniate, erect, white crest.—*Germ* three-celled, &c. as in the genus.

The following species added by Mr. N. Wallich.

Dried specimens of the following two plants were sent to me from the mountains North-East of Bengal under the *Khasee* name, *Kattia jorung*. Flowering time the rainy season.

6. K. *secunda. Wall.*

Caulescent. Leaves petioled, ovate, sub-falcate. Spike terminal, gently curved, unilateral. Crest of the filament entire.

The *Root* appears to send a number of simple fibres from its upper part.—*Stems* simple, slightly ascendent, entirely invested with long, membranaceous, striate, rather loose *sheaths*; 8 or 10 inches long, like the leaves and bractes, smooth.—*Leaves* acute at the base, tapering from the middle into a very long linear point; finely striated, from 3 to 4 inches long.—*Petioles* short, channelled; within their base an oblong, membranaceous slightly hairy, entire *ligula*.—*Spike* scarcely longer than the uppermost leaf, few-flowered.—*Exterior bractes* large, two or three-flowered, imbricating at the back of the spike; *interior*, two, small, embracing the base of the flowers.—Flowers about as large as those of K. Galanga, expanding successively. —*Calyx* short, cylindric, hairy.—*Exterior segments* of the *corol* shorter than the interior; all are lanceolate, acute, spreading. *Nectary* or inferior segment of the interior border, large, obcordate.—*Filament* short, linear, expanding above the anther into a lanceolate, acute crest.—*Anther* two-lobed, linear, scarcely marginated by the base of the filament.—*Germ, style, stigma* and *glands* at the base of the tube as in the genus.

7. K. *linearis*. *Wall.*

Caulescent. Leaves sessile, linear-lanceolate, stipuled. Flowers terminal, few. Crest entire.

Differs from the preceding species in being a slenderer plant. *Leaves* scarcely one-third of an inch broad, tapering into a filiform point.—*Ligulæ* divided to their base into two linear spreading segments, which give the leaves the appearance of being stipuled.—*Sheaths* close.

Obs. In the structure of their extremely thin and membranaceous flowers these two plants are so much like *Kæmpferia*, that I have not considered myself warranted in separating them; though the character of that genus as it now stands differs in the shape of the crest, which in our plants is perfectly entire, and on this account, resembling that of *Costus*. Their general habit partakes of that of *Globba*. If future examinations, especially of living specimens, should confirm them as species of *Kæmpferia*, the generic character of this genus, as well as of Costus, might perhaps be advantageously modified in respect to the shape of the filament, and the species of the former arranged in the following manner.

* Stemless, with a divided crest.

1.—K. Galanga.

2.—ovalifolia.

3.—rotunda.

4.—pandurata.

5.—angustifolia.

** Caulescent, with an entire crest.

6.—secunda.

7.—linearis.

CURCUMA. *Linn. gen. pl. ed. Schreb. No.* 8.

Corolla with both borders three-parted. *Anther* double, base bicalcarate. *Capsule* three-celled; *Seed* numerous, arilled. *Embryo* simple, and furnished with both perisperm and vitellus.

Sect. 1. *Spikes lateral.*

1. C. *Zerumbet*. *Roxb. Ind. pl.* 3. *N.* 201.

Bulbs and palmate tubers pale straw-coloured throughout. Leaves green, petioled, broad-lanceolar, a dark purple cloud down the middle. Flowers shorter than their bractes.

Sans. शटी, Sh*u*tee, गन्धमूली, G*u*ndh*a*-moolee, षड्ग्रंथिका, Sh*ud*-gr*u*nthh*i*ka, कर्वूरा, K*u*rvoora, कर्चूरा, K*u*rchoora, and पलाश, P*u*lasha.

Hind. and *Teling.* K*u*choor*a*.

Beng. Sh*u*thee, Kuchoora.

Pers. Zerumbad.

Kua. *Rheed. mal. vol.* 11. *p.* 13. *t.* 7.

Zerumbed. *Rumph. amb.* 5. *p.* 168. *t.* 68.

Amomum *Zerumbeth. Kön. in Retz. obs.* 3. 55.

Zerumbet, or Cachora of *Garcias.*

The plants from which the following description was taken, were sent by Dr. F. Buchanan, from *Chittagong,* where they are indigenous, to the Botanic Garden at *Calcutta,* in 1798, where they grow freely, and blossom in the month of April. Others have since been procured from thence under the Bengalee name K*u*ch*oora*. From that place the native druggists in *Calcutta,* are chiefly supplied with the root or drug.

Root consists, as in all the other species of this genus, of ovate bulbs, giving support to the parts above ground; from their opposite sides spring the palmate tubers; these two sorts are inwardly of a pale yellow, or straw colour, and possess an agreeable camphoraceous smell, and warm bitterish, spicy taste. From the bulbs chiefly, the proper fibrous roots issue: some of these are thicker than others, penetrate deeper into the soil, and end in an oblong, pearl-coloured, solid tuber; which is more spongy and less fragrant when cut, than the bulbs and palmate tubers.—*Stems* no other than the united sheaths of the leaves, surrounded by two or three obtuse, smooth, green, faintly striated, appressed scales. Height of the whole plant about three feet, or three and a half.—*Leaves* from four to six together, sub-bifarious; in general a pretty long, somewhat winged petiole intervenes between each and its stem-forming base. The leaf itself is broad-lanceolar, fine-pointed, and smooth on both sides; constantly a dark purple cloud runs down the centre; veins numerous, fine and parallel; length from 1 to 2 feet.—*Scape* rises distinct from the leafy stems, and rather before or with them, stout, 5 or 6 inches long, and surrounded with a few, obtuse, lax, green sheaths, of various length.—*Spike* tufted, 4 or 5 inches long, (so that its apex is elevated nearly a foot above the surface of the earth ;) covered with imbricated, oblong, concave bractes, connected by the lower half of their inner margins to the backs of those immediately above, forming as many sacks, or pouches as there are bractes; the lower half of these are broad, shorter, scarcely tinged with red, containing each about three or four, beautiful, yellow, sessile flowers, which expand in succession, and are embraced laterally by their own proper, smaller, short, colourless bractes. The superior half of the large coloured ones, which form the tufts, are generally sterile, and of a deeper crimson, or purple colour, than those which contain flowers.—*Calyx* superior; scarcely one third the length of the corol, irregularly three-toothed, pellucid.—*Corol* funnel-shaped. *Tube* a little curved. *Border* double; *exterior* three-parted, the two lateral segments equal; the third or upper one vaulted and crowned with a subulate point; colour a very faint yellow. *Interior* three-parted; lower segment, or lip, broad, deep yellow, projecting, recurved, bifid; upper or lateral segments obovate, equal, pale yellow, nearly as long as the lip.—*Filament* short. *Anther* double, the lower end of each lobe terminating in a long, sharp, subulate spur.—*Germ* beneath, hairy, three-celled, with many ovula in each, attached to the inner angle of the cell. *Style* slender at the base, embraced by the two nectarial filaments. *Stigma* two-lobed, crowning the anther.—*Capsule* oval, smooth, of a pale straw-colour, thin and nearly pellucid; three-celled, but without any regular division into valves : when the seeds are ripe the elasticity of the segments of the arils bursts the vertex into various portions, from whence the seeds are soon expelled.—*Seeds* several in each cell, arilled; shape various, but the most prevailing is oblong. *Aril* cut to the very base into several slender, unequal, white, fleshy segments, which are united to the seed round the umbilicus. *Integuments* two; *exterior* spongy, with highly polished slippery, light-brown surface; *interior* membranaceous. *Perisperm* pure white, hard but friable, and occupies the lower half of the seed. *Vitellus* clearer, but less white and of a tougher consistence than the albumen; it forms or occupies the upper half of the seed. *Embryo* nearly as long as the seed, tending to be clavate, both ends truncate, the upper half lodged in the *vitellus,* and the lower in the *albumen,* or *perisperm.*

Obs. The dry root pôwdered and mixed with the powdered wood of the *Cæsalpinia Sappan* makes the red powder called *Abeer* by the Hindoos, and *Phag* by the Bengalees. It is copiously thrown about by the natives during the *Hooli,* or Hindoo holidays in the month of March. The root is also used medicinally amongst the natives.

In 1805, I gave some of the sliced and dried bulbous, and palmate tuberous roots of this plant to Sir Joseph Banks, which he gave to Dr. Comb, who found that it was the real *Zedoaria* of our *Materia Medica,* and by the same means ascertained that the root of my *Curcuma Zedoaria,* is *Zedoaria rotunda* of the shops.

2. C. *Zedoaria. R.*

Bulbs small, and, with the long palmate tubers, inwardly yellow. Leaves broad-lanceolar, sessile on their sheaths, sericeous underneath; the whole plant of a uniform green.

Amomum *Zedoaria. Linn. sp. plant. ed. Willd.* i. *p.* 7.

Sans. वनहरिद्रा, V*u*na-h*u*ridra, शटी, Sholee, वनारिष्टा, V*u*narista, शटिका, Sho*l*ika.

Beng. Junglee, or B*u*n H*u*ldi, (wild turmerick.)

Arab. Jedwar, or Zadwar.

This beautiful species is a native, not only of *Bengal,* (and common in gardens about *Calcutta,*) but is also a native of *China,* and various other parts of *Asia,* and the *Asiatic* islands. Flowering time the hot season, the leaves appear about the same period, or rather after; for it is not uncommon to find the beautiful, large, rosy, tufted spikes rising from the naked earth before a single leaf is to be seen.

Root biennial, tuberous, &c. as in the last, and inwardly of a pretty deep yellow colour, approaching to that of *turmeric.*—*Stem* no other than the sheaths of the leaves.—*Leaves* petioled, broad-lanceolar, entire, underneath covered with soft sericeous down, which is particularly conspicuous when the leaves are dry.—The *Scape* rises distinct from the leaves during the dry season, and often not only at some distance, but also sometime before them; it is round, as thick as a man's fore finger; length a few inches, and embraced by its own short, proper, green sheaths.—*Spike* from 6 to 12 inches long, &c. exactly as in the last species.—*Bractes* and Calyx as in the before-mentioned species.—*Corol* one-petalled; at the base tubular. *Tube* short, widening a little; its mouth shut with fine yellow pubescent glands; *Throat* ample. *Border* double; *exterior* of three pale pink-coloured, erect, oblong divisions; the upper one more pointed, and incumbent over the anther and part of the two upper divisions of the inner border. *Interior* somewhat ringent, three-parted, fleshy, yellow; the lower lobe or lip longest, obovate, entire, projecting with a crown and erect margin, while a broad elevation of a deep yellow colour runs along its middle; upper two divisions obovate, and with the upper segment of the exterior border, forming a complete covering or dome over the anther.—*Filament* short, inserted between the two upper divisions of the inner border of the corol. *Anther* two-lobed, each ending in a long, sharp spur at the base.—*Germ* inferior, &c. as in the last. *Style* filiform. *Stigma* three-lobed, with a perforation in the centre.

Obs. The plant when in flower is highly ornamental, few surpassing it in beauty; at the same time it possesses a considerable degree of delicate aromatic fragrance.

Dr. Woodville's observation and quotations, at page 362 of his *Medical Botany*, induce me to think that the roots of this plant are not only the longer species of the *Zedoaria* found in the shops; but also the shorter or roundish, their form and size depending on the manner in which they may have been cut for drying: and I have lately learned from Sir J. Banks, that the dried root of this species agrees well with *Zedoaria rotunda*.

The Hindoos use the roots as a perfume, and also medicinally; they are sold in their shops, under the Arabic name *Judwar*, or the Bengalee one of *Bun Huldi*. When fresh as well as when dry, they are of a pale yellow colour; and possess at all times an agreeable fragrant smell, and warm, bitterish, aromatic taste, whereas the roots of *Kæmpferia rotunda* possess little or nothing of these sensible qualities, and are withal too rare in every part of *India,* where I have been, to allow me to think, any quantity of them can ever be collected for exportation.

3. C. *zanthorrhiza. R.*

Bulbs ovate, with both palmate and pendulous tubers deep yellow. Leaves sessile on their green sheaths, broad lanceolar, a permanent purple cloud down the middle.

From *Amboyna,* this remarkable species was brought to the Botanic Garden at *Calcutta,* in the year 1798, and only flowered for the first time in April and May, 1810. It is the only kind I have yet met with, which has its pendulous tubers inwardly of a deep yellow colour, like that of the bulb and palmate tubers. The *sheaths* of the scapes and leaves are green. The coma large, and of a deep purple, or crimson colour. *Exterior border* of the *corol* red. *Leaves* broad-lanceolar, and oblong; there is a narrow purple cloud down the middle of them, which penetrates to the underside; the whole height about four feet.

4. C. *elata. R.*

Bulbs oblong, obovate, and with the numerous, large, incurvate, palmate tubers, inwardly of a pale yellow. Leaves sessile on their sheaths, oblong, villous underneath; the whole of an uniform green.

A native of *Burma,* and from thence introduced by Dr. W. Carey, into the Botanic Garden at *Calcutta,* where it flowers in March and April.

A full description of this species is unnecessary, as it resembles the other species already described, in every respect, except in such as are pointed out in the definition, viz. its immensely large and numerous *bulbs* and *tubers,* which are inwardly of a pale yellow colour; the latter run very near the surface of the ground, and are much incurved. The *pendulous tubers*, as in every other species, except *C. zanthorrhiza*, are inwardly pearl-coloured; every part above ground is of a uniform green, the spikes excepted. The *leaves* appear when the flowers begin to decay, and are larger than in any other species I have yet seen; being from two to three, or even four feet long, by one broad, and villous underneath; the whole height is about six feet. The *spikes* are large, with a deep rosy, or even crimson coma. Its external appearance resembles most *C. Zedoaria*, the roots however differ much, independent of their size. This plant is altogether a larger species than that.

5. C. *cæsia*. *R.*

Bulbs ovate, and with the palmate tubers inwardly bluish. Leaves lanceolar, petioled; a deep ferruginous purple cloud down the middle, which penetrates to the under surface; every other part green.

Beng. Neel-*k*unt'h*a*, or Kala H*u*ldi.

Tommon-itam, *Rumph. amb.* 5. *p.* 169.

This elegant, strongly marked species is a native of Bengal, where it blossoms in May, while the greatest heat prevails, just before the periodical rains begin: about the same time the leaves begin to appear. In the deep ferruginous purple cloud down the middle of the leaves, it resembles *C. Zerumbet*, but differs widely in the colour of the root.

Bulbs ovate. *Palmate tubers* many, and generally simple; these and the bulbs are inwardly pale blue, verging towards grey, (cæsius of Willdenow's scale;) in some the blue is pretty deep. *Pendulous tubers* ovate, oblong, inwardly of a pale watery pearl-colour, and less fragrant than in the former two species. This indeed is the case with all the other species I have examined.—*Leaves* bifarious, petioled, lanceolar, smooth on both sides, a deep ferruginous cloud runs the whole length of the rib, and penetrates through to the underside; every other part, petioles, and their sheaths, of a clear deep green colour; length about eighteen inches, by five broad. *Petioles* and their sheaths together, about as long as the leaves, making the whole height of the plant about four feet.—*Spikes* lateral, appearing with, or rather before the leaves, about as long as their scapes, together about a foot long. *Scales*, lower fertile; *bractes* green, with a ferruginous tinge. *Coma* deep bright red, tending to crimson.—*Corol*, exterior border red; inner yellow.

6. C. *æruginosa*. *R.*

Bulbs ovate, and with the numerous palmate tubers, inwardly æruginous. Leaves petioled, broad-lanceolar, smooth, above the middle a faint ferruginous evanescent cloud on the upper surface only; every other part green.

A very stately species, introduced from *Burma* by the Rev. Dr. W. Carey, into *Bengal*, where it blossoms in May, immediately after which the leaves begin to appear. It is distinguished from all our other *Curcumas* by the internal æruginous, or verdigrease-like colour of its ovate *bulbs*, and its short thick palmate *tubers;* while the oval pendulous tubers, with which it abounds, are inwardly of a clear pale pearl colour. The comose lateral *spike* is in this, as it is in the other species. The exterior border of the *corol* is rose-coloured; the interior deep yellow. The *leaves* are broad-lanceolar, of a firm texture, and smooth on both sides; length from two to three feet, and from six to nine inches broad. *Petioles* and their sheathing portions together from two to three feet long; which gives to the whole plant a height of from four to six feet.

7. C. *ferruginea*. *R.*

Bulbs and palmate tubers copious, inwardly pale yellow. Sheaths of the scapes, and leaves ferruginous-red, with a faint mark of the same colour on the middle of the upper surface of the leaves.

A native of *Bengal*. Flowers in April and May. By attending to the pale yellow very fragrant roots, and to the rusty reddish sheaths of the scapes and leaves, this species will be readily known. The *leaves*, as in most of the others, are broad-lanceolar, (that is, they taper equally toward each end;) including the sheaths and petioles, they are four feet high, or more; a very faint ferruginous tinge runs down the middle, which is very conspicuous in those that appear first in the season; whereas it is scarcely to be observed in the latter. The *flowers* are large, few in number, with the *exterior* border red, and the *interior* deep yellow. The *bractes* to the fertile part of the spike ferruginous, those of the *coma* few, and of a pretty bright crimson colour.

8. C. *rubescens. R.*

Bulbs ovate-lanceolate, and with the palmate tubers inwardly pearl-coloured. Leaves broad lanceolate, on red winged petioles, above the sheaths; rib red. Scape and scales red. Flowers longer than their bractes.

A native of Bengal. Flowering time the months of April and May, soon after which the leaves appear; and decay about the beginning of the cool season, in November. Every part has a strong, but pleasant aromatic smell when bruised, particularly the root.

Root of several, erect, solid, conical, pale straw or pearl-coloured, powerfully aromatic bulbs, which gave support to the former years foliage, and are strongly marked with the circular scars thereof; from their opposite sides the scapes, and stems of the succeeding year spring, which form similar new bulbs when those of the former year decay; but during their existence, there issues round their lower half, a number of strong fleshy fibres, many of which end in ovate or sub-cylindrical, pale white, slightly aromatic tubers, which also perish with the original parent bulb.—*Stems,* as in the other species, no other than the united sheaths of the leaves, which like them, decay annually about the month of October, and appear again when the flowers begin to perish in April.—*Leaves* bifarious, (six or eight of them forming the above-mentioned stems, of about three, or four feet in height, leaves included,) petioled on their sheathing base, broad-lanceolate, cuspidate, smooth, strongly marked with parallel veins; of an uniform dark-green, with the nerves or ribs red, from twelve to twenty-four inches long, by five or six broad.—*Petioles,* and *sheaths* channelled, smooth, and of a deep red colour; a projecting process on the inside marks, in all this natural order, the limit of the sheath, and the beginning of the petiole.—*Scape,* radical, lateral, cylindric, about six inches long, invested in several, dark reddish sheaths. —*Spike* tufted, five or six inches long, erect; *Coma* less deeply coloured than in Zerumbet.—*Bractes,* or *scales* of the spike exactly as in the other species, each embracing four or five flowers, which expand in succession.—*Flowers* small, bright yellow, rather longer than their bractes, fragrant.—*Tube* of the corol slender, its mouth completely shut with three villous, yellow glands.—*Fruit* not seen in a ripe state.

Obs. The ovate oblong, pendulous tubers of several species of *Curcuma,* yield a very beautiful, pure starch, like the *Arrow-root,* produced from *Maranta arundinacea* and *Tacca pinnatifida,* which the natives of the countries where the plants grow, prepare and eat. In *Travancore,* where some of these species abound, this flour, or starch, forms, I am told, a large part of the diet of the inhabitants. My *C. angustifolia* is another species which yields the same substance; and I have no doubt but the pendulous tubers of this species yield it also, and equally good.

Since writing the above, I have received from *Chittagong, Bhagulpore,* and *Mirzapore,* plants of three other sorts, all employed for the same purpose.

9. C. *comosa. R.*

Bulbs large, oval, inwardly pale ochraceous. Spikes clavate; fertile bractes of a pale pink colour; coma copious and rosy. Leaves broad-lanceolar; a faint ferruginous cloud above the middle of the earliest ones; every other part green.

A native of *Burma.* From *Rangoon* Mr. Felix Carey sent plants to the Botanic Garden at *Calcutta,* where they thrive well, and blossom in May, at which time it is by far the most beautiful, and the largest of the genus, I have yet seen.

Bulbs very large, oval, inwardly of a pale ochraceous colour. *Palmate tubers* scarcely any. *Pendulous tubers* lanceolar, and connected to the bulb by uncommonly long, thick, fibrous cords.—*Leaves* bifarious, petioled on their sheaths, lanceolar, smooth, and green in every part, except in those which appear first in the season, which have a faint ferruginous cloud on the centre of the upper surface beyond the middle; length two to three feet; the petioles with their sheathing bases measuring about as much more, making the whole height of the plant five or six feet.—*Spikes* lateral, appearing rather before the leaves, short-scaped, large, clavate. *Fertile bractes*, of a beautiful pale pink. *Barren ones* of the *coma* a deep, lively, rosy red; on account of the beauty and size of this part, I have given it the trivial name *comosa*.—*Corol* with the *exterior* border pink; *inner* pale yellow.

10. C. *leucorrhiza*. *R.*

Bulbs ovate; palmate tubers long and far-spreading, like the former they are inwardly of a pale straw colour. Leaves petioled, broad-lanceolar, smooth, of an uniform green in every part. Spikes few-flowered, with a coma as long as the fertile portion.

A native of the forests of *Bahar*. From *Bhagulpore* Mr. John Glass, the surgeon of that station, sent roots to this Garden, under the name *Tikor*; and observed that the plant is not cultivated, but grows in the jungles, to the southward of that place. The process, he says, for obtaining the starchy substance called Tikor, is as follows. " The root is dug up, and rubbed on a stone, or beat in a mortar, and afterwards rubbed in water with the hand, and strained through a cloth; the fecula having subsided, the water is poured off, and the Tikor, (fecula,) dried for use."

This plant grows freely in the Botanic Garden, and flowers in May. Here the *palmate*, or *horizontal tubers*, are particularly straight, and often near a foot long; they run deep in the earth, and far. Their inward colour, as well as that of the bulbs, is a very pale ochraceous yellow, almost pure white. *Pendulous tubers* numerous, oblong, inwardly pearl white. *Leaves* broad-lanceolar, uniformly green, about two feet long; whole height of the plant from three to four feet. *Spikes* lateral, the inferior fertile portion not longer than the rosy coma. *Flowers* the length of the bractes; *exterior* border slightly tinged with pink; *inner* yellow. In the structure of the fruit it agrees perfectly with that of *C. Zerumbet.*

11. C. *angustifolia. R.*

Bulb oblong, with pale, oblong, pendulous tubers only. Leaves petioled, narrow-lanceolar. Flowers longer than the bractes.

Tikor of the Hindoos.

Found by H. T. Colebrooke, Esq. in the forests from the banks of the *Sona* to *Nagpore,* and by him introduced into the Botanic Garden at *Calcutta,* where it blossoms in July. The leaves make their appearance at the same time with those of the other species, and decay about the beginning of the cool season, in November.

Root, from a fusiform biennial crown issue many fleshy fibres, which end in smooth, oval, succulent tubers.—*Stem,* consisting of a few pale-coloured, leafless, surrounding sheaths.—*Leaves* petioled, narrow-lanceolar, most acute, striated with fine parallel veins, smooth on both sides; length from one to three feet, (petiole and sheath included,) which is in fact the whole height of the plant.—*Petioles* from six to twelve inches long, the lower half or more of which, expands into a sheath to embrace those within; the upper half or proper petioles slender, and channelled.—*Spike* radical, from four to six inches long, separate from the leaves, crowned with a tuft of oval, lively purple, neuter bractes.—*Bractes common* below the coloured neuter ones, ovate-cordate, obtuse, expanding recurvedly, each embracing three or four flowers, which open in succession.—*Bractes proper* boat-shaped, embracing the proper perianth, and germ of each flower.—*Flowers* large, longer than their bractes, bright yellow, expanding at sun-rise, and decaying at sunset of the same day.—*Calyx* above, three-toothed, somewhat inflated.—*Corol, tube* somewhat gibbous, contracted at the mouth, and there shut with short hairs; *throat* campanulate; *exterior* border pale-yellow, consisting of one large, vaulted, upper segment, and two lower, oblong, smaller, concave ones; *inner border* three-parted; lower segment or *lip* with its exterior part sub-orbicular emarginate, or bifid; *upper segments* obovate, cuneate, with the filament, or short antheriferous segment, between them.—*Anther* as in the genus, double, with a spur from the lower end of each lobe.—*Germ* beneath, villous, three-celled, with many seeds in each. *Style* slender, at its insertion are the usual two nectarial bodies. *Stigma* globular, gaping on the anterior side.

Obs. From the oval pendulous tubers, Mr. Colebrooke observes, a sort of *Starch* or *Arrow-root-like fecula* is prepared, which is sold in the markets of Benares, and is eaten by the natives.

Sect. 2. *Spikes central.*

12. C. *longa*. *Linn. sp. pl. ed. Willd.* 1. 14. *Retz. obs.* 3. 72.

Bulbs small, and with the numerous, long, palmate tubers, inwardly of a deep orange yellow. Leaves long-petioled, broad-lanceolar, of an uniform green.

Amomum *Curcuma*. *Gmelin.* and *Jacq. Hort. Vind.* 3. *t.* 4.

Manjella-kua. *Rheed. mal.* 11. *p.* 21. *t.* 11.

Curcuma domestica major. *Rumph. amb.* 5. *p.* 162. *t.* 67.

Carcumaa of Avicenna.

Dar-zard of Garcias.

Sans. हरिद्रा, H*u*ridr*a*, हलदी, H*u*ldee, पीता, Peeta, काञ्चनी, Kanch*u*nee, निशा, N*i*sha, वरवर्णिनी, V*uru*v*u*rn*i*nee, कृमिघ्ना, Kr*i*m*i*ghna, योषित्प्रिया, Yosh*i*tpr*i*ya, हरिविलासिनी, H*uri*v*i*las*i*nee.

Hind. and *Beng.* H*u*ldee, H*u*l*u*dee, P*i*t-ras.

Heb. כרכם, Kurkum.

Arab. Urukus-sufr, Urukus-saboghin.

Teling. Pampee.

Pers. زردچوب, Zerd-chob.

H*u*l*u*d, or H*u*l*u*dee. Mahrat.

Is much cultivated about *Calcutta*, and in all parts of *Bengal*. Koenig's description published by Retzius, is so very exact and complete, that there is nothing left for me to add.

Cultivation.

The ground must be rich, friable, and so high as not to be overflowed during the rainy season, such as the Bengalees about Calcutta call Danga. It is often planted on land where sugar-cane grew the preceding year, and is deemed a meliorating crop. The soil must be well ploughed and cleared of weeds, &c. It is then raised, in April and May, according as the rains begin to fall, into ridges, nine or ten inches high, and eighteen or twenty broad, with intervening trenches nine or ten inches broad. The cuttings or sets, viz. small portions of the fresh root, are planted on the tops of the ridges, at about eighteen inches, or two feet asunder. One acre requires about 900 such sets, and yields in December and January, about 2000lb. weight of the fresh root.

13. C. *Amada. R.*

Bulbs conic, and with the palmate tubers, inwardly pale-yellow. Leaves long-petioled, broad-lanceolate, smooth. Spikes scanty, few-flowered. The whole plant uniformly green.

Amada of the Bengalees, which means *mango-ginger*, the fresh root possessing the peculiar smell of a green mango. It is used medicinally by the natives. It is a native of *Bengal.* Flowering time the latter part of the rains.

Root consisting of horizontal, palmate, sessile *tubers* united to the sides of an ovate-conic bulb of the same colour, which gives existence and support to the leaves, and spike: from its sides and base long fleshy fibres issue, which penetrate deep into the soil; some of which end in oblong, paler (pendulous) tubers.—*Stems* no other than the united sheaths of the leaves.—*Leaves* radical, bifarious, petioled above their sheaths, ventricose, lanceolar, cuspidate, smooth on both sides; length from six to eighteen inches, by three to six broad; they decay about the end of October, and appear again with the first showers of April or May.—The *petioles* I divide into two portions: the lower or sheathing part from six to twelve inches long; and the upper or proper petioles, from three to nine; the whole measuring in general, from one to two feet.—*Scape* central, about six inches long, invested in several alternate sheaths.—*Spike* shorter than the scape, cylindric, crowned with a tuft of loose, coloured, pale rosy, abortive bractes.—*Bractes* oblong, imbricated, the lower half of their margins united to the backs of the two next above, forming a pouch, for a fascicle of four or five, rather small, yellow flowers, which expand in succession; these have each a proper bracte, which embrace the tube of the flower only, and may be called an inferior calyx.—*Calyx* superior, unequally three-toothed.—*Corol* with a slender *tube;* its mouth shut with three yellow hairy glands; *faux* sub-campanulate. *Border* double, both three-cleft, &c. as in the other species.—*Germ* oval, hairy, three-celled, with several seeds in each attached to their central receptacles.

14. C. *viridiflora. R.*

Bulbs oblong, and with the palmate tubers inwardly of a deep yellow. Leaves long-petioled, broad-lanceolar. The whole plant, (even the spike and coma,) uniformly green.

Tammon Giring seu Giri. *Rumph. amb.* 5. 169.

A native of *Sumatra,* and the other eastern islands; from *Bencoolen* Dr. Charles Campbell sent plants to this garden, where they grow luxuriantly, and blossom in July and August.

Root consisting, like that of the other species, of *bulbs,* and palmate, pendulous *tubers;* aromatic, and bitter, and employed by the *Malays* of *Sumatra* to dye with.—*Leaves* petioled, broad-lanceolar, smooth; length from one to three feet; the petioles, and sheaths thereof about as long.—*Spike* central, large.—*Bractes,* even those of the *coma,* uniformly green; the latter only paler.—*Flowers* small, very pale yellow.

15. C. *montana. Corom. pl.* 2. *N.* 151.

Bulbs conic, with pale-yellow palmate tubers. Leaves short, petioled, oblong. The whole plant uniformly green, except the rosy coma of the spike.

It grows in moist places, amongst the *Circar* mountains.—Flowering time the rainy season.

Root perennial, consisting of conical bulbs enveloped in the remaining sheaths of the leaves, and also of horizontal tubers, as in ginger, &c. with large fleshy fibres from their base.—*Leaves* radical, three or four, petioled, their petioles from eight to twelve inches long, enveloped in a few sheaths, forming something like a stem; the leaves themselves are oval, pointed, beautifully veined, smooth, from eight to twelve inches long, and from five to six broad.—*Spikes* rise from the centre of the petioles; their peduncle is of their length, and involved in its proper sheaths.—*Bractes* numerous; the inferior roundish, the superior oblong; the lower half of the base of the two inner are united to the margins of the next without, forming a sack, which contains from two to four flowers, each of which has a smaller wedge-formed membranaceous proper bracte. The *coma* or *superior bractes* large, waved, rose-coloured, and (generally) without flowers.—*Calyx* as in the two former species.—*Tube* of the *corol* widening, somewhat campanulate; *border* double.—*Exterior* three-parted; *divisions* equal, entire. *Interior* two-lipped; *upper lip* erect, three-cleft; *under lip* entire, roundish, border waved, and a little notched.—*Filament* no other than the middle segment of the upper lip of the interior border of the corol; anther two-lobed, linear, the under point of each lobe ending in a long sharp spur.—*Germ* below. Base of the *style* embraced with the usual two nectarial bodies, which in this species are larger; *stigma* large, globular, two-lipped, somewhat marked.

16. C. *reclinata*. *R.*

Bulbs fusiform, with scarcely any palmate tubers, but many straw-coloured, oval, pendulous ones. Leaves petioled, oblong, reclinate.

Discovered by H. T. Colebrooke, Esq. in the interior parts of *India*, where it blossoms during the rains.

Root a bulbous head, from which descend many fleshy fibres, ending in large, smooth, oval tubers.—*Stem* none, but the height of the whole plant (viz. the spike) is about six inches; the leaves, being reclined, add nothing to its height.—*Leaves* radical; (in the only plant at present in my possession, from which this description is taken), four in number, two on each side of the spike, embracing it closely, with the lower sheathing part of their petioles reclined, oblong, acute; margins slightly waved; smooth on both sides; length, exclusive of the petiole, about nine inches, and about four or five broad.—*Petioles*, including the sheathing base, from three to four inches long, deeply channelled, and smooth.—*Spike* from the centre of the leaves, its short scape entirely hid by their sheaths, the rest loosely imbricated, with large, ovate, obtuse, expanded coloured bractes: the superior sterile, and more deeply coloured as in the other species of this family; but here, and in no other that I have yet seen, are the whole deeply coloured.—*Flowers* small, of a dull dark pink colour, except the lower segment of the inner border, which is tinged with yellow.—*Calyx* inflated, obscurely three-toothed, scarcely half the length of the tube of the corol.—*Tube* of the *corol* sub-cylindric; *throat* campanulate. The *border, stamen, and pistil,* as well as the *nectarial scales* of Koenig, as in the other species, already described.

17. C. *petiolata. R.*

Bulbs and palmate tubers few and small; inward colour pale yellow; tubers numerous, pendulous on short fusiform fibres. Leaves long petioled, oblong, ovate with the base rounded, (this is the only species, known to me, in which they have this form.) The whole plant uniformly green, except the lilac-coloured coma of the spike.

A native of *Pegu*, and from thence sent by Mr. F. Carey to the Botanic Garden at *Calcutta*, where it blossoms in August.

Obs. This species differs conspicuously from all my other *Curcumæ* in the cordate shape of the *leaves*. The spike is central, as in the other late flowering species, its *coma* small, and of a lilac colour; the *bractes* very perfectly united almost to the broad, reniform apex, forming uncommonly deep pouches for the small yellow *flowers*.

AMOMUM. *Schreb. gen. plant.* N. 3.

Corolla with the interior border unilabiate. *Anther* double, surmounted with an entire, or lobate crest. *Capsule* three-celled, three-valved. *Seeds* many, arilled. *Embryo* simple, and furnished with both perisperm, and vitellus.

1. A. *cardamomum. Linn. spec. pl. ed. Willd.* 1. 8.

Leaves short-petioled, lanceolar. Spikes half immersed in the earth, loosely imbricated with villous, lanceolate, acute, one-flowered bractes. Lip with the anterior margin three-lobed. Crest three-lobed.

Cardamomum minus. *Rumph. amb.* 5. *p.* 152. *t.* 65. *f.* 1.

A native of *Sumatra,* and other islands to the eastward of the Bay of *Bengal,* and sent by Dr. Charles Campbell from *Bencoolen,* to the Botanic Garden at *Calcutta,* where it blossoms in April.

Root perennial, creeping under the surface of the soil, like that of *ginger;* but smaller, less fleshy, more ligneous, and white; from which descend, and spread, many fleshy fibres.—*Stems* about biennial, several rising obliquely to the height of from two to four feet, about as thick as a stout ratan, invested in the smooth, deep green sheaths of the leaves.—*Leaves* alternate, bifarious, short-petioled on their smooth stem-clasping sheaths; from broad-lanceolate below, to narrow-lanceolate at top, entire and smooth on both sides; point long, and very fine; length from six to twelve inches.—*Spikes* radical, sessile, oblong, appearing amongst the stems, half immersed in the earth, loosely imbricated with one-flowered, lanceolate, acute, villous, nervous, scariose, ash-coloured bractes; when old their brittle tops are often broken off.—*Bractes.* Besides the exterior one-flowered ones, just mentioned, there is an inner, striated, downy, scariose, two-toothed, tubular one, (which I have sometimes considered an exterior and inferior *calyx,* and which Koenig sometimes called an involucre,) inserted round the base of the germ.—*Flowers* opening in succession, and not very conspicuous.—*Calyx,* clavate, tubular, downy, three-toothed, length of the tube of the corol.—*Tube* of the *corol,* slender, and slightly incurved. *Exterior border* of three sub-equal pellucid divisions. *Lip,* or inner border, rather longer than the exterior great border, somewhat three-lobed, with a crenate, curled margin; middle lobe yellow, with two rosy lines leading up to it from the mouth of the tube.—*Filament* scarcely half so long as the border of the corol, incurved over the mouth of the tube. There is a slender subulate horn on each side of the base of the filament, and nearly its length. *Anther* double, large, fleshy, with a large, three-lobed, concave crest; the infundibuliform stigma rising through a deep groove between the two polliniferous lobes.—*Germ* beneath, downy, and crowned with the two nectarial scales, within the base of the tube of the corol; in this species they are short, and truncated.

Obs. The seeds are to the taste agreeably aromatic, and are used by the *Malays* as a substitute for the true *Cardamomum* of *Malabar.*

2. A. *angustifolium. Linn. spec. pl. ed. Willd.* 1. *p.* 8.

Leaves broad-lanceolate. Spikes elevated, linear-oblong, compact; clavate bractes oblong, rather pointed. Lip obovate, cuneate, entire. Crest three-toothed.

Amomum *angustifolium. Sonnerat it.* 2. *f.* 137.

A native of *Madagascar.* From the island of *Mauritius* Captain Tennent brought plants thereof to the Botanic Garden at *Calcutta,* in 1802; where they grow freely, and blossom at various times of the year. Every part of the plant, when bruised, or wounded, diffuses a strong, but pleasant aromatic smell.

Root perennial, tuberous, horizontal; strong fleshy fibres, together with runners, issue from the crown, or bulb; by these runners the plant soon spreads, and by these it is easily multiplied.—*Stems* numerous, straight, round, smooth, rising to the height of from six to ten feet.—*Leaves* alternate, bifarious, sub-sessile on their sheaths, broad-lanceolate, fine-pointed, smooth on both sides; length from twelve to eighteen inches, and from about four to six broad.—

Sheaths embrace about three quarters of the stem, are smooth, and rise in an obtuse process, (ligula,) above the insertion of the leaves. —*Scapes* from the bulb, or base of the stem, where it unites with the root, rather slender, from six to twelve inches long, and closely imbricated with many, mucronate, smooth, sheathing scales.—*Spikes* linear-oblong, about the size of a man's thumb, imbricated with similar, but larger scales than those of the scape ; the inner more deeply coloured.—*Flowers* few, pretty large, and possessed of a strong spicy fragrance.—*Bractes ; proper* two to each flower, linear-oblong, inserted round the base of the germ, like an inferior perianth.—*Calyx* superior, deep blood red, smooth on both sides, slit two thirds on the outside, apex obscurely three-toothed.—*Corol. Tube* clavate, about as long as the bractes. *Exterior border red,* consisting of one large, erect, oblong, concave segment, which protects the anther and stigma ; and of two inferior, very narrow segments. *Lip* obovate-cuneate, entire, or slightly three-lobed, yellow, and about as long as the exterior border.—*Filament* broad, particularly toward the base, where it has a small, coloured hornlet on each side. *Anther* double, surmounted with a large, three-toothed coronet.—*Germ* clavate, obscurely three-sided, three-celled, with many *ovula* in each, attached to the partitions. *Stigma* funnel-shaped, somewhat bilabiate. *Nectarial bodies of Koenig* are here larger, and longer than in the other species.

3. A. *aculeatum. R.*

Leaves sub-sessile, cordate-lanceolate. Spikes obovate, lax, even with the ground. Lip oval, ventricose. Crest amply three-lobed. Capsules echinate.

A native of the *Malay Archipelago.* It was raised from seed sent to the Botanic Garden at *Calcutta* by Mr. W. Roxburgh, in 1802, where the plants blossom in April and May, and ripen their seed in October and November.

Root perennial, tuberous like ginger, with many, thick, fleshy fibres.—*Stems* a little recurvate, the thickness of a walking cane, slightly pubescent; from five to ten feet high.—*Leaves* alternate, bifarious, very short-petioled, cordate-lanceolate, mucronate, smooth, dark green; general length from twelve to eighteen inches, and from three to four inches broad.—*Sheaths* half stem-clasping, villous on the outside. *Ligula* short, and of a semi-lunar shape.—*Scapes* issuing from the base of the stem, under the surface of the ground, where it joins the root, diverging a little way under the surface of the earth, invested in oval, fleshy scales; those next to the spike longer, and more pointed.—*Spikes* obovate, rising obliquely among, or near the stems, just so much as to elevate its flower-bearing apex above the surface of the earth.—*Flowers* numerous, and expand in succession; prevailing colour deep orange.—*Bractes, exterior* one-flowered, linear-oblong, length of the tube of the corol; *inner* tubular with two-toothed margin, and inserted round the base of the germ, like an inferior perianth.—*Calyx* above, tubular, two or three-toothed.—*Corol, tube* slightly incurved. *Exterior border* three-parted; *segments* unequal, erect, oblong, obtuse, the middle one larger, concave and incumbent over the fissure of the inner border, together forming a campanulate tube. *Lip*, sub-rotund, with crenulate, curled margin, and marked with beautiful crimson spots down the middle, on the inside.—*Filament* erect, with a hornlet on each side of its base. *Anther* two-lobed; *lobes* linear-oblong, and surmounted with a broad, three-lobed crest.—*Germ* inferior, three-celled, with many *ovula* in each, attached to the inner angle. *Stigma* funnel-shaped.—*Capsule* the size of a large nutmeg, oval, (not sulcate as in Koenig's *Amomum or Globba crispa, Retz. obs.* 3. *p.* 50.) but thickly beset with numerous, stout prickles, three-celled, three-valved; *valves* fleshy (not nuciform as in *Globba crispa, Rumph. amb.* 6. *p.* 137). *Partitions* membranaceous.—*Seeds*, many in each cell, oval, smooth, brown; surrounded with a pale, gelatinous, sweetish-acid pulp or aril, intermixed with fine woolly filaments, and attached to a central receptacle by very slender cords, which are enveloped in the same gelatinous pulp which surrounded the seed.—*Perisperm*, white, cartilaginous.—*Vitellus*, nearly spherical, lodged about the middle of the perisperm, and of a clearer colour, and harder texture.—*Embryo* sub-cylindric, the inner end, or cotyledon, lodged in the vitellus, which in some seeds it passes entirely through. *Radicle* pointing directly to the umbilicus.

4. A. *maximum.* R.

Leaves petioled, lanceolar, villous underneath. Spikes oval, even with the earth. Bractes lanceolate. Lip elliptic. Coronet of one semi-lunar lobe. Capsules round, nine-winged.

This plant is now growing in the Company's Botanic Garden near *Calcutta*; it was introduced from the *Malay* islands by the late Colonel Kyd. Flowering time the hot season; the seeds ripen three or four months afterwards.

Root tuberous, as in ginger, but much larger, and with large, conical bulb-like bases to the oldest stems.—*Stems* nearly erect, round, and smooth; from six to eight or more feet high, invested in the smooth, long sheaths of the leaves.—*Leaves* alternate, petioled, lanceolar, and linear-lanceolar, acute, smooth above; villous underneath, with many large parallel veins; from two to three feet long, and about six inches broad; when dry the underside appears to be more closely covered, with most beautiful, soft, sericeous down, than while fresh.—*Petioles,* the sheathing part long, between each petiole and its leaf there is a short, somewhat triangular portion, with a stipulary process, or *ligula* at the inside of its base.—*Spikes* radical, short, their flower-bearing apex being only just above the earth; below the flowers it is covered with large, oval, concave scales.—*Bractes* solitary, one-flowered, lanceolate, membranaceous, rather longer than the tubes of the flowers.—*Calyx* tubular, a little gibbous, length of the tube of the corol; mouth three-toothed.—*Corol*; *Tube* funnel-shaped. *Exterior border* three-parted, as usual; but here they all converge upward into one helmet over the anther and stigma; as in the *Orchideæ*. *Lip* entire, expanding, oblong, marked with a yellow stripe down the middle; margins waved, and crenulate.—*Filament* short, from the mouth of the tube of the corol, opposite to it's interior border, on each side of its insertion, is a small hornlet. *Anther* double or two-lobed, and crowned with a beautiful, broad, semi-lunar crest. *Nectarial* bodies of König linear, fluted, and truncate.—*Germ* short-pedicelled, striated, three-celled, with numerous *ovula* in each, attached to the inner angle, *(axis)*. *Style* slender. *Stigma* funnel-shaped, but shut transversely, except during anthesis.—*Capsule* almost globular, size of a gooseberry, three-celled, three-valved, ornamented with nine, firm, short, ragged (when old, and dry) membranaceous wings.

Obs. The seeds possess a warm, pungent, aromatic taste, not unlike that of *Cardamoms*, but by no means so grateful.

5. A. *dealbatum*. *R.*

Leaves broad-lanceolar, villous underneath. Spikes radical. bractes lanceolate, obtuse, (very perishable.) Exterior border vaulted. Lip round-oval. Crest of one, broad, truncated lobe. Capsules ovate-oblong, nine-winged.

A stately species, a native of *Chittagong*, and other countries which form the eastern frontier of *Bengal;* from the former it has been introduced into the Botanic Garden near *Calcutta,* where it grows luxuriantly, and flowers freely during the months of March and April; and the seeds ripen in September and October.

Roots perennial, tuberous, crooked, jointed, creeping, like common ginger, with strong, fleshy fibres; inward colour of the tubers pale dull yellow, and when newly cut considerably aromatic.—*Stems* numerous, erect in the centre, and oblique in the circumference of the clump, from three to five feet high.—*Leaves* sub-bifarious, the lower sessile on their sheaths and oblong, the superior short-petioled and lanceolar; all are very villous, and paler underneath, length from six to twenty-four inches, breadth from three to six.—*Sheaths* smooth, rising in a bifid ligula above the mouth.—*Spikes* radical, roundish-obovate. *Scape* short and imbricated with scales, &c.—*Bractes* from ovate to lanceolate, obtuse, very perishable, pale, one-flowered. In this species there is no *inner bracte,* like an *interior calyx* attached to the base of the germ, so general in this natural order.—*Flowers* numerous, large, white.—*Calyx* superior, rather longer than the tube of the corol, three-cleft, and like the bractes very perishable.—*Corol*; *Tube* short, the three exterior segments of the border form a vault over the anther, they are linear-oblong, and nearly equal. *Lip* broad-oval, emarginate; *margins* curled; white, with a little yellow in the centre.—*Filament* short. *Anther* linear, crowned with an entire, broad, truncate, pure white crest.—*Germ* short-pedicelled, three-celled, &c. as in the genus.—*Capsules* ovate-oblong, size of a nutmeg, nine-winged, three-celled, but I have not observed them separate into valves.—*Seeds* many, shape various, but oval and ovate are most common. *Integuments* besides the aril, or its dried remains, a single, very hard, dark coloured, brittle one.—*Perisperm* conform to the seed, white, and cartilaginous.—*Vitellus* lodged in the upper part of the perisperm, of a roundish shape, hard, sub-hyaline.—*Embryo* sub-cylindric, its apex immersed in the vitellus, and its base at the umbilicus.

Obs. The seeds of this plant are insipid, whereas those of *Amomum maximum*, which is nearly allied to it, are aromatic, and pass for a sort of *Cardamom*.

6. A. *subulatum.* R.

Leaves lanceolar smooth. Spikes obovate, echinated with the long subulate points of the bractes, calyx, and upper segment of the exterior border of the corol. Lip oblong.

A native of the lower range of mountains which skirt the plains of *Bengal* on the north, where it grows on the borders of the rills of water between the hills. Flowering time in the Botanic Garden near *Calcutta*, March and April; and the seeds ripen during the rains.

Root tuberous, creeping at a small distance under the surface of the soil, &c. as in the genus, and amply furnished with numerous, strong fibres.—*Stems* erect about three or four feet high, and about as thick as the little finger.—*Leaves* alternate, bifarious, sessile on their sheaths, narrow-lanceolar, finely acuminate, deep green, and smooth on both sides; from twelve to fifteen inches long, by three or four broad. *Sheaths* smooth, margins coloured and rising above the mouth into an emarginate ligula.—*Spikes* radical, compact, a little elevated above the soil, obovate, size of a goose-egg, echinate with the long, subulate, pale yellow points of the bractes, inner calyx and upper segment of the exterior border of the corol.—*Bractes* numerous, the *exterior* imbricated upward, obovate, dark-red, smooth, long-subulate, pointed, one-flowered; *the inner, or inferior* calyx, spathiform, opening on one side.—*Flowers* large, yellow.—*Calyx* superior, three-cleft, segments subulate, and much longer than the tube of the corol.—*Corol*; exterior border of three nearly equal segments, the upper one ends subulately. *Lip* oblong, often emarginate; margins a little curled, deeper yellow than the exterior segments. —*Filament* flat. *Anther* two-lobed and crowned with an entire crest.—*Germ* three-celled, with many *ovula* in each, attached to the thick partitions, a little removed from the axis. *Style, stigma, and nectarial scales of König* as in the genus.

7. A. *aromaticum. R.*

Beng. Morung Elachi.

A native of the vallies on the eastern frontier of *Bengal*, where, as well as in the Botanic Garden near *Calcutta*, it blossoms during the hot-season, before the periodical rains begin; and the fruit ripens in September; the capsules are then carefully gathered by the natives, and sold to the druggists, who dispose of them for medicinal and other purposes, where such spices are wanted, under the name of *Morung elachi*, or *cardamon*; though the seed vessel of this species differs in form from all hitherto described sorts of this drug; however the seeds are similar in their shape and spicy flavour.

Root tuberous, running for a small extent at a small depth under the surface of the soil, &c. as in the order.—*Stems* in tufts, erect, or more or less oblique according to their situation in the tuft, those in the centre being nearly erect, while on the outside they bend out considerably, all are covered with the sheaths of the leaves, general height two to three feet; duration from one to two years.—*Leaves* bifarious, lanceolar, acuminate, smooth; from ten to twelve inches long, and from two to four broad.—*Spikes* radical, imbricated, at first clavate, afterwards becoming roundish as the fruit advances. *Scape* from one to three inches long, hid under the surface of the earth, and embraced by numerous imbricated, sheathing scales.—*Flowers* of a middling size, pale yellow.—*Bractes; exterior* one-flowered, oblong, concave, smooth. *Interior,* or inferior perianth, tubular, sheathing the germ, and lower half of the proper perianth.—*Calyx* superior, cylindric, entire, or dentate, villous—*Corol* with a long, slender tube; segments of the *exterior border* sub-lanceolar, obtuse, the upper one somewhat vaulted over the stamina, and stigma. *Lip* nearly round, and undivided, tinged with red down the middle; no spurs, or projections between it and the base of the filament.—*Filament* linear, incurved. *Anther* crowned with a three-lobed crest.—*Germ* villous, three-celled, with many *ovula* in each, attached to the axis.—*Capsule* ovate, size of a large nutmeg, rather fleshy; surface somewhat rugose, three-valved, three-celled.—*Seeds* several in each cell; oval, and obovate are the most prevailing forms.—*Perisperm* conform to the seed.—*Vitellus* nearly round, occupying a considerable portion of the upper part of the perisperm.—*Embryo* clavate, length of the seed, its apex passing through the centre of the vitellus.

8. A. *sericeum.* R.

Leaves lanceolar, sericeous underneath. Racemes radical, sub-globular, ramous. Lip obovate, base calcarate.

A native of the mountains on the North-East border of Bengal, where it is also called *Dooi Kesha.* It is a stately plant, with *leaves* about two feet long, and of the colour and appearance of highly polished silver underneath. The inflorescence a completely compound, sub-globular raceme with the short, villous, branches thereof two or three-flowered, each *flower* pretty long-pedicelled and imbraced during their early period with villous, oblong, membranaceous *bractes,* which soon perish; the calyx is more permanent, tubular, villous, and three-toothed.—*Flowers* large, white, centre of the lip variegated with pink veins on a yellow ground; *exterior border* of three, nearly equal, oblong, white segments, the upper one ends subulately from an obtuse apex; two large spurs between the base of the lip and that of the rather long filament; *anther* of two linear lobes, crowned with a very large, curled, semi-lunar crest.—*Germ* villous, nine-ribbed, three-celled, &c. as in the genus. *Capsule* long, narrow, ovate-oblong, three-celled, three-valved, surface even and smooth, colour a brownish red, inside deeper and brighter.—*Seeds* many, obovate, black, and pretty smooth, aril white and fleshy. *Embryo* simple, erect, length of the seed, and furnished with both perisperm and vitellus.

ZINGIBER.

Corolla with the interior border unilabiate. *Anther* double, crowned with a single, horn-shaped, curved beak. *Capsule* three-celled, three-valved. *Seeds* many, arilled. *Embryo* simple, and furnished with both perisperm and vitellus.

Sect. i. *Spikes radical.*

1. Z. *officinale.* *Roscoe in Trans. of Linn. Soc.* viii. 348.

Leaves sub-sessile, linear-lanceolate, smooth. Spikes elevated, oblong; bractes acute. Lip three-lobed.

Amomum *Zingiber. Linn. sp. pl. ed. Willd.* 1. 6.

Sans. आर्द्रकं, Ardr*u*k*u*m, शृङ्गवेर, Shr*i*ng*u*ver*u*m.

Arab. Zenjabil.

Beng. and *Hind.* Adrak*a*, or Ada. The dry root Sooth*a*.

Teling. Allam.

Inschi, sel Inschi kua. *Rheed. Mal.* 11. *p.* 21. *t.* 12.

Zingiber majus. *Rumph. Amb.* 5. *p.* 156. *t.* 66. *fig.* 1.

This plant is universally known, and cultivated over all the warmer parts of *Asia*; where to be found indigenous I know not. Flowering time the rainy season; but it is very rarely found in that state; nor have I ever met with the seeds.

Root tuberous, biennial.—*Stems* erect, and oblique, invested in the smooth sheaths of the leaves; generally three or four feet high, and annual.—*Leaves* sub-sessile on their long, vaginated sheaths, bifarious, linear-lanceolate, very smooth above, and nearly so underneath.—*Sheaths* smooth, crowned with a bifid stipulary ligula.—*Peduncles* radical, solitary, a little removed from the stems, from six to twelve inches long, enveloped in a few obtuse sheaths, the uppermost of which sometimes end in tolerably long leaves.—*Spikes*, oblong, the size of a man's thumb.—*Bractes*. *Exterior* imbricated, one-flowered, obovate; smooth, membranaceous-bordered, faintly striated lengthways. *Interior* solitary, enveloping the germ, perianth, and the greater part of the tube of the corol.—*Flowers* rather small, when compared with the rest of this natural order.—*Calyx* superior, one-leaved, opening on one side, apex three-toothed.—*Corol*, with a double border; *exterior* of three, nearly equal, oblong segments. *Lip* three-lobed, and of a dark purple colour. *Nectarial* filaments of Koenig large, subulate, fine-pointed.—*Filament* short. *Anther* oblong, double, crowned with a long, curved, tapering, grooved hornlet.—*Germ* oval, three-celled, with many ovula in each. *Style* filiform. *Stigma* funnel-shaped, ciliate, lodged just under the apex of the hornlet of the anther. This famous plant is too well known to require any further description.

2. Z. *zerumbet*. *Roscoe in Trans. of Linn. Soc.* viii. 348.

Stems declinate. Leaves bifarious, sessile, lanceolar. Spike long-peduncled, oval, compact, obtuse. Bractes broad-obovate, obtuse, margins coloured. Lip three-lobed.

Amomum *Zerumbet*. *Linn. Sp. Pl. ed. Willd.* 1. 6.

Katou-Inschi-Kua. *Rheed. Mal.* 11. *p.* 27. *tab.* 13.

Lampujum. *Rumph. Amb.* 5. *p.* 148. *tab.* 64. *fig.* 1

Zinziber *spurium*. *Kön. in Retz. obs.* 3. 60.

Beng. Butch, or mahaburee-butch.

Found wild in the woods about *Calcutta*. Flowering time the rainy season; the seeds ripen in November and December.

Root biennial, tuberous, as in ginger, externally whitish; internally pale yellow, and of a bitter aromatic taste.—*Stems* annual, oblique about three or four feet high, completely invested in the sheaths of the leaves.—*Leaves* bifarious, sessile, broad-lanceolate, entire, smooth, waved; veins numerous, slender, parallel; deep-green above, pale underneath.—*Sheaths* crowned with a large, broad membranaceous, emarginate ligula.—*Peduncle* solitary, a little removed from the stems, erect, from twelve to twenty-four inches long, enveloped in imbricated sheaths.—*Spikes* oval, obtuse. or with a round excavation on their apex, about the size of a goose-egg.—*Bractes* (or scales of the spike) the *exterior* one-flowered, imbricated, broad-obovate, obtuse, with a broad membranaceous border. *Interior* (or proper) solitary. one-flowered, enveloping the germ, perianth, and most of the tube of the corol.—*Flowers* large, pale, of a sulphur colour.—*Calyx* superior, one-leaved, sheathing the lower half of the tube of the corol.—*Corol*; *exterior border* three-parted, &c. as in the genus. *Interior*. or *lip* broad, with the lateral lobes thereof particularly large; middle division two-parted; they are all beautifully curled, and of a pale yellow colour. *Nectarial filaments* of *König* as in the order.—*Filament* scarcely any. *Anther* large, double, with its long, curved horn projecting over the lip. *Stigma* funnel-shaped, fringed with long, distinct hairs.

3. Z. *Cassumunar*. *R.*

Straight. Leaves sessile, linear-lanceolate. Spikes radical, strobiliform, oblong, acute. Bractes obovate, acuminate. Lip three-lobed.

Sans. वन आर्द्रकं, *Vuna Ardrukum.*

Hind. and *Beng.* Bun Ada, (wild ginger.)

Teling. Karpushpoo, (small turmerick.)

A native of *Coromandel*, *Bengal*, and *Bahar*. Flowering time July and August, seed ripe in November and December.

Root perennial. tuberous, furnished with long, white, fleshy fibres, and jointed like ginger, but much larger; when fresh of a deep yellow; possessing a strong, not very agreeable, camphoraceous smell, and warm, spicy, bitterish taste. *Stems* annual, erect, round, from three to five feet high, entirely enveloped in the imbricated sheaths of the leaves.—*Leaves* bifarious, approximate, sessile on their sheaths, linear-lanceolate, deep green above; villous, and paler underneath, where the nerve also is hairy like the sheaths; length from one to two feet, and about three inches broad.—*Sheaths* clothed with exceedingly pungent hairs on the outside; their mouths ending in a stipulary process on each side of the insertion of the leaf.—*Scapes* radical, from six to twelve inches long, invested in several, oblong, downy sheaths.—*Spikes* oblong, strobiliform, compact, sharp-pointed, closely imbricated with numerous obovate, acuminate, greenish ferruginous (where exposed), villous bractes.—*Flowers* solitary, (that is, one to each bracte of the spike, and blossoming in succession,) large, every part of a pale, uniform, sulphur colour.—*Calyx* double; the *exterior*, smooth, spathiform, three-dentate one, may be considered a bracte; it is inserted round the base of the germ, which, as well as the inner, or proper *perianth*, it entirely envelopes; when forcibly expanded, oval. *Inner* or *proper perianth*, superior, one-leaved, somewhat gibbous, membranaceous, three-toothed, on the outside divided nearly half way down.—*Corol. Tube* slender, as long as the scale of the spike, *upper division* of the exterior border boat-shaped, projecting over the stamen; *under* two narrower and reflexed under the lip. *Lip* three-lobed; lateral lobes obliquely obovate, erect; middle lobe nearly round, with the apex emarginate, and the border slightly curled.—*Stam. Anther* ending in a long, subulate, recurved beak.

4. Z. *roseum. Roscoe in Trans. of Linn. Soc.* 8. 348.

Leaves short-petioled, lanceolate. *Spikes* lax, half immersed in the earth. *Bractes* lanceolate, coloured. *Lip* entire.

Amomum *roseum. Corom. pl.* 2. *N.* 126.

Teling. Bumacatchicay.

Is a native of the northern *Circars*, where it flowers in the rainy season.

Root biennial, horizontal, jointed, and white; with many large fibres.

Stems several, invested in the sheaths of the leaves, &c. as in the other scitamineous plants, about two or three feet high, leaves included.—*Leaves* somewhat petioled, bifarious, lanceolate, acute, tending to one side, smooth, finely veined; about twelve inches long, and from three to four broad.—*Spikes* remote from the stems just appearing above the ground when in flower, imbricated with lax, lanceolate, reddish, one-flowered bractes.—*Flowers* solitary, colour a mixture of red and yellow.—*Bractes*, besides the exterior one already mentioned, there is an interior one for each flower between it and the rachis, which is linear, with an obtuse, notched apex.—*Calyx* one-leaved, tubular, above, membranaceous; *mouth* toothed.—*Corol*; *Tube* cylindric, erect, considerably longer than the bractes. *Border* double. *Exterior* three-parted; in this plant the two approximated divisions are above, linear, pointed, and recurved, sometimes revolute; the lower division stands single, and is like the superior two, linear and revolute; all are of a bright red colour. *Interior*, or *lip* single, broad, obtuse, reflexed over the superior two divisions of the exterior border.—*Filament*, and *anthers* as in the other species.—*Style* rather longer than the horn of the anther. *Stigma* perforated, and ciliate. Two similar awl-shaped nectarial bodies at the base of the style, are also present here.

Obs. I have not found that the natives make any use of this plant.

5. Z. *ligulatum*. *R.*

Leaves approximate, sessile, lanceolate. *Spikes* lax, more than half hid in the earth, obovate. *Bractes* cuneiform. *Lip* sub-hastate.

A native of *Hindoosthan*. Flowering time in the Botanic Garden the rainy season; the seeds ripen in November and December.

Root jointed, running horizontally at some depth under the surface of the earth, and furnished with long fleshy fibres from the joints; from these joints shoots rise, and by them the plant is readily propagated.—*Stems* about two feet high, bending considerably to one side.—*Leaves* alternate, approximate, bifarious, sessile on their sheaths, from cordate below to lanceolate toward the top of the plant, both sides smooth, and marked with parallel veins, the under side paler green, length from three to twelve inches, and breadth about four.—*Sheaths* smooth, from the mouth of each rises a remarkably large long strap, or ligula, which by the growth of the plant becomes bifid, and by age scariose.—*Spikes* radical, half hid in the earth, oblong, loosely imbricated with cuneiform, pink-coloured, one-flowered, exterior scales or bractes; and the same number of inner colourless, tridentate shorter bractes; or, as they may be called, inferior perianths.—*Calyx* superior, one-leaved, spathiform, about half the length of the tube of the corol, irregularly tridentate.—*Corol. Tube* slender, the length of the exterior scales of the spike, incurved. *Exterior border* of three, nearly equal, linear, acute, smooth, reddish divisions. *Lip* oblong, obtuse; margins much curled, with two expanding, more or less acute lobes at the very base; colour a pale yellow.—*Filament* short; below the two-lobed *anther*, linear, ending above in the usual long, taper, curved, grooved beak.—*Germ* oval, villous, three-celled, each containing many *ovula* attached to a central receptacle. *Style* very slender, and so long as to elevate the funnel-shaped, ciliate *stigma* to an equal height with the apex of the filament.—*Nectarial* scales long and slender, embracing the base of the style within the bottom of the tube of the corol.—*Capsule* ovate, size of a large olive, three-sided, three-celled, three-valved, opening from the apex down the angles; inside of the cells crimson.—*Calyx* leathery, striated, pale, dull yellowish straw colour.—*Seeds* many, oval, blackish brown; a little rugose, arilled. *Aril* white, nearly complete, and ragged at the upper end.—*Perisperm* conform to the seed, cinereous.—*Embryo* cylindric, central, nearly as long as the perisperm.

6. Z. *rubens*. R.

Leaves lanceolar. *Spikes* lax, half immersed in the earth. *Bractes* linear-lanceolate, straight. *Lip* narrow-ovate, entire.

A native of the interior of *Bengal;* from the district of *Rungpore* Dr. Buchanan sent the plants to the Botanic Garden, where they blossom in August.

Root tuberous &c. as in the other species.—*Stems* herbaceous, somewhat declinate, from two to four feet high.—*Leaves* alternate, bifarious, sessile on their sheaths, smooth on both sides; general length about twelve inches, and from four to five broad.—*Sheaths* with a large, membranaceous, stipulary process at the mouth.—*Spikes* several from the base of the stems, where they join the root, half immersed in the earth, sub-obovate, fastigiate, laxly imbricated with numerous straight, erect, linear-lanceolate, acute, involute, red, slightly villous, *exterior bractes*. *Inner bractes*, or inferior perianth the length of the tube, but shorter than the exterior bractes, and irregularly tri-dentate.—*Calyx* superior, sub-cylindric, membranaceous, pellucid, most slightly villous; *mouth*, three-toothed.—*Corol* with a long, slender, cylindric tube; segments of the exterior border linear-lanceolate, acute, red. *Lip* ovate-oblong, entire, speckled with red and yellow.—*Filament* scarcely any. *Anther* of two long lobes, crowned with the long, characteristic, curved horn, which is incumbent on and reaches to the apex of the lip.—*Nectarial filaments* of *König* oblong, and obtuse.—*Germ* hairy, three-celled, with many *ovula* in each, attached to the axis. *Stigma* funnel-shaped, and ciliate.—*Capsule* ......

7. Z. *squarrosum. R.*

Leaves lanceolar. *Spikes* squarrose, half immersed in the earth. *Bractes* linear, with long, taper, waved, recurved apices. *Lip* three-lobed, apex bifid.

A native of *Burma*, where it ripens its seeds in December. From *Rangoon* Mr. F. Carey sent the fresh roots, entire capsules, and ripe seeds to this garden, where the plants from both the seeds and roots, grow freely, and the latter blossomed in August.

Root tuberous, as in ginger, &c.—*Stems* herbaceous, considerably recurvate, from two to three feet high, entirely hid in the sheaths of the leaves.—*Leaves* bifarious, sessile, lanceolar, smooth above, slightly villous underneath, cuspidate.—*Sheaths* with two large conical processes rising from their mouths, one on each side.—*Spikes* from the base of the stems, three fourths hid in the earth, ovate, laxly imbricated, size of a goose-egg.—*Bractes* green; the *exterior* solitary, one-flowered, linear, concave, long, taper, waved, recurved. *Inner*, one or two, (generally one), as long as the tube of the corol, with a three-toothed apex.—*Flowers* pretty large, inodorous.—*Calyx* superior, &c. as in the other species.—*Corol* with a slender tube. *Exterior border* of three, pink-coloured, lanceolate, acute segments. *Inner*, or *lip* two-lobed at the base; apex bifid, colour a speckled mixture of purple, red, and yellow. *Filament* short, *anther* and beak together nearly as long as the lip.—*Germ* villous, three-celled, with many ovula in each. *Style* and *stigma* as in the genus.—*Capsule* short-pedicelled, ovate, conic, three-celled, three-valved, opening from the apex, inside of the valves bright scarlet while fresh. (Those sent from *Rangoon* by Mr. F. Carey, were, when they reached me, perfectly dry, and then measured three inches in length, and one in diameter.)—*Seeds* many in each cell, arilled, arranged in two rows, and attached to the inner edge of the partitions. *Aril* complete, fleshy, white.—*Perisperm* conform to the seed. *Vitellus* covers every part of the *embryo*, which is clavate, and nearly as long as the seed.

8. Z. *panduratum*. *R.*

Leaves petioled above their sheaths, broad-lanceolar, smooth; *ligula* large and scariose. *Spikes* radical, half immersed in the earth, lax. *Bractes* lanceolate. *Lip* panduriform, with oval, emarginate lamina.

This middling sized species is a native of the country about *Rangoon*. From thence Mr. F. Carey sent seeds to the Botanic Garden in 1808, and in July, 1810, the plants raised from them blossomed abundantly.

Root; numerous, long, fleshy, fibrous fibres, some of which swell into lanceolar tubers, as in *Curcuma*; no palmate tubers, as in that genus. It possesses but little taste and no fragrance.—*Stalks* erect, herbaceous, about three feet high.—*Leaves* bifarious, petioled above their sheaths, broad-lanceolar, acute, smooth on both sides; from six to twelve inches long, by three to four broad.—*Sheaths* smooth above their respective leaves; each ends in a very long, scariose, smooth ligula.—*Spikes* radical, the lower half immersed in the soil, oblong, and rather loosely imbricated with lanceolate, coloured, concave, one-flowered, interior and exterior bractes.—*Flowers* rather small, *exterior border* of three, nearly equal, lanceolate, acuminate, pale red segments; *inner* or *lip* very pale yellow, panduriform, that is with two, pretty large, rounded *lobes* at the base, and the *lamina* nearly oval, retuse, truncate, or emarginate.—*Filament* broad, and short. *Anther* double, and crowned with the usual beak.—*Germ* three-celled, with many *ovula* in each, attached to the inner angle of the cell. *Stigma* infundibuliform.

Observation by N. W. The Burman name is Toukta.—Capsule ripening in November and December, ovate, triangular, acute, with flattened sides, smooth, pale, retuse at the base, and surrounded with numerous withered bractes, as large as a walnut.—Valves thick and fleshy, three-cornered, inner surface waved, shining, scarlet; skin leathery; flesh spongy, pallid.—Dissepiment oblong, ventricose, triangular, scarlet. at the bursting of the capsule adhering to the inner angle of one of the valves.—Seeds obovate, horizontal, in each cell about twelve, attached by small white fleshy placentas to the sides of the dissepiment, faintly aromatic.—Arillus milk-white, leathery, loosely surrounding the seed, perforated at its top with a small toothed opening.—Integument simple, dark-purple, smooth and shining, porous within and replete with a violet-coloured juice: it enters the umbilicated base of the seed and receives the apex of the radicula.—Albumen white, amygdaloid, conform to the seed and occupying chiefly its lower half, sub-obliterate above.—Vitellus white and fleshy, turbinate, constituting almost entirely the upper part of the seed, ending below in a very thin sheath which covers the substance of the embryo.—Embryo cylindric-ventricose, as long as the seed, erect.—Cotyledon sub-cylindric, almost naked.—Radicula thick, somewhat enlarged at its apex, centripetal.

Sect. 2. Spikes terminal.

9. *Z. capitatum. R.*

Herbaceous, *leaves* linear lanceolate, stem-clasping *Spike* terminal, erect, strobiliform, *bractes* lanceolate.

A native of the forests of *Rohilcund*, from thence Colonel Thomas Hardwicke sent plants to the Botanic Garden, under the vernacular name *Junglee Adruk*. Here it flowers during the rains; and the seeds ripen in September and October.

Root composed of many, long peduncled, oval, smooth, pale-coloured, insipid, watery tubers, issuing from the base and sides of the bulb, and palmate tubers, which are of a pale yellow colour, and somewhat spicy taste.—*Stems* erect, round, enveloped in the sheaths of the leaves, which are often a little hairy; height of the whole plant when in blossom, from two to three feet.—*Leaves* sub-bifarious, alternate, sessile on their sheaths, narrow-lanceolate, acute, smooth above, a little hairy underneath, about twelve inches long and about two broad.—*Spike* terminal, sessile on the mouth of the sheaths of the superior leaves, linear-oblong, somewhat strobiliform, many-flowered.—*Flowers* sessile, large, pure yellow in every part.—*Bractes* in pairs, one-flowered, permanent; *exterior* lanceolate, concave, erect, longer than the tube of the corol. *Inner* ovate, shorter, immediately enveloping the germ, calyx, and nearly the whole of the tube of the corol.—*Perianth* superior, one-leaved, about half the length of the tube of the corol, membranaceous, white, two or three-toothed.—*Corol. Tube* cylindric, narrow. *Lip* with two, oblong, very distinct, expanding, lateral lobes; *middle lobe* very large, and nearly round. *Anther* with a very long curved beak.—*Stigma* (as in most plants of the order), funnel-shaped, with ciliate margins.—*Nectarial* scales (of *König*), or hornlet conical.—*Capsule* oval, three-sided, with the angles rounded, pretty smooth, size of a small olive, three-celled, three-valved; each cell containing several, black, shining seeds, enveloped in a deeply jagged, white aril.—*Perisperm* conform to the lower half of the seed which it occupies, pure white, cartilaginous. *Vitellus* above the perisperm, occupying nearly the upper half of the seed.—*Embryo* vertical, passing through the centre of both perisperm and vitellus, sub-cylindric, both ends o[illegible]se, &c. in every respect so exactly like *Z. Cassumunar*, that the same drawings and description may well serve for both.

10. Z. *marginatum*. *R.*

Leaves sessile, lanceolate; *ligula* large and membranaceous. *Spike* terminal. *Bractes* obovate, with transparent, membranaceous margin. *Nectary* three-lobed.

11. Z. *elatum*. *R.*

Leaves bifarious, linear, recurved. *Spikes* terminal, lanceolar, compactly imbricated with lanceolate, one-flowered bractes.

A native of the interior of *Bengal*; in the Botanic Garden it blossoms in July and August, and the seeds ripen in September and October.

Root tuberous, as in common ginger, but inwardly yellow; from the tubers proceed the proper roots, many of which end in oblong, paler coloured tubers, as in *Curcuma*.—*Stems* herbaceous, straight, from four to five feet high, including the spike.—*Leaves* bifarious, linear, recurved, smooth above; with many adpressed, short, soft, white hairs underneath; from twelve to eighteen inches long, and only one inch broad. *Sheaths* a little hairy; *mouth* crowned with a very narrow, membranaceous process.—*Spikes* terminal, solitary, of a narrow lanceolar shape, six inches long, and three in circumference, every part compactly imbricated with lanceolate, one-flowered bractes, which are an inch and a half long, and a little hairy on the outside.—*Bractes*; *interior*, or *inferior perianth* shorter and broader than the *exterior*, but like them stiff, and a little hairy.—*Calyx* superior, white, smooth; *mouth* unequally toothed, little more than half the length of the bractes.—*Corol* exactly as in the other species, only considerably longer; the colour a tolerable bright, lively yellow.—*Stamina* with a short filament, and very long curved beak crowning the linear two-lobed *anther*.—*Nectarial* bodies subulate.—*Germ* oval, hairy, three-celled, with numerous *ovula* in each, attached to the axis. *Stigma* ciliate.—*Capsule* oval, size of a small olive, three-celled, three-valved.—*Seeds* numerous, lucid, black, enclosed in a large white ragged aril.—*Embryo* simple sub-lanceolate, base truncate, embraced by the perisperm; the upper half by the *vitellus*.

COSTUS. Schreb. gen. No. 4.

Corolla with the interior border of one, campanulate, back-cleft lobe. *Filament* lanceolate, with the *double anther* attached far below its apex. *Capsule* three-valved, three-celled. *Seeds* numerous, arilled. *Embryo* simple, and furnished with both perisperm and vitellus.

1. C. *speciosus. Smith in Trans. of Linn. Soc.* 1. 240. *Linn. Sp. Pl. ed. Willd.* 1. 11.

Leaves sub sessile, spirally arranged, oblong, cuspidate, villous underneath.

Tjana-Kua. *Rheed. mal.* 11. *p.* 15. *f.* 8.

Tsiana *speciosa. Gmel.* 9.

Herba spiralis hirsuta. *Rumph. amb.* 6. *p.* 143. *t.* 64. *f.* 1.

Sans. केमुक :, Kemooka.

Hind. and *Beng.* Keoo.

Banksia *speciosa of Dr. König, see Retz. obs.* 4. 75.

This is one of the most elegant looking plants of the natural order *Scitaminea.* It is a native of moist shady places, such as the low banks of water courses, &c. Flowers during the wet season.

Obs. König's description agrees so well with my plant, that I have scarcely any thing to add: the chief differences are,

1*st.* Mine is generally from three to four feet high, ascending spirally, with the leaves spirally disposed up and round the stem.

2*d.* In this the spike is at all times erect, and rigid; from about three to six inches long.

3*d.* There is nothing like a crown to the *anthers;* and

4*th.* The stigma consists of two, broad, glutinous lips, which gape wide while the pollen is shedding; between them is evidently seen with the naked eye, the mouth of the perforated *style;* on the back of the upper lip are seen the two white glands mentioned by König. I impute his not taking notice of the two large lips of the *stigma,* to his having examined the flowers sometime after they were gathered, when they coalesce.

The fresh roots are almost insipid. The natives make a preserve of them, which they deem wholesome, and nutritious; the dry root does not at all resemble the *Costus arabicus* of the shops.

ALPINIA. *Schreb. gen. No.* 5.

Corolla with the interior border unilabiate. *Anther* double, naked. *Capsule* berried, three-celled. *Seeds* few, or many, arilled. *Embryo* simple, and furnished with both perisperm and vitellus.

Sect. I. *Inflorescence terminal.*

1. A. *Galanga. Linn. Sp. Pl. ed. Willd.* 1. 12. *Roscoe in Trans. of Linn. Soc.* 8. 345.

Perennial. *Leaves* sessile, broad lanceolar. *Panicle* terminal. *Lip* oblong, unguiculate; *apex* bifid. *Capsule* obovate, smooth, seeds few.

Galanga *major. Rumph. amb.* 5. *t.* 63.

Sans. कुलंयोगः, Koolunyog*a*, धमूलः, Dhumool*a*, तीक्ष्णमूलः, Teekshn*a*-mool*a*, कुलञ्जनः, *Koolunjuna*, सुगन्धा, Soog*u*ndha, महाभराबचा, M*u*habh*ura*-*vu*cha.

Hind. Koolinjan.

Arab. Kholinjan or Khoolunjan.

Dr. Charles Campbell, at *Bencoolen*, sent the roots of these species from thence to the Botanic Garden near *Calcutta*, where the plants thrive remarkably well, and are in blossom during the latter half of the hot season, April and May. Seeds ripen, though rarely, in November.

Root perennial, tuberous as in ginger, &c. possessing a faint aromatic smell, and strong pungent taste, like a mixture of pepper and ginger, accompanied with some degree of bitterness, and now ascertained to be the real *Galanga major* of the shops.—*Stems* perennial, or at least more durable than those of herbaceous plants, nearly erect, round, smooth; general length, when in flower, about six or seven feet; and as thick as a slender walking cane. *Leaves* occupy from the middle of the plant upward; the lower half being inverted in leafless sheaths only: they are short-petioled, bifarious, lanceolar, margins white and somewhat callous, both surfaces smooth; length from twelve to twenty-four inches, and from four to six broad.—*Sheaths* smooth, with a short, rounded, ciliate, stipulary process above the insertion of the leaf.—*Panicle* terminal, erect, oblong, crowned with numerous, spreading, simple dichotomous branches, each supporting from two to five, or six, pale greenish white, somewhat fragrant flowers.—*Involucre* of König, in our plants generally two-leaved, lanceolate, first withering, then falling before, or while in blossom.—*Bractes* concave, solitary at the divisions, and subdivisions of the panicle.--*Calyx* superior, sub-cylindric, smooth, white, scarcely the length of the tube of the corol, *mouth* one toothed.—*Corol. Exterior border* of three, nearly equal, linear, recurved, smooth, pale greenish divisions. *Inner border*, or lip unguiculate, somewhat ascending, so as nearly to meet the slightly declining anther; oval or ovate oblong, concave, apex divided into two lobes by a pretty deep, straight fissure; margins incurved, curled, minutely laciniate, white, with here and there a small reddish speck, and two recurved fleshy, coloured hornlets at the base of the claw.—*Filament* rather longer than the claw of the lip. *Anther* double, apex deeply emarginate, projecting horizontally over the middle of the lip.—*Germ* beneath, smooth, oval, three-sided, three-celled, with only two *ovula* in each cell, attached to the middle of the partition. *Style* filiform. *Stigma* funnel-shaped, fringed; while the flower is in vigorous blossom erect between the lobes of the anther, afterwards recurved under the anther.—*Nectarial* bodies (of König) nearly square, fluted.—*Germ* oblong, smooth, three-celled, with two *ovula* in each, attached to the axis.—*Capsule* the size of a small cherry, obovate, smooth, deep orange red, three-celled, but entire, as no separation seems to exist, the thin dry cortex crumbles, or breaks to pieces, without any tendency to separate into valves.—*Seed* one, rarely two in each cell, much compressed, size of a grain of black pepper. *Aril* nearly complete, there being only an opening at the apex, thin, rather fibrous, and of a dull white color. *Integuments* two, or three; *exterior*, tough, fibrous, polished, of a dark chesnut colour, a little wrinkled, *inner* hard, dark brown; between the two is a spongy lamina of the same colour, and I cannot well say to which it most properly belongs.—*Perisperm* white, friable, and very hard.—*Embryo* simple, dull white, (canus,) roundish in the body of the perisperm, and with the narrow, conic *radicle,* passing through it and pointing to the umbilicus.

Note by Mr. Colebrooke. "The root of this plant being ascer-"tained to be the *Galanga major* of the druggists, comformably with "what has been said of it by *Rumphius,* there is no doubt of its being "*Kholinjan* of the Arabs, termed in *Hindee Culinjan.* In Sanskrit it "is called *Culanjana,* according to one authority, (the *Rajnighantu* ;) "but *Sugandha*-vacha, sweet-scented Acorus; as also *Mahabhari*-"*vacha,* according to another Indian treatise, (the *Bhava-pracasa.*) "If the first name be genuine Sanskrit, which is however doubtful, "the similar names in other languages, including the European, must "be derived directly, or immediately from it."

2. A. *Allughas. Roscoe in Trans. Linn. Soc.* 8. 346.

Leaves lanceolar, polished. *Panicle* terminal. *Lip* bifid ; lobes retuse. *Capsules* spherical, (black,) one-valved, many-seeded.

Hellenia *Allughas. Linn. Sp. Pl. ed. Willd.* 1. 4.

Mala-inschi-kua. *Rhéed. mal.* 11. *p.* 29. *t.* 14.

Sans. तारक :, Tar*u*k*a*.

Beng. Tar*u*k*a*, or Tara.

This plant is common near *Calcutta,* in low moist places, among brush wood. Flowering time the beginning of the rainy season.

Root tuberous, biennial, or perennial, horizontal, brownish, with many pretty thick fibres, considerably aromatic.—*Stem* erect, slightly compressed, entirely covered with the sheaths of the leaves; from three to six feet high.—*Leaves* alternate, bifarious, petioled, oblong, ending in a fine twisted point ; both sides smooth, and glossy; the under side paler colour; veins numerous, straight, exceedingly fine, going off at an acute angle ; generally about a foot long, and four inches broad. —*Petioles* sheathing with a stipulary rounded process rising above the insertion of the leaf.—*Panicle* terminal, bending slightly to one side. —*Flowers* numerous, in succession, large, of a beautiful rose colour, inodorous.—*Calyx* superior, gibbous, fleshy, generally two or three-toothed.—*Coral* one petalled. *Tube* short. *Border* double; the *exterior* three-parted; *divisions* nearly equal, outwardly downy, 'concave, particularly near the ends, pale greenish rose colour; the superior one more remote. *Interior* border, or *lip* two-cleft, with two fleshy protuberances near the base; *divisions* with two-toothed apices and waved margins, colour a beautiful mixture of deep, and pale rose.—*Nectary* (of *König and Retz.*) two oblong, compressed, fleshy bodies, embracing the base of the style.—*Filament* short, broad, much flattened. *Anther* double, each lobe ending in an obtuse hornlet.—*Germ* inferior, globular. *Style* lodged in the grove of the filament. *Stigma* clubbed, somewhat three-sided, obtuse, with a crescent-like fissure, and ciliate margin.—*Capsule* globular, inclining to be three-sided, smooth, when ripe black, and friable, not opening by sutures, but crumbling to pieces, crowned with the permanent, withered calyx, and coral, three-celled.

3. A. *bracteata. R.*

Leaves lanceolar, villous underneath. *Raceme* terminal, simple; pedicels, one-flowered. *Bractes* green.

A native of the Eastern parts of *Bengal.* In the Botanic Garden near *Calcutta* it flowers in March. This is one of the smallest of our East India *Alpinias;* this circumstance, and the green bractes, immediately distinguish it from *malaccensis,* which is not so easily done in a specific definition.

Root biennial, or perhaps, perennial.—*Stems* erect, about three feet high, entirely invested in the smooth sheaths of the leaves.—*Leaves* bifarious, petioled, from ovate-oblong below, to lanceolar above, fine-pointed; the upper surface smooth, polished, deep green, villous and paler underneath; length from six to eighteen inches.—*Petioles* proper about an inch long, and deeply channelled. *Sheaths* smooth, rising a little above the insertion of the proper petioles on the inside.—*Raceme* terminal, erect, simple, about a span long, slightly villous; before expansion embraced, by one or two caducous sheaths.—*Pedicels* throughout one-flowered, round, short, and villous.—*Bractes* (or inferior *perianth*) solitary, one-flowered, more permanent than in the other species, and of a dull green colour, hiding the calyx.—*Calyx* shorter than the bracte, open on the underside, bidentate.—*Corol; exterior border* of three oval, nearly equal, white segments; *lip* somewhat three-lobed; lateral lobes large and incurved into a tube round the stamina; exterior narrower, with a bifid apex, colour a beautiful mixture of crimson and yellow, with a paler belt on each side.—*Filament* short, with two curved spurs at its base. *Anther* double, linear.—*Germ* oval, three-celled ; with many ovula in each, attached to the inner angle. *Stigma* funnel-shaped, sub-ciliate.

4. A. *malaccensis. Roscoe in Trans. Linn. Soc.* 8. 345.

Leaves lanceolate, petioled, villous underneath. *Raceme* terminal, simple. *Lip* obscurely three-lobed, with two spurs at the base, lateral lobes incurved. *Capsules* obliquely spheriodal.

Maranta *malaccensis. Linn. Sp. Pl. ed. Willd.* 1. 14.

Galanga *malaccensis. Rumph. amb.* 5. *p.* 176. *t.* 71. *f.* 1.

A most beautiful and stately plant, a native of *Chittagong*, and from thence sent, by Mr. W. M. Maddern, to the Botanic Garden near *Calcutta*, where it blossoms in April and May, the very hottest time of the year; and ripens its seed in the rains.

Root perennial.—*Stems* numerous, (within four years, a single root encreased so much, as to produce about fifty,) from six to ten feet in height; those in the centre erect, in the circumference bending out; in two, three, or four years, the stoutest blossom, ripen their seed and decay.—*Leaves* bifarious, petioled, lanceolate, acute; margins often waved, and slightly fringed with short brown hairs; upper surface smooth; under downy; length from two to three feet, and from three to eight or nine inches broad.—*Sheaths* smooth, embracing the stems completely, *ligula* ovate, obtuse, entire, villous. *Petioles* (I mean the space between the ligula and leaf), about three inches long and channelled.—*Racemes* terminal, solitary, erect, always simple; from six to twelve inches long. *Peduncle* round, and villous. *Pedicels* short, villous, one-flowered, the largest sometimes droop a little from the weight of the flowers.—*Involucre* of two, or three, caducous, boat-shaped leaves embracing the raceme.—*Flowers* very large, a pure smooth shining white, except the inner border, or labium, which is a beautiful mixture of orange and crimson.—*Bractes,* (or inferior calyx), gibbous, bursting on one side to the base, and partially on the other.—*Calyx* superior, one-leafed, gibbous, length of the bracte, splitting into various portions.—*Corol; exterior border* of three, oblong, obtuse, entire segments, of which the upper is largest. *Inner border* or *labium,* very large; when expanded, three inches broad, and two and three quarters long, somewhat three-lobed, the two lateral semi-circular and incurved into a tube; the middle one smaller, more projected, curled, and sometimes bifid; from the base on each side an incurved, sharp spur.—*Filament* about as long as the large, broad, deeply bifid, obtuse, two-lobed *anther.*—*Nectary* of König; in most of the other plants of this order, I have found this part to consist of two distinct bodies here it is of one, very thick, and short with its truncated, apex tuberculate.—*Germ* inferior, obovate, hairy, three-celled, with many *ovala* in each, attached to central receptacles. *Style* toward the apex hairy. *Stigma* rising erect through the bifid apex of the anther, funnel-shaped, hairy.—*Capsule* obovate, spheroidal, size of a very large gooseberry, armed with short, sharp, stiff, slender bristles; when ripe nearly dry, and yellow, three-celled, three-valved.—*Seed* numerous, ovate, oval, or obovate, immersed in a large portion of a soft pulpy matter, which in the immature state form distinct arils; on each side, and round the apex a groove. *Integuments* two, brown, and rather thin.—*Perisperm* cartilaginous, cloven at both ends by a brown spongy fissure, but deepest from the apex.—*Vitellus,* the centre of each of the superior limbs of the perisperm, is occupied by a small, thin, oblong body of a clearer colour; and widely distinct from both embryo and perisperm. —*Embryo,* the upper part crescent-shaped, with its horns projecting into the limb of the perisperm, and in some degree into the vitellus. *Radicle* inferior, projecting straight from the convex, or under side of the crescent, with its base to the umbilicus.

Obs. This is the most beautiful plant of the whole genus, even surpassing *A. nutans,* which I believe the Botanical Repository considers a species of *Renealmia.*

5. A. *nutans. Roscoe in Trans. of Linn. Soc.* 8. 346.

Leaves lanceolar, short-petioled, smooth. *Racemes* compound by the lower pedicels being two or three-flowered, drooping. *Lip* broad, three-lobed, the lateral incurved into a tube; the exterior curled and bifid. *Capsules* spherical, opening on the sides. *Seeds* few.

Beng. Poonnag-champa.*

Renealmia *nutans. Andr. Bot. Repos.* 5. *t.* 360.

Globba *nutans. Linn. Sp. Pl. ed. Willd.* 1. 15. 3.

This very elegant plant, a native of the Eastern Islands, was introduced by the Rev. Dr. Carey, into the Botanic Garden near *Calcutta,* where it blossoms in April and May.

Root perennial, tuberous, &c. as in the other species.—*Stems* more or less bent to one side, (a direction common to most of the plants of this order;) from four to six feet high.—*Leaves* bifarious, subsessile on their sheaths, occupying the upper half of the stems; the lower are broad lanceolar, the upper much narrower, all are entire, and remarkably smooth on both sides; while young the margins villous; length from one to three feet.—*Sheaths* of the leaves smooth, with a stipulary bearded process, rising above the insertion of the leaf.—*Racemes* terminal, compound, variously curved, villous, about a foot long, drooping under the leaves.—*Involucre* of generally two, caducous, lanceolate, boat shaped sheaths.—*Flowers* numerous, large.—*Pedicels* two or three-flowered.—*Bractes* large, concave, pure white, with a pink-coloured apex, smooth; a common one to the two or three flowers, and a proper, smaller one to the second and third

* From this name, which Dr. Roxburgh mentions as the Bengal name, it is natural to suppose this plant to be the *Poonnaga* of the *Sungskrita* Materia Medica. If so the following *Sungskrita* Synomina should stand before the Bengalee name, পুন্নাগ, *Poonnaga,* পুরুষঃ, *Pooroosha.* তুঙ্গঃ, *Toonga,* পুমান্, *Pooman,* পাটলঃ, *Patula,* রক্তপুষ্প, *Rukta-pooshpa,* পুন্নামা, Poonnama, রক্তরেণু, *Rukta-renoo,* অরুণঃ, *Uroona.* It is however doubtful whether this plant is the *Poonnaga. Ed.*

flowers, which expand in succession for a long time after the first. —*Calyx* superior, colour of the bractes, three-toothed.—*Corol* with a double border; *exterior* three-parted, pale pink, with deeper rosy apices; upper division oval, incumbent; lower two linear-oblong, revolute, with the apices rounded. *Inner* lip large, ovate-cordate, obscurely three-lobed, lateral lobes incurved into a large companulate tube; apex of the interior lobe bifid, margins curled; at the base are two spurs which embrace the base of the style, colour of the whole a mixture of deep orange yellow, with a dark crimson ground, and veins.—*Germ* beneath, oval, hairy, three-celled, with many *ovula* in each, attached to the inner edge of the partitions. *Stigma* on the first expansion of the flower erect through the fissure of the apex of the anther; afterwards, as the *style* lengthens, its apex becomes involute, and throws the *stigma* under the anther.—*Nectaries* (of König) two, square, fluted, fleshy bodies within the tube, on the apex of the germ, embracing the base of the style.

Obs. This is no doubt one of the most gaudy of our East *Indian Scitamineæ.* It multiplies itself by the root very fast, grows quickly, and continues to produce its numerous, large, elegant, drooping panicles, of beautiful flowers for two months, and more or less the whole year round.

6. A. *mutica. R.*

Leaves short-petioled, narrow-lanceolar, polished. *Racemes* erect, compound. *Lip* three-lobed, base spurless. *Capsule* buried. *Seeds* numerous, angular. *Aril* evanescent.

Found indigenous in the forests of *Prince of Wales* Island by Mr. W. Roxburgh, and from thence introduced into the Botanic Garden near *Calcutta,* where it flowers about the beginning of the rains, and ripens its seed from October till January.

Stems numerous, erect, involved in the sheaths of the leaves, from four to six feet high, generally biennial, as they do not perish till after the seed is ripe.—*Leaves* alternate, bifarious, short-petioled, narrow-lanceolar, acute, polished, and entire, from ten to twenty-four inches long. *Sheaths* smooth, ending in a rounded ligula, above the insertion of the leaf.—*Racemes* terminal, erect, compound, conical, many-flowered: peduncles a little downy. *Pedicels* alternate, short, villous, two, three, or even four-flowered.—*Bractes* of the pedicels, and pedicelli solitary, oblong, caducous.—*Flowers* numerous, large, drooping.—*Calyx* longer than the tube of the corol, widening from the base; mouth irregularly three-toothed, with the lower fissure deepest, pure white. *Corol. Tube* much recurved, much shorter than the calyx; border double. *Exterior* three-parted; *superior divisions* incumbent, oval, concave, *inferior* two smaller, linear oblong, narrower than the *superior* one; all are obtuse and of a pure glossy white colour. *Lip* or *inner border*, large, elegantly variegated with crimson and yellow, surrounded with an orange-coloured edging; three-lobed; the two lateral lobes simi-lunar and incurved into a large sub-campanulate tube under the upper segment of the *exterior* border; exterior lobe smaller, curled, and bifid. In this species there are no spurs between its base and the filament, as in most of the other species of this genus.—*Filament* nearly as long as the broad double *anther*, which projects over the middle of the lip.—*Germ* round, villous, three-celled, with many *ovula* in each attached to the partitions near the axis. *Style* filiform, embraced at the base by a single, truncate, dentate, glandular body, (nectary of König, Retzius, and Willdenow.) *Stigma* funnel-shaped, ciliate, rising with a curve through the bifid apex of the anther.—*Capsule* berried, size of a large gooseberry, deep orange, or yellow, according to their state of maturity; not opening spontaneously as in A. nutans, when dry, wrinkled, and a little hairy.—*Seeds* numerous, angular, arilled. *Aril* complete, but thin, and when perfectly dry scarcely traceable. *Integuments* two; exterior thin, inner dark brown, and spongy.—*Perisperm* cartilaginous, a deep pit at the base, and a deep cleft from the apex down to near the middle. *Vitellus* as in *Alpinia malaccensis*, in the centre of each of the superior lobes of the perisperm, is a somewhat clearer, distinct body, as if a continuation of the horns of the crescent of the embryo, but perfectly distinct, and readily separable from them, and the perisperm.—*Embryo* an invested crescent, with the cylindric radicle descending from the convex side through the pit at the base of the perisperm to the umbilicus.

Obs. This is also a very elegant species, and holds a middle rank between *nutans* and *calcarata*.

7. A. *calcarata*. *Roscoe in Trans. of Linn. Soc.* 8. 347.

Leaves linear-lanceolar, polished. *Spike* compound, erect. *Lip* ovate-oblong, apex curled, and bifid.

Renealmia *calcarata*. *Andr. Bot. Repos.* 6. *t.* 421.

A native of *China*, and from thence introduced by Captain J. Garnault, into the Botanic Garden near *Calcutta* in 1799; where it thrives luxuriantly in the common soil of the garden, and blossoms in March, April, and May, but more or less all the year.

Root perennial, horizontal, somewhat woolly, crooked, pale-coloured, stoloniferous, fragrant, enveloped while young in short, alternate sheaths; when old marked with their cicatrices; and furnished with many fleshy ramous fibres.—*Stems* oblique, smooth, from two to four feet high.—*Leaves* alternate, bifarious, short-petioled on their sheaths, narrow-lanceolar, fine-pointed, entire, smooth on both sides; length frequently twelve inches, and the breadth from one to two.—*Sheaths* of the leaves stem-clasping, smooth, with a stipulary process projecting above the insertion of the leaf.—*Racemes* terminal, solitary, erect, compound.—*Flowers* numerous, large, generally in pairs or more, but expanding at different periods.—*Bractes* solitary, boat-shaped, white, one-flowered.—*Perianth* superior, three-toothed, white, smooth.—*Corol* with double border. *Exterior* three-cleft; *divisions* equal, linear, expanding, pellucid, white. *Interior*, or *lip* ovate-oblong, large, deeply coloured with dark purple veins on a yellow ground, apex two-lobed, lobes rounded and curled; there are no lateral lobes in this species, nor are the sides much incurved; but two, incurved, sharp, subulate horns project from its two posterior angles and embrace the base of the filament.—*Filament* single, opposite to the lip of the corol, projecting, &c. as in the genus.—*Germ* spherical, hairy, three-celled, with many *ovula* in each, attached to the inner angle of the cells. *Stigma* funnel-shaped, ciliate.—*Nectarial* scales nearly four-sided, with apices crenulate.

8. A. *spicata*. *R.*

Spike terminal, oblong, compactly imbricated with narrow lanceolate, acute bractes.

A native of *Sumatra*, from thence introduced by Mr. Roxburgh into the *Calcutta* Botanic Garden in 1803. Is the slowest growing and smallest of the genus in this garden. Flowering time the close of the rains.

Root as in the other species.—*Stems* arising in an oblique direction to the height of about two feet, when from two to three years old the flowers appear, and after their functions are performed, the stems decay.—*Leaves* bifarious, sessile on their sheaths, broad lanceolar, fine-pointed, surfaces and margins smooth, length from six to twelve inches.—*Sheaths* of the leaves smooth, ending in a long, bifid, or simple ligula.—*Spikes* terminal, strobiliform, compact.—*Bractes*, the *exterior* imbricate, ensiform-lanceolate, one-flowered, outside slightly villous. *Interior* or *inferior calyx*, one-leaved, subcylindric, length of the perianth proper, mouth obliquely truncate, outside villous.—*Calyx* superior, like the inner bracte, and of the same length, &c.—*Corol*, *exterior border* of three, very acute, linear lanceolate segments. *Inner* or *lip* uncertain, as the only two spikes I have yet seen did not expand sufficiently, to allow me to ascertain its form.—*Filament* short, anther large, of two, long linear lobes with the style sunk between them.—*Germ* oval, very hairy, three-celled, with numerous *ovula* in each, attached to the axis. *Style* slender. *Stigma* large, consisting of two, gaping unequal lobes.

Sect. 2. Inflorescence radical.

9. A. *cardamomum*. *R.*

Scapes from the base of the stem, compoundly flexuose, procumbent. *Lip* three-lobed, with calcarate base.

Amomum *repens, seu Cardamomum. Wood, v.* 2. *p.* 365. *t.* 131.

Amomum *repens. Roscoe, Willd. and Sonnerat.*

Elettaria. *Rheed. mal.* 11. *p.* 9. *t.* 4 *and* 5.

Cardamomum minus, Pharm. Lond. and Edinburgh.

Sans. पृथ्वीका, Prithweeka, चन्द्रवाला, Chundruvala, एला, Ela, निष्कुटि, Nishkooti, बहुला, Buhoola.

Beng. Elachi, Elaich.

Ailum chedy of the Malabars, of the Malabar Coast.

Hil of the Persians.

Kakulah, also Hal, of the Arabs.

Teling. Sana-yallacci.

Cardomum. *Buch. Journey,* 2. 336—510, 538, *and* 3. 225.

A native of the mountainous part of the *Coast* of *Malabar.*

Root tuberous, with numerous fleshy fibres.—*Stems* pérennial, erect, smooth, joined, enveloped in the spongy sheaths of the leaves; from six to nine feet high.—*Leaves* bifarious, sub-sessile on their sheaths, lanceolate, fine-pointed, somewhat villous above; and sericeous underneath, entire; length from one to two feet.—*Sheaths* slightly villous, with a rounded stipulary process rising above the mouth.—*Scapes* several (three or four) from the base of the stems, resting on the ground, flexuose, jointed, ramous, from one to two feet long. *Branches* or *Racemes* alternate, one from each joint of the scape, sub-erect, two or three inches long.—*Bractes* solitary, oblong, smooth, membranaceous, nerved, sheathing, one to each joint of the scape, which embrace the insertion of the raceme, or branch; and one at each of their joints.—*Flowers* alternate; short-pedicelled, solitary at each joint of the racemes, opening in succession for a length of time, as the racemes lengthen.—*Calyx* above, widening to the three-toothed mouth, about three quarters of an inch long, striated with fine nerves, permanent.—*Corol* withering. *Tube* slender, as long as the calyx. *Border* double, exterior of three, oblong, concave, nearly equal, pale greenish white divisions. *Inner lip* or *nectary* obovate, much longer than the divisions of the exterior border; margins somewhat curled, with the apex slightly three-lobed, marked chiefly in the centre, with purple violet stripes; at each side of its insertion and close by the base of the filament is a small acute hornlet, as in most of the plants of this genus, and in several of the other genera of our other *Indian* Scitamineœ.—*Filament* short, erect. *Anther* double, emarginate.—*Germ* beneath, oval, smooth. *Style* slender. *Stigma* funnel-shaped.—*Nectarial* scales of König, subulate, almost half the length of the tube of the corol.—*Capsule* oval, somewhat three-sided, size of a small nutmeg; three-celled, three-valved.—*Seeds* many, angular.

Obs. The following account of this interesting plant has lately been transmitted to me by Captain Dickson, the commanding officer of the district of *Wynaad* where the *Cardomum* is cultivated, viz.

" The *Cardomum* shrub is found in great abundance among the western mountains of *Wynaad,* and is called by the natives of Malabar *Ailum chedy* (the *Ailum shrub*). I cannot obtain any satisfactory derivation of the true import of the word *Ailum,** unless, as is alleged, it implies in the Sanskrit language, celebrity, and eminence.

" The shrub is said to be produced as follows. Before the commencement of the periodical rains, in June, the cultivators of the *Cardomum* ascend the coldest, and most shady sides of a woody mountain; a tree of uncommon size and weight is then sought after, the adjacent spot is cleared of weeds, and the tree felled close at its root. The earth shaken, and loosened by the force of the fallen tree shoots forth young *Cardomum* plants in about a month's time. I have repeatedly enquired of the natives the means by which the *Cardomum* plants are first produced, and have invariably been told as already stated. They attribute no other effects to the falling tree, which may be selected from any species, than such as are derived from its weight and strength, added to the shade and shelter which its branches afford to the young plants. I have heard it by some asserted that the *Cardomum* like the famed Phœnix, would emerge from the ashes of any large forest tree; but these stories were uttered only by the ignorant, and are too absurd to require comment.

" The shrub continues to grow in this manner until after the early rains of the fourth year, in February, when it has reached its ut-

* It is undoubtedly corrupted from एला, the Sungskrit name of the plant. *Ed.*

most height, which varies from six to nine feet; four or five tendrils are now seen to spring from its stem near the root, and afterwards the fruit is produced, which is gathered the following November; and requires no other preparation, than drying in the sun. The fruit is anually collected in this way, until the seventh year, when it is usual to cut it down, and from the trunk other sprouts arise, in the course of the next monsoon; which grow, flourish, and are cultivated as before. The husk with the seed I have called the fruit, as it corresponds with the native name. The seeds are termed the rice, and the tendrils the threads of the plant.

"I do not hear of any varieties in the species, and now enclose you a specimen of some just given to me by the person who farms the collection, and accounted of an inferior growth. It may not be unnecessary to mention, that this commodity yields to Government from 25, to 30,000 rupees per annum. The inhabitants use it as a general condiment to their food, and it is likewise held in sacred estimation for the purposes of sacrifice. It is constantly chewed with betle, and some medicinal properties are ascribed to it when used in decoction for complaints of the bowels. I am assured, by all, that the seeds of the Cardomum will never produce plants, and that it is only to be propagated as already stated, or by cuttings."

10. A. *punicea. R.*

Leaves lanceolate. *Spikes* radical, compact, imbricated. *Lip* entire, sub-parabolic; base spurless.

A stately species, a native of *Sumatra*, from thence sent by the late Dr. C. Campbell, to the Botanic Garden near *Calcutta*, where it blossoms in March and April.

Roots hard, and bifarious. *Suckers* numerous, running far under the surface of the earth.—*Stems* numerous, erect, or oblique, smooth, swelled near the base; general height about six or seven feet, and as thick as a slender walking cane.—*Leaves* sub-bifarious, short-petioled on their sheaths, lanceolate, rounded at the base, both sides smooth; from six to twenty inches long, by from three to six broad.—

Sheaths smooth on both sides, rising into a short, ciliate ligula, above the insertion of the short, semi-cylindric, proper petioles.—*Spikes* radical, half immersed in the earth, sub-cylindric, pretty compactly imbricated with from oblong to lanceolate, coloured, smooth, one-flowered bractes.—*Flowers* large, colour a bright red, like the finest carmine.—*Bractes interior,* or *inferior perianth,* tubular, length of the tube of the corol; with an unequally two, or three-toothed mouth.—*Perianth proper* superior, tubular, length of the tube of the corol; *mouth* generally bifid.—*Corol* with long, slender *tube* hid within the bractes, and perianths.—*Exterior border* of three, small, paler coloured, lanceolate segments. *Lip* highly coloured, large, entire, horizontal, sub-parabolic, the lower part broad, and encurved round the stamina; *lamina* linear, with entirely rounded, or slightly marginate apex.—*Filament* short, base spurless. *Anther* double, apex bifid, and naked.—*Germ* three-celled, with numerous ovula in each, attached to their respective receptacles, a little removed from the axis. *Base* of the *Style* embraced by two subulate nectaries. *Stigma* a lamellate, ciliated gape.

11. A. *Cardamomum medium. R.*

Leaves lanceolar, villous underneath. *Spikes* radical, lax. *Lip* ovate lanceolate. *Capsules* pedicelled, of an ovate oblong, conical shape, and nine-winged.

A native of the hilly countries in the vicinity of *Silhet,* where the plant is called Do-Keswa. It blossoms about the beginning of the rains, in June, and the fruits ripen in September, when they are gathered, and sold to the dealers in drugs.

Leaves petioled above their sheaths, linear-lanceolar, downy underneath, from two to three feet long, by two to four inches broad.—*Sheaths* villous, rising in an obtuse ligula above the insertion of the leaf.—*Spikes* radical, oblong, laxly imbricated, rising but little above the earth; lower part, or scape hid in the soil, and clothed with shorter scariose bractes.—*Flowers* numerous, large, red, fragrant.—*Bractes exterior,* one under each flower, lanceolate, ribbed, smooth, yellowish pink, *interior*, or *inferior perianth* tubular, length of the proper perianth.—*Calyx* superior, length of the tube of the corol, tubular, with three-toothed, coloured apex.—*Corol; Tube* cylindric, long and slender; *segments* of the border linear-oblong, obtuse. *Lip* with pretty broad cordate base, from thence tapering to its entire obtuse point, much longer than the segment of the exterior border; margins curled.—*Filament, anther, germ, style, stigma, and nectarial bodies,* as in the genus.—*Capsules* pretty long-pedicelled, ovate-oblong, while fresh above an inch and a half long, and nearly one in diameter, somewhat three-lobed, each angle marked with a larger vertical wing, and two smaller on the flatter sides, between the large ones, three-celled.—*Seeds* numerous, obovate, with a groove on one side. *Integuments* two; *exterior* soft, while fresh it may be called the succulent aril; *interior* white and rugose.—*Perisperm* conform to the seed, white, and friable, perforated by a spongy, brown substance above the embryo.—*Vitellus* somewhat hyaline, rising on each side of the perforation like two horns.—*Embryo* subclavate, its small end lodged at the umbilicus.

Obs. The form of the Capsule, which resembles that of Gærtner's Zinziber Ensal, and the acrid, aromatic taste of the seeds, induce to conclude that this is the plant which produces the *Cardamomum medium* of the writers on Materia medica.

12. A. *linguiforme. R.*

Spikes radical, linear, rather open. *Lip* linguiform, bifid; *base* broad, and spurless; *sides* incurved.

A native of the interior parts of *Bengal,* where it blossoms during the rains.

Root perennial, throwing off numerous suckers, which run both above, and under the surface of the earth, by which this species is soon multiplied.—*Stems* many, erect, or nearly so, as thick as a man's fore-finger, smooth, from four to six feet high.—*Leaves* bifarious, sessile on their *sheaths,* lanceolate, smooth on both sides; length, from twelve to twenty-four inches, by from four to five broad *Sheaths* smooth, rising little above the insertion of the leaf.—*Spikes* radical, solitary, the apex only rising above the soil, laxly imbricated with oblong, obtuse smooth, exterior, and interior, one-flowered, green *bractes;* besides the interior, proper, tubular, thin, colourless *bractes* or *inferior calyx.*—*Calyx* superior, tubular, length of the tube of the corol; *mouth* three-toothed, split quite down on one side; colour from the middle upwards deep red.—*Corol; exterior border* of three, sub-lanceolate, obtuse, red segments. *Lip* sub-linguiform, the base broad, and axis incurved, forming an envelope for the stamen, from thence projecting almost horizontally, into a long linear, bifid lamina, with curled margins; colour yellow, with a little red down the centre.—*Stamen* as in the genus.—*Germ* hairy, three-celled; *ovula* many, attached to one fleshy receptacle *in the inner angle.* *Style* embraced at the base by the nectarial bodies, which are in this species bidentate. *Stigma* large, and red.

GLOBBA. Schreb. Gen. Plant. N. 54.

Corolla with the interior border two-lobed, or none. *Filament* very long, base tubular, and winged with a cruciform lip. *Anther* double, with an appendix, or naked. *Capsule* one-celled, three-valved. *Seeds* many attached to three parietal receptacles. *Embryo* simple, and furnished with both perisperm and vitellus.

1. G. *marantina. Willd. Spec. Plant.* 1. 153.

Leaves lanceolar. *Spike* terminal, sub-sessile, strobiliform, bulbiferous. *Anther* four-horned.

G. *marantina. Smith's Exot. Bot.* 2. *p.* 85. *t.* 103. *Roscoe in Trans. of Linn. Soc.* 8 356.*

Lompujum silvestre minus. *Rumph. amb.* 5. 150. *t.* 64. *f.* 2.

A native of the *Moluccas*, and from thence introduced into the Botanic Garden near *Calcutta* by bits of the root found amongst the earth in which nutmeg and other plants were brought from *Amboyna.* Flowering time, in *Bengal*, the rainy reason.

Root perennial, consisting of many, small, oblong, fleshy tubers, united to the base of the stems.—*Stems* annual, about twelve or eighteen inches high, somewhat recurved, inveloped in the sheaths of the leaves, the whole as thick as a rattan, and a little compressed.—

* Ceratanthera amomoides. *Hornem. Hort. Hafn. p.* 5.—N. W.

Leaves bifarious, sessile on their half stem-clasping sheaths, broad-lanceolate, acute, smooth above, somewhat villous and whitish underneath, margins waved; length from one to six inches, the lower being very small, and cordate.—*Sheaths* half stem-clasping, smooth, with semi-lunate, projecting, bearded mouths, rising above the insertion of the leaves.—*Spikes* terminal, sessile, solitary, strobiliform, oblong.—*Bractes* numerous, ovate-cordate, the lower half, or more, and sometimes even the whole, have each a single, small, brownish, sessile, pointed, ovate bulb in the axil; the superior are smaller and generally embrace a single flower each.—*Flowers* slender, bright-yellow, and fragrant.—*Calyx* superior, somewhat gibbous, three-toothed.—*Corol* one-petalled. *Tube* long, and very slender. *Border* double; *exterior* three-parted, nearly equal, oblong, concave, and reflexed; *inner* of two lobes, which are longer by one third than those of the exterior border, cuneate, and of a deeper yellow, reflected over the anterior two divisions of the outer border. *Lip* elevated on a continuation of the tube, a little above the former, consisting of two cuneate, reflected segments, united at the middle only. The upper end continued into a very long, slender, incurved, grooved filament. *Anther* double, with two horns from each side.—*Germ* beneath, round, tubercled, one-celled, *ovula* parietal. *Style* slender, as long as the tube of the corol and filament, and lodged in a groove therein. *Stigma* funnel-shaped.—*Pericarp;* I have not found it come to maturity. The little bulbs found in the axills of the lower bractes of the spike, readily vegetate, and thus the plant is abundantly propagated.

2. G. *bulbifera*. *R.*

Stems bulbiferous. *Leaves* oblong. *Racemes* terminal, erect, shorter than the leaves, bulbiferous.

Teling. Conda pooshpoo.

A native of the northern *Circars.*

3. G. *orixensis*. *R.*

Leaves oblong. *Panicle* terminal. *Anther* naked. *Capsule* verrucose.

Is a native of the moist vallies in the northern *Circars*, and flowers during the rainy season.

Roots biennial.—*Stems* annual, bending obliquely to one side, as in the case with most plants of the natural order, about two or three feet long; lower part envolved in a few simple sheaths, above with the sheathing petioles of the leaves.—*Leaves* alternate, bifarious, very short-petioled, oblong, ending in a tapering filiform point, entire, smooth, finely veined; from eight to ten inches long, and from two to four broad.—*Petioles* short. *Sheaths* opening on the side opposite to the leaf.—*Panicles* terminal, erect, oblong, thin, rigid.—*Flowers* sessile, small, deep orange yellow.—*Bractes* solitary, one-flowered, oval, falling.—*Calyx* one-leaved, *mouth* three-toothed.—*Corol; Tube* a little curved, widening gradually to the mouth, twice the length of the calyx. *Border* double; *exterior* three-parted; upper division largest, sub-erect, concave; the other two inferior, and ovate. *Interior* of two, smaller, opposite, ovate lobes.—*Filament* very long, much curved, base tubular, and ornamented with a large, orange-coloured, two-lobed lip, or apron. *Anther* at the extremity of the filament, naked, double, or two-lobed.—*Style* embraced at the base by the two subulate bodies (*nectaries of König*); above the tube of the corol, it takes a curved direction like the filament, but is shorter, so that the curve is less; its end passes between the lobes of the anther. *Stigma* clubbed, half immersed between the lobes of the anther.—*Capsule* sub-globular, watery, fleshy, one-celled.—*Seeds* many.

4. G. *Hura*. *R.*

Leaves ovate-oblong. *Racemes* terminal, erect; pedicels tern, three-flowered. *Anther* with a membranaceous crest.

Hura siamensium. *Kön. in Retz. obs.* 3. 49.*

Nat. of the peninsula of *Malacca*.

5 G. *pendula*. *R.*

Leaves lanceolate. *Racemes* terminal, compound, much longer than the leaves, pendulous. *Anther* bicalcarate.

* G. *versicolor*. Smith Exot Bot. ii. 116. tab. 117. *a, b, c.*—N. W.

Found by Mr. William Roxburgh, on the Island of *Pulu-Pinang*. Flowering time the rainy season.

Root fibrous.—*Leaves* lanceolate, smooth, fine-pointed. *Sheaths* a little hairy on the outside.—*Racemes* terminal, solitary, compound, very long, pendulous.—*Pedicels* remote, diverging, generally three-flowered.—*Bractes* caducous; those of the pedicels solitary, oblong, ciliate; those of the flowers oval, ciliate.—*Calyx* campanulate, three-toothed.—*Corol; Tube* ascending, long and slender; *Border* double. *Exterior* three-parted, of these the upper one is concave; the other two obliquely obovate, expanding, and flat. *Interior* border of two, opposite, lanceolate, expanding lobes.—*Filament* very long, curved; base tubular and winged, with a bifid, cuneiform, yellow lip or apron, above it is grooved for the reception of the style. *Anther* two-lobed; each lobe bending in a long curved spur.—*Germ* beneath, oval. *Style* filiform. *Stigma* funnel-shaped, mouth ciliate.—I have not seen the fruit.

6. G. *Careyana. R.*

Leaves ovate-lanceolate, villous underneath. *Racemes* terminal, compound, bulbiferous.—*Corol* with the two segments of the inner border as long as the lip. *Anther* naked, suborbicular.

From *Pegue* this pretty little plant was introduced by Mr. F. Carey into the Botanic Garden near *Calcutta,* where it flowers in August.

Root tuberous, about biennial, perishing at one end, and shooting forth at the other. This may be considered the general habit of all the species.—*Stems* numerous, oblong, as thick as a slender ratan, about eighteen inches high, and invested in the sheaths of the leaves.—*Leaves* sessile on their sheaths, alternate, bifarious, from broad-lanceolate to ovate-oblong, very finely acuminate, villous underneath; about six inches long, by two broad.—*Sheaths* villous on the outside, scarcely ending in any ligula, but rising a little on each side.—*Racemes* terminal, erect, composed of little, alternate fascicles, of about three flowers each, supported on very short, common pedicels; each of the lowermost of them produces a small, obovate bulb, and two or three more between the racemes and the leaves. *Flowers* numerous, small, of a deep orange-colour.—*Bractes* concave, acute, one to each fascicle; and a smaller coloured one to each flower; all caducous.—*Calyx* superior, tubular, three-toothed, coloured.—*Corol; Exterior border* of three, nearly equal, oval lobes; *Interior* of two linear lobes, which are as long as the lip; soon after expansion they are reflected back over the exterior border, but become involute before the flower drops. *Lip* a little elevated above the five segments of the exterior border, bifid at both ends.—*Filament* long, incurved, &c. as in the genus. *Anther* broad, ovate, or nearly round, with its two polliniferous cells within, projecting at the sides a little beyond its periphery, but without any other kind of appendage, as in most of the other species.—*Germ* round, rugose, one-celled, containing many *ovula*, attached to three parietal receptacles. *Style* length of the corol and filament, filiform, taking a more direct course to the anther than the groove of the filament. *Stigma* funnel-shaped, ciliate.—*Capsule* globular, size of a large marrow-fat pea, warted, one-celled, three-valved.—*Seeds* several, attached to the middle of the valves of the capsule, round-ovate, resting on a large spongy pedicel.—*Integuments* two; *exterior* firm, light brown, a little villous; *interior* membranaceous.—*Perisperm* conform to the seed, hard, and friable, even while fresh. *Vitellus* in small quantity, clear, lining the inside of the upper part of the cavity of the *perisperm* so as to embrace the upper half of the clavate *embryo*.*

7. G. *subulata. R.*†

Panicles radical. *Filament* with a subulate wing on each side near the base; *anthers* semi-lunar, winged.

Found wild in the forests of *Chittagong* by Mr. W. Roxburgh, and by him introduced into the Botanic Garden near *Càlcutta*, in 1801; where it blossoms in April, May, and June, during which period the new foliage also appears above the ground. There is a constant suc-

* *G. racemosa.* Sm. Exot. Bot. ii. 115. t. 117. seems to come very near to this species and to G. orixensis.—N. W.

† *G. purpurea.* Andr. Repos. x. 615. Mantisia saltatoria. Curt. Mag. xxxii. 1320. —N. W.

cession of flowers from the extremities of the lengthening branches of the panicle, for nearly two months, which renders this lovely plant one of the most charming of the whole natural order that I have yet met with. The specific name is taken from the shape of the wings of the filament.

Root perennial, composed of numerous, diverging, fleshy fibres, some fusiform, some columnar.—*Stems* annual, bending a little to one side; from one to two feet high.—*Leaves* alternate, bifarious, sessile on their sheaths, lanceolate, with a long, withered, filiform point; villous on both sides; length from six to nine inches, and about one and a half broad. *Sheaths* smooth, keeled; *ligula*, or stipulary process truncated.—*Scapes* radical, from three to four inches long, often several arise from the same cluster of roots, erect, enveloped in three or four, alternate, smooth, short sheaths.—*Panicle* erect, about a foot high, including the scape, and expanding nearly as much, the whole a very lively blueish purple. *Ramifications* alternate, long, simple, diverging, bearing in succession numerous, alternate, sessile, small, lovely purple and yellow flowers.—*Bractes* solitary, light purple, smooth; those of the ramifications oblong; those of the flowers cordate.—*Flowers* solitary, sessile in the axils of their proper bractes.—*Calyx* superior, clavate, smooth, light purple; mouth three-toothed, permanent, with the lower half of the tube of the corol.—*Corol*; *Tube* long and very slender; the lower half, which is inclosed in the tube of the calyx, projecting horizontally; the upper half erect; at the angle, which those portions of the tube make with each other, and which is nearly a right one, the flower drops, leaving the lower half to wither. *Border* single and three-parted; upper division erect, pressing on the back of the base of the filament, oblong, boat-shaped; lower two divisions obliquely cordate, and reflexed, all the three are of a light purple colour. *Lip* cuneate, deeply two-parted, much larger than the divisions of the exterior border, deep orange yellow, forming a pretty contrast with the lively purple of every other part of the panicle.—*Filament* very long and very slender, on the first day much incurved, afterwards revolute, finely grooved on the underside for the reception of the style; near the base are two, very long, slender, flat, subulate, diverging, recurved wings. In my other six species with terminal inflorescence, these are entirely wanting, but the inner two-lobed border is present. *Anther* double, and enlarged at the sides with two, broad, reniform wings.—*Germ* beneath oval, with nine longitudinal ridges, a transverse section exhibits one cell only, with numerous seeds, affixed to three parietal receptacles. *Style* most slender. *Stigma* funnel-shaped.—*Capsule* globular, purple, size of a small cherry, six-ridged, one-celled, three-valved, opening spontaneously from the apex. *Receptacle* a longitudinal ridge on the inside of each valve.—*Seeds* many, minute, oblong, villous, connected by a large, white, spongy membrane to the parietal receptacles.

8. G. *spathulata*. *R.*

Panicles radical. *Filaments* with a spatulate wing on each side near the base ; *anthers* simi-lunar, winged.

This second, beautiful species, with large azure coloured, radical inflorescence, is a native of the vallies on the north and eastern boundaries of the province of *Silhet*, and from thence was introduced by Mr. M. R. Smith into the Botanic Garden near *Calcutta*, where it flowers in April, at which time the foliage begins to appear. At the beginning of the cold season it perishes down to the root.

Panicles radical, erect, oblong, from eight to twelve inches high, composed of many, diverging, simple, lengthening branches, every part, bracte included, of a pretty, light azure colour.—*Bractes* oval, and oblong, concave, a larger one under each branch of the panicle, and one to each flower.—*Flowers* numerous, produced in a continued succession for several weeks on the same panicle.—*Perianth* superior, widening to its three-toothed mouth.—*Corol ; Tube* long, slender, curved, and villous. *Border* three-lobed ; *lobes* nearly equal and equally arranged, shape oblong, and concave. *Lip* as in G. *subulata*, but less divided ; yellow, as in that species.—*Filament* long, slender, recurved, &c. as in the genus ; near the base, instead of an inner border to the corol, there is a spatulate, diverging wing on each side. From its shape the specific name is taken, and readily distinguishes this species from *subulate*, which it very exactly resembles when in flower. *Anther* with a similunar wing on each side.—*Germ* one-celled, containing many *ovula,* attached to three *parietal* receptacles.*

The following plant, which was omitted through an oversight in its proper place, viz. next after Hedychium coronarium, is here inserted at the end of the natural order of Scitamineæ.

Hedychium flavum. R.

Leaves broad-lanceolar. *Spikes* terminal, imbricated. *Bractes* about four-flowered; interior two-segments of the corol linear; lip obcordate, with apex deeply retuse.

A native of the vallies amongst the hills near *Silhet,* where it is called Kattia-rityam by the natives. Flowering time the rainy season.

Root tuberous, like that of most other *Scitamineæ.*—*Stems* erect, two or three feet high, and rather stout —*Leaves* lanceolar, very fine pointed, smooth on both sides. *Sheaths* smooth, with a very large, smooth membranaceous stipula rising far above the mouth.—*Spike* terminal, solitary, erect, oblong, imbricated; length six or eight inches.—*Flowers* numerous, large, yellow, resembling those of *H. coronarium,* fragrant.—*Bractes; exterior* between oval and oblong, rather obtuse, smooth, finely nerved, three or four-flowered, about two or two and half inches long and nearly one and a half broad. *Interior,* one to each flower and much smaller.—*Calyx* superior, nearly as long as the tube of the corol, membranaceous, inflated, obliquely two or three-toothed.—*Corol; Tube* slender, length of the exterior bractes, all the five *proper* segments of the border linear, the inner two more deeply coloured and obtuse. *Lip* or sixth segment, obcordate, deeply retuse, placed on the enterior or upper side opposite to the filament.—*Filament* linear, and together with the two-lobed *anther* about as long as the lip.—*Germ* hairy, three-celled, with numerous *ovula* in each, attached to the inner angle. *Style* length of the co-

* This beautiful plant is a second species of the genus *Mantisia,* which has been established by Dr. Sims, and confirmed by the illustrious author of the article *Globba,* in Rees's New Cyclopædia.—N. W.

rol. *Stigma* large, and villous.—*Capsule* turbinate, size of an olive, smooth, when ripe yellow, three-celled, three-valved.—*Seeds* a few, smooth, bright deep red, with a partial ragged aril of the same colour. —*Embryo* simple, sub-lanceolate, furnished with both perisperm and vitellus.*

SALICORNIA. *Schreb. gen. N.* 14.

Calyx gibbous like an aril lining the inside of the cavities of the fructification in the joints. *Corol* none. *Seed* one.

1. S. *brachiata. R.*

Perennial, erect; branches numerous, decussate, joints clubbed. *Spikes* cylindric. *Flowers* conspicuous.

Teling. Quoiloo.

This seems an undescribed plant; nor can I collect what Dr. König thought it was. It is a small, generally erect, perennial species, growing plentifully on such low, wet, salt ground as is overflowed by the spring tides. Flowers all the year round. It should be compared with *S. herbacea.*

Stems generally erect, woody, about as thick as a goose-quill, though sometimes at the base as thick as the little finger; from twelve to eighteen inches high. *Branches* decussate, ascending, very numer-

* In January 1816 I received roots of this most charming species from Mr. Smith at Silhet, which produced blossoms in the Botanic Garden in October the following year. It differs from H. coronarium, in regard to its flowers, which are about one third smaller, having the inner segments of the corol linear-clavate, the fissure of the lip narrow with straight sides, and its base contracted into a linear claw. They partake not only of the yellow colour of those of *Michelia Champaca*, but possess even the peculiar fragrance of the latter, only in a less powerful and therefore more grateful degree. In stature and leaves both species are alike.

The following quotation should be added to *H. angustifolium*, p. 11.—*Edwards's Bot. Regist.* n. 157.—Notwithstanding the opinion which this distinguished botanist has expressed I am led to suspect that *H. coccineum of Sir J. E. Smith* (*Rees's new Cyclopædia*, in loco) is a distinct species, of which I have examined the original drawing, left by my inestimable friend and predecessor Dr. F. Buchanan at the Botanic Garden at Calcutta.—N. W.

ous, divided, and subdivided in the same manner; their extremities are the succulent, cylindric jointed spikes.—*Joints* of the spikes, or leaves, thick, succulent, annular, cap-like portions, that closely surround the extremities of the filiform, woody branchlets, like beads strung on a thread; above they are concave, with two opposite elevations, the lower convex end of the next leaf above enters this concavity; these elevations, like the branches, are decussated with those of the next above and below, with a ring.—*Flowers* conspicuous, three-fold, (three on each side of each joint,) opposite.—*Calyx,* a flask-like fleshy substance, with a longitudinal slit for the stamen, and style; they adhere firmly to one another, and to the substance of the leaf, till the seed is ripe.—*Corol* none.—*Stamens* constantly one. *Anther* sagittate.—*Stigma* two-cleft.—*Capsule,* no other than the calyx now dry, which separates, and drops with its single seed.

2. S. *indica. Vahl. enum. pl.* 1. 10.
Perennial, diffuse; joints gibbous. *Spikes* cylindric; *flowers* hid.
Beng. Jodoo Palung.
Teling. Quoia-pipaly.
This is also very common, and grows with the former on similar ground.

Stems and large *branches* woody, diffuse over an extent of some feet; the latter sometimes alternate, sometimes opposite, those that are young, and covered with the leaves are ascending, or nearly erect.—*Joints,* or leaves longer than in the former species, but in other respects the same. *Floral joints* or leaves, very short.—*Spikes* terminal, cylindric, erect, obtuse.—*Flowers* inconspicuous, opposite, three-fold, hid by the upper margin of the next floral leaf below.—*Calyx* as in the former.—*Stamen;* in this species I have never been able to discover even the rudiments of one.—*Pistil,* &c. as in the former.

Obs. Cattle will not eat these plants, although their taste is simply saltish. These two species are very plentiful on the coast of *Coro-*

mandel, where large quantities of the fossil alkali might no doubt be made at a low rate. Such manufactures deserve much encouragement, particularly here, where there are many more labouring people* than can possibly find employment. Large quantities of this substance are annually imported into England from the Mediterranean, for making soap, glass, &c. It is worth about twenty-eight or thirty shillings per cwt. and as labour is cheap on this coast, the alkali might be made, I think, at so low a rate as to admit of its being sent to *England*, and paying a freight of 10£. the ton, or more; provided it could be there imported duty free.

* This was written before the famine of 1791, 2, and 3; the case is now (1794) different, as fully half, if not more of the labouring classes were carried off by that deplorable calamity.

Class 2d.

DIANDRIA MONOGYNIA.

NYCTANTHES. *Schreb. gen. N.* 21.

Calyx campanulate. *Corol* salver-shaped. *Capsule* superior, obcordate, compressed, two-celled, two-valved. *Seeds* solitary. *Embryo* erect without perisperm.

1. N. *arbor tristis. Linn. Spec. Plant. ed. Willd.* 1. 35.
Parlium arbor tristis. *Gært. Sem.* 1. 234. *t.* 51.
Scabrita scabra. *Vahl. Symb.* 2. 28. *Gært. Sem.* 2. 265. *t.* 138.
Scabrita triflora. *Linn. Mant.* 1. *p.* 37.
Manja pumeram. *Rheed. mal.* 1. *p.* 35. *t.* 21.
Sephalica. *As. Res.* 4. 244.
Sans. Hind. and Beng. Singahar, Sewly, Nibari, Har-singhar, &c.*

Of what country this is a native I know not, for I have never found it but in a cultivated state; and it is always raised from seed, which may be the reason we have no varieties of this most delightfully fragrant plant. In our gardens it is found in the state of a large shrub, or small tree. Flowers, on the Coast, nearly the year round; in *Bengal* only during the rains. The seeds ripen in the cold season.

Trunk erect. *Bark* scabrous. *Branches* numerous, spreading in every direction; the young shoots alone are four-sided; four ligneous, cord-like nerves, under the bark, forming the angles.—*Leaves* opposite, short-petioled, cordate, those near the flowers oblong, pointed, sometimes entire, sometimes very grossly serrate, and sometimes the

* At Napaul, Baleeja Soah; Soa, Soang and Swa being synonymous terms in that country for *Flower*.—N. W.

lower parts are angular; scabrous; from three to five inches long, and from one to three broad.—*Inflorescence* a large, terminal, leafy, cross-armed, *panicle,* composed of small, five-flowered, terminal *umbellets.*—*Flowers* numerous, middle-sized, exquisitely fragrant, open at sun-set, and fall off at sun-rise; the tube orange-coloured, and the border white.—*Involucre* to the umbellets four-leaved, leaflets inverse-cordate, opposite, sessile.—*Calyx* campanulate; mouth a little contracted, and lightly five-notched, downy, withering.—*Corol.* *Tube* cylindric, length of the calyx. *Border* spreading, from five to eight, but generally six or seven parted, contorted; *segments* triangular, or obliquely lobed, exterior margin scolloped.—*Filaments* scarcely any. *Anthers* two-lobed, sitting within the tube.—*Germ* above, compressed. *Style* length of the tube. *Stigma* headed, glandular.—*Capsule* size of a man's thumb nail, obcordate, or nearly orbicular, compressed, two-celled, two-valved, opening transversely from the apex.—*Seeds* one in each cell, compressed, &c. as described and figured by Gœrtner; I have not, however, been able to detect any thing like a perisperm.

Obs. The flowers of this tree are exquisitely fragrant, partaking of the smell of fresh honey, and on that account the plant is much esteemed; for when destitute of flowers it has but an indifferent appearance.—The orange tubes of the flowers dye a most beautiful buff or orange colour, with the various shades between them, according to the preparation, and mode of conducting the operation; but unfortunately, no way has yet been discovered of rendering this elegant colour durable.

JASMINUM. *Schreb. gen. N.* 22.

Corol salver shaped. *Germ* two-celled, one seeded.* *Berries* one or two, superior. *Seed* solitary. *Embryo* erect, without perisperm.

* In all the species hereafter mentioned, there is a two-lobed, two-celled germ, with one seed in each, attached to the upper end of the partition; and succeeded by two one-seeded berries, when both come to maturity.

1. J. *Zambac. Willd.* 1. 35.

Shrubby, twining. *Leaves* opposite, sub-sessile from cordate to oblong, acute, or obtuse. *Segments* of the calyx subulate. *Berries* globular.

Magorium Sambac. *Lamarck. Encycl.* iv. 210.

Of this we have three varieties, viz.

1*st. Single flowered Arabian Jasmine,* which I suppose to be Jasminum undulatum. *Willd.* 1. 36.

Tsjiregam-Mulla. *Rheed. mal.* 6. *p.* 97. *t.* 55.

Mallica, and Asphota, *As. Res.* 4. *p.* 245, are I suspect both this variety. The first is found in gardens, though a small flower, and the second wild in hedges, &c.

Beng. B*u*n-m*u*ll*i*ka; or Desh*i*-m*u*l*i*ka.

2*d. Double flowered Arabian Jasmine.*

Nalla-mulla. *Rheed. mal.* 6. *p.* 87. *t.* 50.

Jasminum Limonii folio conjugato. *Burm. zeyl.* 128. *t.* 58. *f.* 2.

Flos Manoræ. *Rumph. amb.* 5. *p.* 52. *t.* 30.

Sans. S*u*ptala, N*uva*-mull*i*ka. *See As. Res. l. c.*

Beng. Bela, or m*u*ll*i*ka.

3*d. Great double Arabian* or *Tuscan Jasmine.*

Beng. Mootta-bela; B*u*ra-bela, or Patna m*u*ll*i*ka.

Kudda-mulla. *Rheed. mal.* 6. *p.* 89. *t.* 51.

Flos Manoræ plenus. *Rumph. ibid.*

Septala. *As. Res. l. c.*

All the varieties flower during the rains chiefly.

The single variety of this plant, (which is the one described,) is one of the most common in every forest on the *Coast.* Birds eat the berries, and drop the seeds, which vegetate. In this way, I account for its being so general. Flowers during the hot season.

Stem and *branches* woody, twining. *Bark* scabrous on old plants.—*Leaves* opposite, short-petioled, cordate-ovate, or oblong-waved, sometimes scolloped, pointed, smooth, except when young; in the axils of the veins on the back of the leaves there are tufts of down, as in some species of *Gardenia.*—*Flowers* terminal, generally forming small, trichotomous umbellets, white, fragrant.—*Calyx* bordered with from five to nine, long, subulate, permanent segments. *Stigma* clubbed. *Germ* two-lobed, one *ovulum* in each, attached to the lower part of the partition.—*Berries* twin; each round, smooth, juicy, one-seeded, when ripe of a beautiful shining black.—*Seed* conform to the berry; integument single.—*Perisperm* none.—*Embryo* erect.—*Cotyledons* conform to the seed.—*Radicle* small, inferior.

Obs. From this species we have the common double; and rose, or great flowered, and full *Moogaries,* (*Arabian,* or *Tuscan Jasmines;*) these are always raised from layers, and have been improved by culture into their present state as above noted.

The plant that bears the common double sort still retains its twining habits, but it is lost in the great rose, or full flowered sort: here the branches are erect, or spreading, but never twining.

I am inclined to think that *Nyctanthes undulata, Sp Pl.* 2. *p.* 8. is only a variety of this species, arising from soil, culture, &c. for with the utmost attention, I could never discover any thing like a fixed specific character to distinguish it from *Zambac,* as the number of petals or divisions of the corol are inconstant, and at most mark only the variety.

2. J. *scandens. Willd.* 1. 37.

Shrubby, twining, and scandent, smooth. *Leaves* opposite, simple, cordate-oblong, entire, taper-pointed. *Corymbs* terminal, globular, trichotomous. *Calycine* segments six or seven, bristle-shaped, of the corol six to eight, lanceolate, very acute, nearly the length of the clavate tube. *Stigma* bilobate.

Nyctanthes *scandens. Retz. obs. fasc.* 5. *p.* 9.

A native of the southern parts of *Bengal.* The plants in the Botanic Garden near *Calcutta* were reared from seed sent from *Chittagong*; which after they were two years old began to blossom in January and February, and continue so to do, but have not produced good seed.

Trunk none, but many, long, ligneous, at first scandent, but finally twining, round, smooth branches of considerable extent.—*Leaves* simple, opposite, short-petioled, cordate-oblong, tapering from near the somewhat re-entering base, fine-pointed, entire, smooth, and rather lucid on both sides ; length from one to six inches, and from half an inch to two inches broad.—*Petioles* flattened, channelled.—*Corymbs* terminal, sub-globular, dense, that which terminates the larger branches more compound. *Peduncles* short, with the *pedicles* still shorter, all are villous.—*Flowers* numerous, pure white, delightfully fragrant.—*Calyx* villous, *border* of six, or seven, bristle-shaped, spreading, villous segments.—*Tube* of the corol widening to the mouth, and more than twice the length of the segments of the calyx. *Border* of from six to eight, lanceolate, acute segments, which are rather shorter than the tube.—*Germ* turbinate, two-celled, with one seed in each, attached to the thickened middle of the partition.

3. J. *elongatum. Linn. Spec. Pl. ed. Willd.* 1. 37.

Scandent. *Leaves* opposite, and alternate, lanceolate, villous on both sides. *Corymbs* terminal. *Corol* eight or twelve-cleft; *segments* linear. *Stigma* bifid.

Nyctanthes *elongata. Linn. suppl.* 82.

A native of the forests about *Hidgelee,* near the mouth of the Hoogly in *Bengal.* Flowering time the beginning of the hot season.

Stem woody, straggling to a considerable extent, or climbing. *Bark* of the woody parts light ash colour ; of the *young shoots* very villous.—*Leaves* opposite, and alternate, short-petioled, from lanceolate to ovate-lanceolate ; very soft with fine down in both sides ; length from one to five inches.—*Flowers* terminal, from simply tern to a large compound corymb, pretty large, pure white, fragrant.—*Pedicels* clavate, villous.—*Bractes* short, subulate, villous.—*Calyx* villous, five or six-cleft ; *divisions* short, and subulate.—*Corol. Tube* swelling towards the mouth, greatly much than the calyx. *Border* of from six to twelve, linear, spreading divisions with margins revolute.—*Filaments* short. *Anthers* lodged just within the mouth of the tube of the corol.—*Germ* turbinate. *Style* so long as to elevate the two-lobed *stigma* so as to make it level with the anthers.

Obs. The species differs from my *arborescens* in being scandent, and in the shape of the leaves, as well as in the pubescence; here they are lanceolate, and very villous; there cordate and smooth.

4. J. *pubescens*. *Linn. Spec. Pl. ed. Willd.* 1. 37.
Leaves cordate, downy. *Umbels* terminal, sessile many-flowered.
Katu-Tsjiregam-Mulla. *Rheed. mal.* 6. *p.* 95. 1. 54.
Nyctanthes *multiflora Burm. Flor. Ind. p.* 5. *t.* 3. *f.* 1.
Nyctanthes pubescens. *Retz. obs.* 5. 9.*
Sans. माघ्यं, Maghyun, कुन्दं, Koondum. *See As. Res.* 4. 244.
Beng. Koonda.

A very ramous shrub, brought originally from *China* into the Company's Botanic Garden. It is also indigenous in *Bengal*. Is in flower during the rains chiefly.

Stem scarcely any; *branches* numerous, covered with ash-coloured bark; the young parts very downy.—*Leaves* opposite, short-petioled, cordate, entire, downy, particularly while young; the terminal ones crowded (involucre-like) round the umbels.—*Umbels* terminal, sessile, tending to be trichotomous, containing from ten to thirty, sessile flowers; downy in all parts.—*Flowers* large, pure white, fragrant.—*Bractes* few, lanceolate.—*Calyx* six-cleft; border of the corol from six to nine-parted; *divisions* lanceolate.—*Stamens* within the tube. *Anthers* beaked.—*Germ* two-lobed, two-celled, with one *ovula* in each, attached to the upper part of the partition. *Stigma* entire, clubbed.

Obs. This is a goodly looking plant, its numerous, large, pure white, fragrant flower, opening in succession make it very desirable.

* J. hirsutum. Sm. Exot. Bot. ii. 117. t. 118. J. multiflorum. An'r. Repos. viii. 496.—N. W.

5. J. *laurifolium.* R.

Scandent, polished. *Leaves* opposite, ovate-lanceolate, lucid, three-nerved. *Flowers* axillary, and terminal, from one to five, long-peduncled. *Calycine segments* six or seven, filiform; those of the corol from nine to twelve, linear, length of the tube.

A very elegant, climbing and twining, highly polished, shrubby species; a native of the mountainous countries east of Bengal, where it flowers during the cool season.

Leaves opposite; subsessile, lanceolate, firm and glossy; margins entire, somewhat thick; faintly three-nerved; from two to four inches long, and from one to two broad.—*Peduncles* terminal and axillary, sometimes single, sometimes several together, and sometimes, when terminal, sub-racemed; long and slender.—*Bractes* opposite, subulate.—*Calyx* from five to seven cleft; *segments* long, slender, subulate, and smooth.—*Corol.* *Tube* twice the length of the calyx, sub-clavate; *border* of from nine to twelve, ensiform, spreading *segments* which are as long as the tube.—*Stamina* within the tube.—*Germ* two-celled, with one *ovula* in each, attached to the middle of the partition.—*Style* and *stigma* clavate, scarcely reaching to the anthers.

6. J. *coarctatum.* R.

Shrubby. *Leaves* oblong, smooth, acute. *Corymbs* terminal, peduncled, crowded with ternate, subsessile flowers, and amply bracted. *Calyx* five-cleft.

A very ramous shrub, without any tendency to ramble or climb. It is the first simple-leaved *Indian Jasmine* I have found which has not that habit. Is a native of the hills near *Chittagong*, where it flowers in April and May; may be readily known, without any other mark, by the great number of flowers which form the little dense corymbs.

7. J. *trinerve.* *Vahl. Symb.* 3. *p.* 2.

Scandent, polished. *Leaves* ovate-lanceolate, long-acuminate, three-nerved, polished. *Flowers* axillary and terminal, solitary. *Calycine segments* six or seven, subulate; those of the corol six or eight, sub-filiform and longer than the long tube.

A native of the forests near *Silhet*, where it blossoms in March and April, and the berries ripen in June.

Stem, and *branches* together, some fathoms in extent, running up, and over large shrubs and trees. *Bark* of the old woody parts ash-coloured and rather rough; *tender shoots* polished deep green. —*Leaves* opposite, petioled, ovate-lanceolate, long, fine-pointed, firm and glossy, three-nerved, three or four inches long, and one or one and a half broad.—*Petioles* short, with a remarkable joint near the middle.—*Flowers* terminal and axillary, solitary, sub-sessile, very large, white and sweetly fragrant.—*Calyx* with a short, smooth tube, border of about six subulate segments which are nearly half the length of the tube of the corol.—*Corol; Tube* long, widening above the middle for the stamina; *Border* of six to eight slender, spreading segments, which are an inch and a half in length.—*Filament* short. *Anthers* linear, lodged about the middle of the tube of the corol.—*Germ* oval, two-lobed, two-celled, with one *ovulum* in each, attached to the partition.—*Style* length of the tube of the corol. *Stigma* bifid.

Obs. The flowers were solitary the first year the plants were in the Botanic Garden, but during the rainy season of the second year there were usually nine.

8. J. *bracteatum*. *R*.

Twining. *Leaves* opposite, ovate-oblong, acute, villous. *Flowers* in terminal bracted fascicles of three, five, or eleven flowers. *Calycine segments* from five to seven, subulate; those of the corol from five to eight, oblong, with rounded acuminate apices. *Style* longer than the tube. *Stigma* entire.

A native of the island of *Sumatra*. From *Bencoolen* Dr. C. Campbell sent plants to the Botanic Garden, under the name *Coast*

Jasmine. In Bengal they blossom during the cool, and the beginning of the hot season; the seed ripen in April.

Stem scarcely any; *branches* numerous, ligneous, climbing and twining, extending up and over trees of considerable size; tender parts round, and clothed with much, very soft, short, pale green down.—*Leaves* simple, opposite, short-petioled, ovate-oblong, acute, entire, downy, two or three inches long, and one or two broad.—*Flowers* numerous, pretty large, white, fragrant, in terminal fascicles of generally three, or five, subsessile ones, or growing on short, axillary branchlets; embraced by one, or two pair of large bractes, and by a greater number, when the fascicle terminates a primary branchlet; the additions in this, as in most of the other *Indian* species, proceed in opposite pairs of three-flowered peduncles.—*Bractes* opposite, always two, generally unequal pairs, of an ovate-cordate shape, under each fascicle, and some small subulate ones within them, all villous, like every other tender part of the plant.—*Calyx* downy. *Tube* short. *Segments* five, six, or seven, subulate.—*Corol* smooth. *Tube* twice the length of the calyx, sub-cylindric. *Border* of from five to eight, oblong segments, rounded at the apex, with a small point in the middle.—*Germ* two-celled, one ovula in each. *Style* longer than the tube of the corol. *Stigma* entire, linear, with a groove on each side.—*Berries* one or two, united from the middle downwards, size of a large pea, oval, smooth, succulent, and of a shining black colour when ripe; one-celled.—*Seed* single, oval. *Integuments* two; *exterior* composed of woolly fibres; *inner* thinner and more tender, both dark, brownish black.—*Perisperm* none.—*Embryo* erect. *Cotyledons* conform to the seed, green. *Plumula* two-lobed. *Radicle* patelliform, inferior.

9. J. *latifolium*. *R.*

Shrubby, twining. *Leaves* opposite, petioled, cordate. *Corymbs* terminal; *calycine segments* from five to seven, subulate; those of the corol from ten to twelve, linear, and cuspidate. *Berries* kidney-shaped.

Teling. Adivi-mulli.

To the best of my knowledge, this species is a native of the mountainous parts of the *Circars* only. It is much larger than the other species, though like them it has a woody twining, or climbing stem, but its branches are in general straight and spread. *Branchlets* opposite smooth.—*Leaves* opposite petioled, varying from oblong, to broad-cordate, pointed, entire, smooth; size various.—*Corymbs* terminal, divided by threes, diffuse.—*Flowers* large, white, fragrant.—*Calyx* with a short tube, and seldom more than five-cleft border.—*Corol* tubular; border from eight to twelve-cleft; segments slender, and patent.—*Berries* one or two; when single oblong; when two kidney-shaped.

10. J. *arborescens.* R.

Arborescent. *Leaves* opposite, and three-fold, oblong, downy. *Flowers* terminal, numerous, corymbiform, border from ten to twelve-cleft. *Stigma* two-lobed.

Sans. सप्तला, Suptula, नवमल्लिका, Nuva-mullika.

Beng. Bura-koonda, Nuva-mullika.

Nyctanthes *grandiflora. Loureir. Cochinch.* 26.

A native of the more elevated parts of *Bengal,* where it blossoms about the beginning of the hot season.

Trunk scarcely any; *branches* many, stout, woody, sub-erect, covered with smooth, ash-coloured bark; *young shoots* round, and somewhat villous. Height of the whole tree ten or twelve feet.—*Leaves* opposite, rarely three-fold, or alternate, short-petioled, ovate-cordate, acuminate; margins entire, and often waved, smooth on both sides; length from two to four inches, and from one to three broad.—*Petioles* channelled, jointed near the middle.—*Flowers* on terminal, trichotomous, three-flowered peduncles, often forming large, corymbiform panicles; they are large, white, and very fragrant.—*Bractes* subulate, villous.—*Calyx* crowned with five or six subulate, incurved, villous divisions.—*Corol. Tube* longer than the calyx. *Border* of ten or twelve, somewhat acute, linear divisions

—*Anthers* ovate, within the tube.—*Germ* turbinate, two-lobed, two-celled, with one *ovula* in each, attached to the upper part of the partition.—*Stigma* two-lobed, about half way up the tube of the corol.

Obs. This species is nearly allied to my *J. latifolium*, and indeed the only specific difference I have yet been able to observe, is, that this species has no tendency to twine or climb; that is twining.

11. J. *angustifolium. Vahl. En. Pl.* 1. 29. *Linn. Spec. Pl. ed. Willd.* 1. 36.

Shrubby, twining; polished. *Leaves* opposite, petioled, ovate, smooth, of a shining deep green. *Flowers* terminal, one, two, or three; *corol* eight or nine-parted. *Berries* single, ovate.

Nyctanthes triflora. Burm. Ind. 4. *t.* 2.?

J. vimineum. Willd. loc. cit.

Katu-pitsjegam-Mulla. *Rheed. mal.* 6. *p.* 93. *t.* 53.

This is much like *Burman's* figure quoted above, but does not accord with his character of alternate leaves; it is very common, and at the same time one of the most beautiful species of Jasmine I know. It abounds in every forest on the coast of *Coromandel.* Blossoms during the hot season; but in a state of cultivation flowers are produced the greatest part of the year.

Stem and *branches* woody, twining, smooth.—*Leaves* opposite on every part of the shrub, short-petioled, ovate-oblong, or oblong, sometimes nearly cordate, finely-pointed, smooth, shining, firm, very deep green, veinless.—*Flowers* terminal, generally three-fold, peduncled, large, white, with a faint tinge of red, star-shaped, having a peculiar but very pleasant fragrance.—*Peduncles* clubbed, smooth.—*Calyx; tube* short; *segments* acute.—*Corol; border* generally eight or nine-parted; *divisions* lanceolate.—*Style* short. *Stigma* lanceolate.—*Berries* oblong.

Obs. I have for many years cultivated this species. It grows easily in every soil and situation, is constantly covered with leaves, and their bright, shining deep green colour, renders it always beautiful, and particularly well adapted for screening windows, covering arbours, &c.

12. J. *tubiflorum.* *R.*

Scandent. *Leaves* sub-sessile, ovate-oblong, smooth. *Flowers* terminal, from three to five, short-peduncled. *Calycine segments* from seven to eight, subulate, one-sixth the length of the very long slender tube of the corol. *Berries* ovate.

A native of the Malay *Archipelago,* differs from all the other *Indian* species hitherto examined by me in the great length of its slender tube, which is three times longer than the seven or eight, lanceolate segments of its border.

13. J. *simplicifolium.* *Linn. Spec. Pl. ed. Willd.* 1. 38.*

Shrubby, spreading. *Leaves* oblong, polished. *Flowers* from three to many terminal. *Border* of the corol of from six to eight, linear, acute segments, equalling the tube in length.

Native place, the Friendly islands, and the eastern Archipelago, &c. Flowering time in the Botanic Garden the hot season. It does not ripen seeds in *Bengal.*

Trunk scarcely any; *branches* many, stout, ligneous, round, smooth; *branchlets* spreading in all directions.—*Leaves* simple, opposite, short-petioled, oblong, entire; highly polished on both sides; about two inches long, and about one broad.—*Flowers* terminal, from three to a decussate panicle, with its ramification trifid. *Pedicels* clavate, round, smooth.—*Bractes* minute, subulate.—*Calyx* campanulate, with a five-toothed border.—*Corol; tube,* many times longer than the *calyx.* *Border* from six to eight-parted; *divisions* linear, acute, about as long as the tube.—*Anthers* half erect. *Style* shorter by half than the tube of the corol. *Stigma* two-cleft.

14. J. *auriculatum.* *Linn. Spec. Pl. ed. Willd.* 1. 38.

Shrubby, twining. *Leaves* sub-ternate, leaflets ovate, the pair mi-

* Curt. Bot. Mag. Vol. xxv. 980.—N. W.

nute or wanting. *Border* of the calyx with five, obscure, glandular teeth. *Corol* seven-cleft. *Berries* globular.

Sans. मागधी, Magudhee, गनिका, G*u*nika, युथिका, Yoot'h*i*ka, अम्बष्टा, Umbustha. *See As. Res.* iv. 246.

Beng Jut'hi, or Jui.

This species is a native of various parts of India, but not common. Its habit is twining, with a woody stem, and branches.

Leaves opposite, ternate. *Leaflets,* the pair very small, standing opposite, sometimes wanting; the terminal one at least ten or fifteen times larger; all ovate, entire, with a minute point, smooth, except when young when they are downy —*Corymbs* terminal, always three-forked, dense.—*Flowers* small, white, sweet-smelling, though less fragrant than those of the former species.—*Calyx* five-notched, with a round firm glandular process on the outside of each.—*Corol; border* from five to eight-cleft, divisions oblique, obtuse. *Stigma* clavate.—*Berry* exactly as in Sambac.

15. J. *lanceolaria. R.*

Shrubby, erect. *Leaves* ternate; leaflets lanceolar. *Corymbs* terminal.

A native of the jungles or copses of *Silhet,* where it blossoms in May.

16. J. *paniculatum. R.*

Shrubby, erect, every part polished. *Leaves* ternate; leaflets oval, obtusely acuminate. *Panicles* terminal.

Sam-yeip-sou-hing of the Chinese.

From Canton in China this fine erect, shrubby, highly polished species was received into the Botanic Garden, where it blossoms during the rains. The leaflets are from oval to oblong, with an obtuse point, polished, of a hard texture. The flowers are small and white but numerous, on terminal pretty large open, brachiate panicles.

17. J. *chrysanthemum.* *R.*

Shrubby, erect. *Leaves* alternate, pinnate; leaflets three, five, or seven, lanceolate. *Corymbs* sub-terminal. *Calycine segments* attenuate.

Hemapushpica, or yellow yuthica. *As. Res.* iv. 246.

Sans. हेमपुष्पिका, Hema-pooshpika.*

A native of the mountainous countries north of *Hindoostan.* Colonel Hardwicke found it on his Journey to Sirinagore, (*see As. Res.* vi. 349. Jasminum, No. 3.) and Dr. Buchanan in *Napaul.* In the Botanic Garden it grows freely from cuttings, and becomes a stout, erect ramous shrub, even a small tree, without the smallest tendency to lean, or twine. Flowers more or less the whole year; but, like the other species, the proper season is April and May, at which time it is the most desirable *Jasmine* I have yet seen.

Stem stout and woody, in plants seven or eight years old as thick as a man's leg. *Bark* smooth ash-coloured. *Young shoots* erect, flexuose, angular, smooth; height of the shrub from eight to twelve feet.—*Leaves* alternate, unequally pinnate; at the base of the young shoots ternate or even sometimes solitary. *Leaflets* two, or more generally three pair, and a terminal one; opposite, lanceolate, smooth, dark green, but not glossy; length from one and a half to three inches. —*Corymbs* generally terminal, from ten to twenty-flowered, as in the figure of *Jasminum odoratissimum, in Curtis's Bot. Mag. No.* 285. —*Flowers* large, bright yellow, delightfully fragrant.—*Bractes* ensiform.—*Calycine segments* five, alternate, nearly as the tube of the calyx.—*Segments* of the *corol* five, oval, obtuse, length of the tube.—*Filament* short. *Anthers* linear, with their apices rather above the mouth of the tube of the corol.—*Germ* two-celled, with two ovula in each, attached to the middle of the partition. *Style* half the length of the tube. *Stigma* clavate.

18. J. *grandiflorum. Linn. Sp. Pl. ed. Willd.* 1. 41.

Shrubby, scandent (in India). *Leaves* opposite, pinnate; exterior three or five leaflets confluent. *Peduncles* terminal.

* Agee-Soa in Nepala.—N. W.

Catalonian, or Spanish Jasmine.

Pitsjigam mullo. *Rheed. mal.* 6. *t.* 52.

Sans. सुमाना, Soomuna, मालति, Malutee, जाती, Jat*i.*

Beng. Jatee.

This plant I have only found in gardens, where it flowers greater part of the year.

19. J. *heterophyllum. R.*

Shrubby. *Leaves* alternate, petioled, simple, sometimes ternate, ovate-lanceolate, waved, lucid, finely acuminate.

A stout shrubby species, a native of *Napaul.* In the Botanic Garden its branches are weak and variously bent; the bark roughened with little elevated gray spots. While the plants are young the leaves are for the most part ternate, afterwards they are simple.*

20. J. *dispermum. Wall.*

Shrubby, scandent, smooth. *Leaves* opposite unequally ternate; *leaflets* ovate-oblong, acuminate, five-nerved. *Corymbs* terminal and lateral, brachiate.—*Calyx* campanulate with subulate teeth. *Segments* of the *corol* oblong obtuse. *Berries* twin, two-seeded!

A native of *Napaul,* where it is called *Gooje-Soah.*

It appears to be a ramous rambling shrub, with slender, four-sided, dotted, divaricated branches.—*Leaves* distant.—*Leaflets* membranaceous, rounded and sometimes slightly retuse at the base, tapering into a long acumen; besides the short five-nerves at the base, there are a few remote subtransversal ones from the rib, which all unite into a waved submarginal vein. The terminal leaflet is about four inches long; the others three-times smaller, subsessile.—*Petioles* slender-grooved, curved, an inch long, not articulated.—*Corymbs* ovate, with opposite, four-sided, three-flowered peduncles; and two opposite subulate bractes at their base.—*Flowers* large fragrant, white. *Tube* clavate, about an inch long. *Border* five-parted,

* This shrub has been in the garden about sixteen years, without having as yet produced any flowers.—N. W.

half the length of the tube.—*Berries* about the size of a small olive, of a dark purple colour; their skin separating, transparent.

Obs. In March 1818 I received specimens with flowers and ripe berries from the Honourable Mr. Edward Gardner, the resident at Katmandu in Napaul, to whose unremitting kindness and exertions the Botanic Garden owes a very great accession of the most rare and curious plants, natives of that interesting country. This handsome species is very distinct from the preceding one, but seems to approach to J. didymum, Forst and J. divaricatum, Brown Prodr. i. 521.

PHILLYREA. Schreb. gen. N. 24.

Calyx four-toothed. *Corol* one-petalled, four-cleft. *Germ* two-celled, two-seeded. *Drupe* or *Berry* superior, one or two-seeded. *Embryo* inverse, and furnished with a perisperm.

1. P. *paniculata. R.*

Arboreous. *Leaves* opposite, ovate-oblong, entire, smooth. *Panicle* terminal.

A native of *China*, and from thence introduced into the Botanic Garden near *Calcutta*, before 1793. Now, 1802, they have grown to be beautiful small trees, of from fifteen to twenty feet in height, clothed almost to the ground, with numerous, opposite, spreading, and drooping branches, covered with smooth, ash-coloured bark.

Leaves opposite, petioled, drooping, ovate, and ovate-oblong, sides incurved, entire, leathery, smooth on both sides, above a shining deep green, underneath glaucous, with two or more obscure glands near their base: length four or five inches, breadth from two to three.—*Petioles* short, recurved, channelled.—*Stipules* none.—*Panicles* terminal from the exterior axills, ramifications thereof opposite, and smooth.—*Flowers* very numerous, rather small, pure white, somewhat fragrant.—*Bractes* minute, caducous.—*Calyx* obscurely four-toothed.—*Corol* one-petalled; *tube* short; *segments* linear, revolute, smooth.—*Filaments* opposite, inserted into the bottom of the tube of the corol, and about as long as the segments of its border. *Anthers* incumbent.—*Germ* above, but lodged deep in the bottom of the calyx, two-celled, with two ovula in each. *Style* short. *Stigma* elavate, entire.—*Drupe* size of a small French bean, obliquely obovate, smooth, generally one-celled.—*Nut* solitary, with the rudiment of a second, but I never saw more than one come to maturity; pointed at the base, furrowed on the outside, covered, with a thick, somewhat indurated envelope, though scarcely hard enough to be called a nut. *Embryo* inverse, lodged in a pearl-coloured amygdaline perisperm.

Obs. Ligustrum *Japonicum. Thunb. Flor. Japan. P.* 17. *F.* 1. *is a tolerably good* representation of this tree; but the one-seeded nut, short style, and clavate stigma of my plant, will not allow me to think they can be the same.

2. P. *robusta. R.*

Leaves ovate-oblong, entire, acuminate. *Panicles* terminal, large and spreading. *Berries* sub-cylindric, one-seeded.

Bhooeemoora, the vernacular name in *Silhet,* where it grows to be a very large tree, and furnishes the natives, &c. with very hard durable wood. Flowering time June and July; the seed ripen in January and February.

Young shoots void of pubescence, but dotted with small whitish specks.—*Leaves* opposite, short-petioled, ovate-lanceolate, acuminate, entire, smooth; three or four inches long, and one or one and a half broad.—*Panicles* terminal, large, very ramous; ramifications four, round, and villous.—*Flowers* oblong, acute, villous.—*Calyx* bowl-shaped, four-toothed.—*Corol* campanulate. *Tube* very short; border four-parted.—*Filaments* short, inserted on the little tube of the corol opposite to each other. *Anthers* oblong.—*Germ* superior, depressed, two-celled, with two ovula in each, attached to the top of the partition.—*Style* and *stigma* clavate, apex flat and emarginate.—*Berries* (for they can scarce be called drupes) sub-cylindric, smooth, succulent, dark olive-purple, size of a large berberry, rarely more than one-celled.—*Seed* solitary, conform to the berry. *Integuments* two; *exterior* fibrous, and striated: *interior* rather thick, brown, adhering to the perisperm.—*Perisperm* conform to the seed, oily.—*Embryo* straight, nearly as long as the perisperm, inverse. *Cotyledons* lanceolate. *Radicle* cylindric, superior.

MILLINGTONIA. R.

Calyx three-leaved, calycled. *Corol* three-petalled, nectarial scale on the inside of each. *Germ* two-celled, two-seeded. *Drupe* with one, or two-celled, two-valved, nut seed solitary. *Embryo* curved and folded, with little or no perisperm, and with a curved inferior radicle.

Having found it necessary to deprive our countryman the late Sir Thomas Millington of the genus assigned to his memory by the younger Linneus, (*Supl.* 45 *and* 201) because, on finding the ripe seed vessel of the only species thereof, I ascertained it to be a true *Bignonia* as I suspected; I have therefore restored that respectable name to the system, under a different dress, by giving it to the two trees which at present constitute this strongly marked family, and, which I am inclined to think, have not, until now, been described.*

1. M. *simplicifolia*. *R.*

Leaves alternate, simple, broad-lanceolate.

A large tree, a native of *Silhet*, where it is called *Dantrungee* by the natives; the timber is used for various purpose. Flowers in February and March; seed ripe in July and August.

Leaves alternate, petioled, broad-lanceolar, tapering most toward the base, entire, or very remotely subserrate, in very young plants completely serrate, rather acuminate, smooth; veins simple and parallel; from six to twelve inches long, by three or four broad.—*Panicle* terminal, large, oblong, rather open, composed of many, alternate decompound, long, patent, brown, villous branches.—*Flowers* numerous, sessile, very minute, yellow.—*Bractes* oblong, clothed with

* Millingtonia, Donn. Hort. Cantab. is Roxburgh's *Flemingia*.—N. W.

ferruginous pubescence.—*Calyx* three-leaved, independently of two or three minute, villous bractes, like a *calycle*; leaflets ovate, smooth, permanent.—*Petals* three, broad-ovate, waved, twice the length of the calyx, permanent.—*Nectary* a variously lobed, smooth scale from the base of each petal on the inside, they form a dome over the pistillum, and round the base of the germ is found a flat, triangular body, with its angles bidentate.—*Filaments* two, opposite, incurved, inserted without the interior, three-angular nectary; bifid, the inner lamina supporting on its apex a patelliform receptacle, on which the two-lobed yellow *anther* rests. *Exterior lamina* bifid, *segments* subulate, rising rather higher than the anther.—*Germ* superior, two-celled, with *two ovula* in each, attached to the thickened middle of the partition. *Style* single, short. *Stigma* obscurely two-lobed.—*Drupe* size of a pea, nearly round, succulent, smooth, black, one-celled. *Nut* conform to the drupe, rarely more than one-celled, hard, cell somewhat angular; the second or abortive cell always traceable.—*Seed* solitary, concave on the side next the abortive cell of the nut, hemispheric on the other. *Integument* single, membranaceous.—*Perisperm* none.—*Embryo* curved, and doubled, as in the *Convolvulaceæ* with a curved inferior *radicle.*

2. M. *pinnata. R.*

Leaves alternate, abruptly pinnate; leaflets from six to twelve pair, sub-opposite, lanceolate.

Batiwa, the vernacular name in *Silhet,* where it in indigenous, growing to be a middling sized tree, the wood of which is employed by the natives in constructing their houses. Flowers in April and May. The seeds ripen in September.

Branchlets smooth.—*Leaves* alternate, abruptly-pinnate; from six to eighteen inches long. *Leaflets* from six to twelve pair, sub-opposite, short-petioled, lanceolate, remotely serrate, acuminate, smooth, three to six inches long, by one or one and a half broad.—*Panicles* terminal, very large and very ramous, a little hairy.—*Bractes* minute.—*Flowers* numerous, small, white.—*Calyx* three-leaved with two smaller ones under them, which may be called *bractes* or a *calycle*; or, the calyx may be called five-leaved.—*Nectary*, from the inside of the base of each petal arises a variously lobed, ragged leaflet or scale which is incurved over the germ; besides which round the base of the germ is a small, three-sided cup with bidentate angles.—*Filaments* two, opposite, length of the germ, short, broad, with enlarged apex; from the base on the outside, rises a broad, tridentate scale, of about the length of the filament. *Anthers* twin, each lobe globular, opening transversely on the inside.—*Germ* superior, cordate two-celled, with two ovula in each cell, attached to the partition. *Style* short. *Stigma* small, bidentate.—*Drupe* size of a pea, smooth, black, succulent, one-celled. *Nut* roundish thick for its size and very hard, rugose, one celled, two-valved, with a projection into the cell opposite to the remains of the style on one side near the base, which marks the abortive lobe.—*Seed* solitary, conform to the nut. *Integument* single, membranaceous.—*Perisperm* none, or in small quantity, and blended with the integument.—*Embryo* curved, and folded up nearly as in the *Convolvulaceæ*. *Cotyledons* ovate-oblong. *Radicle* inferior, curved.

OLEA. *Schreb. gen. N.* 25.

Corol and *calyx* four-cleft. *Germ* two-celled, two-seeded. *Drupe* superior, one-seeded. *Embryo* inverse and furnished with a perisperm.

1. O. *fragrans. Linn. Sp. Pl. ed. Willd.* 1. 4. 6.

Shrubby. *Leaves* oblong-lanceolate, slightly serrate. *Peduncles* axillary.

Olea *fragrans. Thunb. Jap.* 18. *t.* 2. good.

A native of *China*, in the Botanic Garden it blossoms during the cool season.

2. O. *paniculata. R.*

Leaves opposite, petioled, oblong, entire, smooth. *Panicles* axillary. *Bractes* deciduous. *Lobes* of the stigma divaricate.

A small tree; a native of the *Circar* mountains. Flowers during the hot season.

Leaves opposite, petioled, oblong, waved, entire, running down the petiole, smooth on both sides, firm, about five inches long, and two broad.—*Petioles* an inch and half long, smooth, flat above. —*Stipules* none.—*Panicles* below the leaves, and also from the axills, small, thin, brachiate.—*Bractes* small, nearly sessile.—*Calyx* four-cleft, small.—*Corol* one-petalled; *tube* shorter; *border* four-cleft; *divisions* oblong, spreading, concave; apices sharp and incurved.—*Filaments* two, opposite, short. *Anthers* large, oblong, erect. —*Germ* superior, round.—*Style* very short. *Stigma* two-cleft; *divisions* entire, acute. Ripe fruit not observed.

3. O. *dioica*. *R.*

Arboreous, dioicous. *Leaves* opposite, short-petioled, oblong, serrate. *Panicles* axillary.

Atta-jam, the vernacular name in *Silhet*, where it is indigenous; also on the hills of *Chittagong*. In both places it grows to be a pretty large tree, the timber of which is reckoned excellent, and put to many uses by the natives. Flowering time the month of March and April; fruit ripe in July.

Bark of the old ligneous parts ash-coloured; of the young shoots smooth and green.—*Leaves* opposite, short-petioled, oblong, tapering equally at each end, remotely, and acutely serrate, acuminate, smooth, from four to eight inches long, by from two to four broad.—*Stipules*, the oval concave scales of the buds only.—*Panicles* axillary, and opposite, below the leaves, thin and not above half their length.—*Male* flowers more numerous, small.—*Calyx* four-toothed.—*Corol* one-petalled, *tube* very short; *border* four-parted.—*Filaments* two, very short, inserted on the short tube of the corol. *Anthers* oval, hid in the corol.—*Germ* none, a small two-lobed gland supplies its place.—*Female* on a separate tree, (probably HERMAPHRODITE, and in that case, all the corols with the stamina, must have fallen off at a very early period; which I have no reason to think was the case, as they were examined at what I consider a proper age, and no traces of the place where a corol grew could be found.)—*Calyx* as in the male. —*Corol* none.—*Germ* superior, ovate, two-celled, with two *ovula* in each, attached to the top of the partition. *Style* short. *Stigma* two-lobed.—*Drupe* nearly round, in size and colour much like the common sloe, Prunus *spinosa*, one-celled. *Nut* conform to the seed, thin and rather brittle than hard, one-celled.—*Seed* solitary, conform to the nut. *Integument* single, thin.—*Perisperm* conform to the seed, firm, pearl-colour.—*Embryo* inverse, shorter than the perisperm, and considerably narrower. *Cotyledons* ovate-lanceolate. *Radicle* superior.

CHIONANTHUS. *Schreb. gen. N. 26.*

Calyx four-parted. *Corol* one-petalled; *segments* long. *Germ* two-celled, two-seeded.—*Drupe* superior, one or two-seeded. *Embryo* inverse without perisperm.

1. C. *ramiflora. R.*

Arboreous. *Leaves* opposite, broad-lanceolar, entire. *Panicle* below the leaves.

A native of the *Moluccas.* Flowering time March and April. Seed ripe in September.

Trunk erect, *bark* smooth, ash-coloured; *branches* opposite; *young shoots* smooth, and round.—*Leaves* opposite, petioled, broad-lanceolar, entire, acute, smooth on both sides; length about six inches.—*Petioles* channelled, smooth, about an inch long.—*Panicles* opposite, from the naked *branches* below the leaf-bearing shoots, ovate, brachiate, much longer than the leaves, smooth in every part.—*Flowers* numerous, small, yellow.—*Bractes* opposite, lanceolate.—*Calyx* four-parted.—*Corol* to the base four-cleft, consequently there is little or no tube, segments of the border lanceolate, with incurved margins.—*Filaments* short and thick, inserted both on the base of the corol and the receptacle. *Anthers* a poleniferous groove on each side of the short, thick filaments.—*Style* scarcely any. *Stigma* four-lobed.—*Drupe* oblong, one-celled.—*Nut* oblong, striped, one-celled. *Perisperm* none. *Embryo* inverse. *Cotyledons* conform to the seed.

2. C. *Zeylanica. Linn. Sp. Pl. ed. Willd.* 1. 47.

Leaves opposite, short-petioled, obovate, entire, smooth. *Racemes* axillary, often compound ; flowers ternary. *Bractes* ovate.

Linociera *purpurea. Vahl. En. Pl.* 1. 47.

Said to be a native of *Ceylon.*

3. C. *dichotoma. R.*

Shrubby, dichotomus. *Leaves* cuneiform-lanceolar, recurved, entire. *Racemes* axillary ; pedicels three-flowered.

Native place *Coromandel.* The Asiatic synonima are uncertain. Flowering time in the Botanic Garden the month of March.

Trunk short ; *bark* dark ash-coloured, with many, elevated, lighter coloured specks. *Branches* dichotomous throughout, very numerous; height of the shrubs in the Botanic Garden, about seven feet in seven years.—*Leaves* opposite, short-petioled, broad-cuneiform lanceolar, entire, rather obtuse; surfaces smooth, texture hard; three inches long by one broad.—*Racemes* axillary, solitary, scarcely half the length of the leaves. *Pedicels* sub-opposite, diverging, three-flowered.—*Flowers* small, white, inodorous.—*Calyx* four-parted; segments ovate, obtuse, villous.—*Corol*, the segments can scarcely be said to be united at the base, except by the filaments, which are inserted on the bases of two, and in that manner form two pair of segments; they are nearly exact, taper a little, and have their margins much incurved.—*Filaments* two short. *Anthers* large, with bifid apex.—*Germ* four-sided, ovate, two-celled, with two ovula in each attached to the middle of the partition. *Style* short. *Stigma* obscurely two-lobed.—*Drupe* superior, oval, or short obovate, size of a black currant, succulent, when ripe purple, one-celled.—*Nut* obovate, rather thin, and tender, marked with four ribs from the apex to the base, with four small, imperfect ones between, one-celled.—*Seed* solitary, conform to the nut. *Integument* single, thin, and tender.—*Perisperm* none. *Embryo* inverse. *Cotyledons* conform to the seed.—*Radicle* patelliform superior.

4. C. ? *smilacifolia. Wall.*

Shrubby, scandent, with four-sided branches. *Leaves* oval, acuminate, three-nerved, gloucous. *Panicle* terminal and axillary, ramous, brachiate.

A native of the forests of *Chittagong*, where it is called *Chunalata*. In the Botanic Garden it blossoms in April.

A rambling ever green ramous shrub, with ash-coloured separating bark.—*Branches* long and slender, acutely and exactly four-sided, as well as the other parts of the plant gloucous, smooth.—*Leaves* opposite, spreading, petioled, five to seven inches long, about three inches broad, acute at the base, firm and leathery, upper surface undulated (bullate) between the principal vessels, nerves and rib much elevated on the pale under surface, and uniting a little above the base; veins a few, reticulated. There are two additional very thin nerves originating at the insertion of the petiole, which together with the preceding larger ones run along the margins of the leaf, forming two arched sub-marginal veins on each side.—*Petioles* scarcely an inch long, round, channelled above, articulated at the base.—*Stipules* none.—*Panicles* short-peduncled, pyramidate, acuminate, ramous, brachiate, about the size of the larger leaves.—*Peduncles* four-sided, much sub-divided, the upper ones sub-alternate; pedicels pubescent, slightly flattened.—*Bractes* lanceolate, very small, deciduous, opposite under each subdivision of the panicle.—*Flowers* small, almost scentless, ternate, sessile, smooth, greenish yellow.—*Calyx* urceolate, inferior, with four lanceolate acute erect, purple-margined teeth.—*Corol* infundibuliform, somewhat fleshy. *Tube* twice as long as the calyx, large, obscurely four-sided. *Limb* four-parted, longer than the tube, spreading; divisions linear clavate, equal, contracted at their base, thickened blunt and slightly vertical at their end. *Mouth* shut up by the anthers.—*Filaments* two subulate, inserted above the base of the tube.—*Anthers* fleshy, with marginal cells, erect, scarcely elevated above the tube, between two opposite fissures of the border.—*Ovarium* ovate, two-celled; *ovula* solitary inserted at their base.—*Styles* (*Stigmas ?*) two, subulate, acute, parallel, very short.

Obs. This elegant shrub was introduced into the Garden in 1811, and commenced blossoming five years afterwards. Not having observed a single fruit I am doubtful of its identity with *Chionanthus*, under which however I have ad interim placed it. The erect position of the ovula seems to indicate that it belongs to the family of *Jasmineæ*, as limited by Mr. Brown in his matchless *Prodromus Floræ Novæ Hollandiæ*; in which case it may perhaps be found to constitute a new genus. The æstivation of the corol is valvate.—N. W.

SCHREBERA. R.

Calyx bilabiate. *Corol* salver-shaped. *Capsule* superior, turbinate, two-celled, two-valved. *Seeds* several, membrane-winged.

S. *swietenioides*. *R. Corom. pl.* 2. *N.* 101.

Teling. Mucaadi.

Tam. Mogalinga marum.

A large timber tree, a native of vallies over the mountainous parts of the *Raja-mundri Circar, Balaghata* mountains, &c. Flowering time the beginning of the hot season.

Trunk erect. *Bark* scabrous. *Branches* numerous, spreading in every direction, so as to form a large beautiful shady head.—*Leaves* nearly opposite, petioled, pinnate with an odd one, about a foot long. *Leaflets, three or four* pair, opposite, short-petioled, the lowermost largest, and obliquely-ovate, or cordate, while those towards the apex become narrower; all are entire, pointed, smooth on both sides, and about three or four inches long.—*Petioles* round, smooth.—*Stipules* none.—*Panicles* terminal, thin, trichotomous.—*Bractes* small, caducous.—*Flowers* rather small, white and brown variegated, very fragrant during the night.—*Calyx* beneath, tubular, somewhat two-lipped; *lips* nearly equal, emarginate, sometimes there are two lateral toothlets, one on each side in the fissures which separate the lips.—*Corol* one-petalled, salver-shaped. *Tube* cylindric, three times longer then the calyx. *Bo der* spreading, divided into five, six or seven cuneate, truncate segments.—*Filaments* two, short, inserted below the middle of the *tube*. *Anthers* oblong, hid within the tube of the corol.—*Germ* above, oval. *Style* a little longer than the tube. *Stigma* bifid.—*Capsule* large, pear-shaped, size of a pullet's egg, scabrous, very hard, two-celled, two-valved, opening from the apex.—*Seed* four in each cell, they are irregularly oval, and compressed, with a long membranaceous wing.

Obs. The wood of this tree is of a grey colour, very close-grained, heavy and durable. It is reckoned less subject to bend or warp than any other hereabouts; on which account it is employed by weavers in many parts of their looms, particularly the beam. It is also used for a great variety of other purposes.

I am inclined to think it would answer well for scales to mathematical instruments; it does not look so well as box, but its not being subject to warp may render it fitter, though not so handsome.

ERANTHEMUM. *Linn. Flor. Zeyl.* 15.

Calyx five-cleft. *Corol* hypocrateriform; border regular, or nearly so. *Filaments* four, two of them sterile. *Capsule* two-celled, two-valved, bursting with elasticity opposite to the partition.

1. E. *montanum*. *R.*

Shrubby, diffuse. *Leaves* oblong, acute. *Spikes* terminal. *Bractes* linear, hairy.

Justica montana. *R. Corom. Pl.* 2. *N.* 176.

A small, straggling, jointed, under shrub; a native of the *Circar* mountains, flowers during the cool season.

Stems many, striated, jointed, smooth, leaning on the ground, and rooting at the joints, young shoots four-sided.—*Leaves* opposite, petioled, oval and oblong, pointed, running down the petioles, from three to four inches long, and about two broad.—*Spikes* many, terminal, erect, slender. *Peduncles* four-sided.—*Bractes* exterior opposite, decussate, linear acute, hairy, one-flowered. *Bracteoles* two to each flower, pressing laterally the calyx, and like the exterior, but much smaller.—*Flowers* opposite, solitary, decussate, very pale blue.—*Calyx* five-parted; *segments* subulate.—*Corol* with a very long, and very slender tube, border of five equal obovate truncate, pale blue segments.—*Filaments* short. *Anthers* single, oblong, within the tube of the corol.—*Style* length of the tube of the corol. *Stigma* unequally bifid, appearing just without the tube.—*Capsule* linear-oblong, compressed, pointed, two-celled, two-valved, opening across the partition, with an elastic jerk.—*Seeds* two in each, compressed.

2. E. *pulchellum. Andrews. Repos. vol.* ii. 88.

Shrubby, erect. *Leaves* ventricose-oblong. *Spikes* terminal, imbricated. *Bractes* ovate-oblong.

Justicia pulchella. Roxb. Corom. Pl. 2. *No.* 177.

Justicia nervosa. Vahl. En. Pl. 1. 164.

Beng. Kalla-Jati.

Hind. Gool-sham.

A most beautiful, stout, flowering shrub; a native of dry, uncultivated, shady places, over various parts of India, flowering time the cold season. It is generally in its full beauty in February.

Stems many, erect, and ascending, or nearly so; two or three feet high. *Branches* many; young shoots four-sided, smooth.—*Leaves* opposite, short-petioled, oblong, waved, pretty smooth; from six to nine inches long, and from three to four broad; they are of a much deeper and brighter green than in E. montana, which this resembles most.—*Spikes* terminal.—*Bractes* disposed as in the former, but larger, broader, ciliate, and much nearer to each other.—*Flowers* disposed as in the former but larger, and of a deep bright blue purple colour.—*Capsule* and seeds as in E. montana.

3. E. *suffruticosum.* R.

Shrubby, erect. *Leaves,* ovate-oblong, smooth. *Spikes* terminal; *bractes* wedge-shaped, naked. *Stamens* within the tube.

A native of the *Andaman* islands, from thence several plants were brought to the Botanic Garden, by Colonel Kyd; and are now (March 1797) in blossom. It is a small shrub, and of slow growth.

Stem and branches sub-erect, jointed; the younger somewhat four sided and a little hairy.—*Leaves* opposite, short-petioled, oblong, and ovate-oblong, somewhat repand, smooth on both sides.—*Spikes* terminal, flowers opposite.—*Bractes* three-fold, one-flowered, the exterior large, wedge-shaped, and naked; the interior daggered.—*Corol; tube* length of the exterior *bractes. Border* regularly five parted, pale blue.—*Stamens* two, abortive filaments issuing from the base of the two fertile ones. *Anthers* half hid in the tube of the corol.

4. E. *diantherum.* R.

Suffruticose, erect. *Leaves* broad-lanceolar, smooth. *Racemes* terminal, in axillary, peduncled umbellets. *Corol* somewhat bilabiate. *Anthers* twin.

A small delicate, perennial plant, brought from the *Moluccas* in 1800, blossomed in the Botanic Garden in September 1801.

Stem curved, simple; *branches* few in our small young plants.—*Leaves* opposite, petioled, ovate-oblong, entire, obtuse, destitute of pubescence, somewhat *bullate;* about two or three inches long, by one broad.—*Flowers* numerous, collected on small, long-peduncled, axillary umbellets; and in opposite fascicles, or solitary, on a long terminal, slender, erect raceme; small, white, with a slight tinge of red, and some minute red spots on the base of the middle divisions of the under lips.—*Bractes,* a single, ensiform one, at the base of each peduncle, with one or two, still smaller, about the middle of each pedicel.—*Calyx* divided almost to the base into five slender, acute divisions.—*Corol; tube* long, slender, and curved. *Border* sub-bilabiate: *upper lip* erect, two-parted; *lower* three-cleft; all the five segments equal.—*Filaments* four, short, in the mouth of the tube, two of them very minute, and abortive. *Anthers* erect, twin.—*Capsule* with two, or four, somewhat scabrous flattened orbicular seeds.

Obs. Is nearly allied to my *E. racemosum,* but differs from it in having long, slender-peduncled fascicles of flowers in the axills of the leaves, as well as in having from one to three or more flowers on each of the lower opposite peduncles of the racemes, while in that there is only one; and lastly, in having the upper lip of the corol only two-parted, which in that is three-cleft.

5. E. *racemosum. R.*

Suffruticose. *Leaves* oblong, ventricose. *Racemes* terminal; *corol* somewhat bilabiate, both lips three-parted.

A small, under shrub, of great beauty, brought into the Botanic Garden from the *Moluccas,* in 1798; and blossomed in November, 1799.

Stem small, with a few, slender branches; height of the whole plant about two feet.—*Leaves* opposite, petioled, oblong, ventricose, and running down on the petiole, entire, and pretty smooth; from two to three inches long, and broad in proportion.—*Racemes* terminal, some of them long, and slender; others shorter.—*Flowers* opposite, pedicelled, approximate, pretty large, colour a pale pink, or white, tinged with red.—*Bractes* subulate, one at the base of each pedicel, and two on it.—*Calyx;* divisions equal, and subulate.—*Corol* infundibuliform; border somewhat bilabiate; both lips three-parted, the under one considerably larger.—*Filaments* four, short, inserted within the mouth of the tube, two of them minute, and abortive. *Anthers* half erect.

6. E. *barlerioides. Roxb.*

Shrubby, erect. *Leaves* ovate oblong, smooth. *Spikes* terminal. *Bractes* wedge-shaped, naked. *Stamens* within the long slender tube, adjoined to them are two abortive filaments.

A native of the *Andaman* Islands, from whence several plants were brought to the Botanic Garden, by Colonel Kyd, which blossom in March. It is a small shrub, and of slow growth.

Stem and *branches* sub-erect, jointed, the younger somewhat four-sided, and a little hairy.—*Leaves* opposite, short-petioled, oblong and ovate-oblong, somewhat repand, smooth on both sides.—*Spikes* terminal; flowers opposite.—*Bractes* three-fold, one-flowered, the exterior large, wedge-shaped, and naked; the interior daggered.—*Corol; tube* length of the exterior bractes. *Border* regularly five-parted, pale blue.—*Stamens,* two abortive filaments issuing from the base of the two fertile ones. *Anthers* half hid in the tube of the corol.

7. E. *strictum. Colebr.*

Suffruticose erect pubescent, with ascending decussate simple branches. *Leaves* lanceolar, obscurely crenulated. *Spikes* terminal, slender, much elongated. *Bractes* opposite, fourfold, remote, one-flowered.

A native of the mountains near *Sylhet,* and from thence introduced into the Botanic Garden in 1813, where it produces its beautiful large blossoms in succession in the months of January, February, and March.

Bengalee name Neel-Vas*ooka*.

A small shrub of about four or five feet in height, slightly covered with short hair.—*Stem* almost round, jointed, sending forth four-sided slender *branches* in remote pairs.—*Leaves* about four inches long, acute at each end; their margins somewhat revolute and crenulate, smooth and shining, of a peculiar greyish-green colour above, very pale, with prominent, hairy, and reticulated nerves and veins below. The uppermost or floral leaves, at the base of the spikes, approach to the size and figure of the bractes.—*Petioles* about an inch long, flattened above, and slightly marginated by the decurrent base of the leaf.—*Spikes* solitary, erect, from one to two feet long.—*Rachis* sharply four-sided, almost four-winged.—*Flowers* large dark-blue, opposite in alternate pairs, which become remote as the spike elongates.—*Bractes* adpressed imbricating, with their ends cuneate-lanceolar, dark green, acute, ciliate, about an inch long; the lowermost barren and becoming floral leaves.—*Interior* two very small, scarcely longer than the five calycine segments, and like them linear, pubescent.—*Tube* of the *corol* slender, pubescent slightly enlarged towards the mouth, about twice the length of the outer bracte.—*Segments* obovate truncate spreading and flat, equal, very pale below.—*Two* filiform barren stamina between the *filaments.* *Anthers* in the mouth of the corol with parallel cells.—N. W.

JUSTICIA. Schreb. gen. N. 35.

Corol one-petalled, irregular. *Capsule* superior, two-celled, two-valved, bursting with elasticity contrary to the partition.

Sect. 1*st. Anthers single.*

1. J. *Ecbolium. Linn. Sp. Pl. ed. Willd.* 1. 85.

Shrubby, erect. *Spikes* terminal, four-sided. *Bractes* oval, ciliate. *Corol* bilabiate, tube filiform; upper lip linear, reflexed, with bipartite apex. *Capsule* two-seeded.

Carim curini. *Rheed. mal.* 2. *t.* 20.

Hind. Oodoo-jati.

A native of our *Indian* forests. *Flowers* the greatest part of the year.

Stems several, woody, straight, round, smooth, jointed, and swelled above the joints; from two to four feet high.—*Leaves* opposite, petioled, oval, tapering a little towards each end, smooth, a little waved.—*Spikes* terminal, four-sided, imbricated with four rows of large, oval, sharply serrated, waved, one-flowered, bractes.—*Flowers* a pale blue or azure colour.—*Corol; tube* very long and slender, *upper lip* linear, reflexed, becoming revolute, apex bifid; under lip very broad, three-parted.—*Anthers* simple, linear, erect.—*Stigma* perforated, ciliate.—*Capsule* two-seeded.

2. J. *thrysiflora*. *R.*

Shrubby, erect, smooth. *Leaves* lanceolar, drooping. *Racemes* terminal, flowers verticilled. *Corol* sub-bilabiate, clavate, ten or twelve-seeded.

This beautiful stately species is a native of the interior part of Bengal, and from thence was introduced by Dr. William Carey, into the Botanic Garden; where it flowers in January and February; and the seeds are ripe in March and April.

Root ramous.—*Stem* straight, woody. *Branches* nearly straight, general height from four to eight feet; woody parts round and covered with smooth ash-coloured bark; young shoots smooth, swelled at the insertion of the leaves, somewhat four-sided, and tinged with dull purple.—*Leaves* opposite, short-petioled, drooping, lanceolar, entire, smooth on both sides, of a firm texture, the upper side deep green, the under side pale; length from four to eight inches, and from one to two broad.--*Racemes* terminal, often solitary, though sometimes two, or even three-fold, thyrsiform, shape of the flame of a candle, crowded with large orange-coloured flowers.—*Pedicels* four-fold, verticilled, approximate, one-flowered, shorter than the divisions of the calyx.—*Bractes* three-fold, subulate; the inner two permanent, and inserted on the base of the pedicels; the exterior single one larger and caducous.—*Calyx* simple, five-cleft, to the base *divisions* equal, straight, subulate, half the length of the tube of the corol.—*Corol* about an inch long. *Tube* incurved, compressed, laterally somewhat villous. *Border* bilabiate; *upper lip* broad, projecting straight forward, deeply emarginate; *under lip* of three, equal, deeply-divided, ovate, acute divisions.—*Anthers* even with the upper lip of the corol, dark brown, oval, with an opening on the under side for the discharge of the white pollen, and studded on the back.

—*Style* scarcely so long as the stamina. *Stigma* entire.—*Capsule* linear-clavate, smooth.—*Seeds* five or six in each cell.

Obs. This is one of the most beautiful species of *Justicia* I have yet met with in *India*, even when only in foliage. It grows readily from cuttings.

3. J. *vitellina*. *R.*

Shrubby, erect. *Leaves* broad-lanceolar, reflexed, smooth. *Racemes* terminal, cylindric ; flowers opposite ; below fascicled, about the middle tern, and above single ; sub-ringent, with ample-campanulate throat. *Capsule* linear, few-seeded.

A native of the hilly parts of *Chittagong*, where it blossoms in February and March.

Stems scarcely any ; but several, erect, stiff, short *branches*, with their ultimate tender divisions four-sided, the angles sharp, and rather rough ; whole height of full grown plants about three feet.—*Leaves* opposite, short-petioled, reflexed, from oblong to lanceolar, entire and pretty smooth ; length from two to four inches, by from one to two broad.—*Racemes* terminal, solitary, rigidly erect, cylindric ; from three to six inches long. *Rachis* four-angled.—*Flowers* rather small, colour yellowish pink, on the lower part of the raceme in opposite fascicles, about its middle in threes on common very short peduncles and pedicels ; above single.—*Bractes* to the single flowers solitary, to the lower three.—*Calyx* single, five-parted ; divisions broad-subulate.—*Corol* with the tube shorter than the calyx, and an ample-campanulate throat ; *border* sub-bilabiate ; upper lip projecting almost horizontally, bifid ; lower lip broad, and three-parted.—*Filaments* much curved, hairy near the base. *Anthers* single, linear, lodged under the upper lip of the corol.—*Germ* two-celled, about four *ovula* in each.—*Stigma* simple, obtuse.—*Capsule* linear, smooth.

4. J. *alba*. *R.*

Shrubby erect, smooth. *Leaves* opposite, ovate-lanceolate. *Spikes* terminal, and axillary. *Flowers* remote. *Corol* with filiform tube, and bilabiate border.

This very pretty plant, a native of the great *Andaman*, was brought from thence to the Botanic Garden, by Capt. Stokoe. Flowering time the beginning of the hot season.

Stem erect, round, smooth, while young swelled above the insertions of the leaves.—*Leaves* opposite, short-petioled, oblong, waved, smooth and shining on both sides.—*Spikes* numerous, terminal, and axillary, spreading, with a considerable recurvature, longer than the leaves, round, and smooth.—*Flowers* opposite, remote, pure white.—*Bractes*, three small ones to each flower, shorter than the calyx.—*Corol*; *tube* long, straight, filiform; *upper lip* of two, erect, wedge-shaped *divisions* united below the middle; *under lip* three-parted; *divisions* broader and pointing downwards.—*Filaments* short. *Anthers* just without the tube.

5. J. *picta*. *Linn. Sp. Pl. ed. Willd.* 1. 88.

Shrubby, erect. *Leaves* broad-lanceolate, pointed. *Racemes*, subglobular. *Tube* of the corol widening, vertically compressed, margins of the segments of the border revolute.

Tsjude-maram. *Rheed. mal.* 6. *p.* 111. *t.* 60.

Folium bracteatum. *Rumph. amb.* 4. *p.* 73. *t.* 30.

A large, elegant, ramous shrub, common in gardens, and one of our finest ornaments. I never saw it wild; is in flower most part of the year.

Leaves opposite, short-petioled, ovate-lanceolate, smooth-pointed, generally variegated with large white spots, though sometimes of a uniform green, and we have a variety with the leaves uniformly ferruginous.—*Racemes* terminal, short, erect, smooth.—*Flowers* large, generally of a beautiful crimson colour.—*Bractes* opposite; below three or four-flowered; above one-flowered.—*Corol*, *throat* compressed, *divisions* of the border soon after they expand becoming spirally revolute, with their inside wrinkled, and beautifully ornamented with small chrystalline specks.

6. J. *paniculata*. *Linn. Sp. Pl. ed. Willd.* 1. 89.

Herbaceous, erect, four-sided, brachiate. *Leaves* sub-sessile, lanceolate. *Racemes* terminal; flowers secund erect. *Corol* bilabiate; lips linear, revolute. *Anthers* coalesced, and bearded at the base.—*Capsules* linear, many-seeded.

Beng. Kal*a*-megh, Kal*u*p-nath, or M*u*ha-t*i*ta; the last name implies *King of bitters*, and a very powerful, much esteemed one it is.

Kreat is the name by which it is known on the *Malabar* coast, &c. where it is much used in medicine, and is the bitter of the famous Nostrum, called *Drogue amere*.

Cara-canirum. *Rheed. mal.* 9. *p.* 109. *t.* 56.

Teling. Nella-vemoo.

A native of dry ground, under the shade of trees, bushes, &c. Flowering and seed time, the wet and cold season.

Stem erect, generally annual, at least down to the root, branchy; slender, four-sided, jointed, smooth, one to two feet high. *Branches* opposite, cross-armed, four-sided, spreading.—*Leaves* opposite, short-petioled, lanceolate, entire, smooth; two or three inches long.—*Racemes* terminal, and from the exterior axills, horizontal, long secund, four-sided.—*Flowers* remote, rose-coloured, long-pedicelled, alternate, erect, downy.—*Bractes*, a large, exterior, opposite pair; within the flower-bearing one there are two smaller, embracing the base of the pedicels.—*Corol*, bilabiate. *Tube* recurved; lips linear, and reflected; the *superior* broadest, three-toothed; the *inferior* two-toothed.—*Filaments* as long as the lips of the corol, projecting, hairy. *Anthers* very flat, obovate, firmly united at the base, and three-bearded.—*Capsule* erect, sub-cylindric.—*Seeds* three or four in each cell, cylindric.

7. J. *echioides*. *Linn. Sp. Pl. ed. Willd.* 1. *p.* 87.

Annual, erect, four-sided, hairy. *Leaves* sessile, decussate, oblong and linear. *Spikes* axillary, recurved over the bosoms of the leaves, secund. *Capsule* oblong, with four, verrucose seeds.

Pee-tumba. *Rheed. mal.* 9. *t.* 46.

A native of *Coromandel,* and generally found on old walls; in flower all the year.

Stem annual, erect, four-sided, hairy, with ascending, opposite, similar branches from the base; whole height about a foot.—*Leaves* opposite, decussate, sessile, from oblong to linear, obtuse, recurved, hairy; from one or two inches long.—*Spikes* axillary, spreading out over the bosom of the leaves, and in general shorter than them; in luxuriant plants compound.--*Bractes* opposite, small, lanceolate, hairy.—*Flowers* alternate, sessile, erect, upper lip purple, the rest white. —*Calyx* of five, long, hairy, ensiform divisions.—*Corol* bilabiate. *Tube* sub-cylindric, *upper lip* erect, broad, and trifid, *under lip* narrower, recurved, and bidentate.—*Filaments* projecting, straight, and about as long as the lips of the corol. *Anthers* simple, united and bearded at the base.—*Capsule* oblong, erect, hairy, length of the calyx.—*Seeds* generally two in each cell, oblong, rough.

8. J. *polysperma. R.*

Annual, creeping, round, jointed. *Leaves* sessile, lanceolate, downy. *Spikes* imbricated. *Bractes* three-fold, one-flowered, woody. *Capsules* with from twenty to thirty seeds.

Beng. Chhota-micheta.

Is a native of *Bengal;* appears during the cold season amongst grass, &c. uncultivated places.

Stems diffuse, or creeping, very ramous, six to twelve inches long, round, jointed, a little downy.—*Leaves* opposite, sessile, lanceolate, entire, downy.—*Spikes* terminal, and axillary, imbricated with opposite, decussated flowers, and bractes.—*Flowers* small, pale blue.—*Bractes* three-fold, one-flowered, the lower one broader a little than the two lateral, all are woolly.—*Calyx* woolly.—*Capsules* linear, containing from twenty to thirty, small roundish seeds.

9. J. *acaulis. Linn. Sp. Pl. ed. Willd.* 1. 84. *R. Corom. pl.* 2. *N.* 127.

Root perennial, stemless. *Leaves* radical, sessile, ovate-oblong, crenate. *Spikes* ascending, imbricated. *Corol* ringent. *Capsules* many-seeded.

Elytraria *crenata. Vahl. Enum. plant.* 1. 106.

Teling. Nella mura.

Grows on pasture ground, generally under the shade of trees. Flowering and seed time the wet, and cold season.

Root perennial.—*Stem* none.—*Leaves* radical, spreading close on the ground, sessile, wedge-form or oblong, crenulated, rugose, woolly on the under side; from six to twelve inches long.—*Spikes* radical, several, as long as, or longer than the leaves, ascending, imbricated from top to bottom, with small, rigid, pointed bractes, each of the superior ones embracing a flower, and two small lateral bractes.—*Flowers* small, white.—*Calyx* five-leaved, the superior leaflets much the largest.—*Corol; tube* cylindric, recurved; *upper lip* vaulted, *under lip* broad, three-parted, with the divisions two-parted.—*Capsules* conical, eight-seeded.—*Seeds* kidney-form.

Sect. 2. Corols bilabiate. Anthers double.

10. J. *nasuta. Vahl. Symb.* 2. 15. *Linn. Sp. Pl. ed. Willd.* 1. 90.

Shrubby, erect. *Leaves* short-petioled, lanceolate. *Panicle* axillary and terminal. *Upper lip* of the corol linear, and bidentate; *under lip* broad, three-lobed.

Pul-colli. *Rheed. mal.* 9. 135. *t.* 69.

Sans. यूथिकापर्णी, Yoothika-purnee.

Beng. Jooi-pana.

Teling. and *Tamul.* Naga-mulli.

This species I have only found in gardens: from whence it originally came, I cannot say. *Flowers* all the year round, but never ripens its seed, owing, I imagine, to its being always propagated by cuttings, layers, or slips. It grows to be a large thin shrub, of five, or more feet in height.

Root woody, ramous.—*Stems* many, erect, ramous, the old woody parts round, and covered with pretty smooth ash-coloured bark; the tender branches, and young shoots jointed, smooth, and obsoletely six-seeded.—*Leaves* opposite, petioled, broad-lanceolate, point obtuse, above smooth, below a little downy, entire; from two to four inches long, and from one to two broad.—*Panicles* corymbiform, axillary, and terminal, always three-cleft, as also the subdivisions.—*Peduncles* and *pedicels* short, round, a little downy.—*Bractes* minute.—*Flowers* small, white.—*Corol* with a long, slender, compressed tube; *under lip* broad, three-cleft, *upper lip* erect, linear, sides reflected, apex bifid.—*Nectary*, a fleshy ring surrounding the base of the germ.—*Anthers* without the tube, twin.—*Capsule;* —I have not seen it.

Obs. Milk boiled on the roots, is reckoned, by the Indian physicians aphrodisiacal, and as such often employed; for no medicines are more sought after by the natives of *India* than those which are imagined to belong to that class.

The roots rubbed with lime juice, and pepper, are used, and often with good effect, to cure the *ring worms,* or Herpes miliaris, which in *India* is a most troublesome disease, and very common.

They are also used for the bite of poisonous snakes; hence the *Telinga* and *Tamul* name *Naga mulli,* or Jasmine of the *Cobra de capella: Naga* is therefore prefixed to the name of most of their antidotes, or medicines supposed to destroy the effect of poison, and occurs very often in their writings. For like the ancient Physicians of *Greece, Rome,* &c. those of *India* at this day, make poisons, and their antidotes, a very chief part of their studies; but from the seeming inactivity of a very large share of them, we may safely conclude, that their practice is equally frivolous with that of the ancients, and ought to be discouraged as much as possible; because at least, they prevent the application of more efficacious remedies.

11. J. *lanceolaria.* R.

Shrubby, erect. *Leaves* petioled, lanceolar, smooth. *Fascicles* terminal, panicled, clammy, from two to four-flowered, embraced by many, long, ensiform, bractes. *Tube* of the corol curved, and twisted; *upper lip* three-dentate, *under lip* entire.

This beautiful shrub, is a native of the *Silhet* district, where it is called Toogee. Flowering time the cold season.

Stems erect, or nearly so, *branches* round and smooth, and while young, swelled just above the insertion of the leaves; whole height from two to four feet.—*Leaves* opposite, petioled, lanceolar, (that is tapering equally at each end) entire, acute, smooth; from three or six inches long by one or two broad.—*Flowers* numerous, large and rosy, collected in numerous, terminal, clammy fasicles of from two to four each, and embraced by numerous, ensiform, clammy bractes, like a common calyx, and longer than the proper perianths; the whole forming a pretty large panicle at the end of each branch.—*Calyx* five-parted; segments ensiform, smooth.—*Corol; tube* long, slender, curved outward, and like my *J. speciosa, latebrosa,* &c. has a half twist; which, as in the *orchideæ,* brings the lower *lip* uppermost. *Border* bilabiate, *lips* nearly equal, oblong; the *superior,* in its inverted state, three-toothed; the *inferior* entire.—*Filaments* from the mouth of the tube of the corol; shorter than the lips, and projecting straight over the under one. *Anthers* double, one at the end of the other, forming one straight line; the exterior half of the posterior one being attached to the apex of the filaments and to the base of the exterior anther.

12. J. *speciosa. R.*

Perennial, erect. *Leaves* ovate-cordate. *Peduncles* axillary and terminal, proliferous, few-flowered. *Bractes* opposite, foliaceous. *Corol* bilabiate, tube curved, and twisted; lips equal; upper one tridentate, lower one bidentate.

A native of the interior parts of Bengal, where it blossoms during the cold season, and is then one of the greatest ornaments of the forests.

Stems erect, jointed. *Branches* few, and erect. *Bark* of the lig-nous parts ash-coloured; of the young shoots smooth and green. —*Leaves* opposite, decussate, petioled, ovate-cordate, entire, rather obtuse, a little wrinkled; from one to four inches long.—*Peduncles* axillary, often proliferous and much longer than the petioles, supporting two, or three flowers.—*Flowers* large, beautiful, bright carmine colour.—*Bractes,* or floral leaves opposite, sessile, from ovate to linear.—*Calyx* simple, divided almost to the base into five ensiform, equal portions.—*Corol; tube* long, slender, curved, villous, and somewhat twisted; *upper lip* linear-oblong, recurved over the tube, apex three-toothed, at the base some dark coloured spots; *under lip* broad lanceolate, pointing directly to the earth, apex slightly two-toothed.—*Filaments* inserted in the mouth of the tube of the corol, nearly as long as its border, and, with the lower lip, pointing downwards. *Anthers* double, linear, purple, with pale coloured pollen.—*Style* as long as the stamens, and beyond the tube taking the same direction with them. *Stigma* slightly emarginate. —*Capules* pedicelled, with two seeds in each cell.

13. J. *tinctoria. R.*

Shrubby. *Leaves* petioled, ovate-oblong. *Flowers* terminal, fascicled, involucred. *Anthers* linear.

Folium tinctorium. Rumph. amb. vi. 51. *tab.* xxii. *f.* 1.

Beng. Batia-rung.

Found in the woods about *Calcutta,* in full flower in October.

Stem scarcely any; *branches* numerous, sub-erect; young shoots jointed and swelled above the joints, round a little downy.—*Leaves* opposite, petioled, oblong, entire, covered with a little harsh pubescence, two to three inches long.—*Peduncles* terminal, forming the exterior axills, each supporting some few fascicles, of four to six sessile flowers.—*Bractes* of the peduncles and pedicels opposite, those of the flowers, or rather, the *involucres* subulate, opposite, sessile, linear oblong, cordate, entire, unequal; the upper one being the largest, and incumbent over the others, and the flowers they contain: of the calyx small.—*Flowers* sessile, large, of a beautiful deep pink colour, opening in succession.—*Calyx* five-toothed.—*Corol, tube* length of the largest involucres, filiform, having a small twist near the apex. *Border* two-parted, both divisions oblong, the upper one three-toothed.—*Nectary* a gaping cup in which the germ rests.—*Stamens* projecting. *Anthers* two on each filament.—*Style* rather longer than the stamens. *Stigma* obtuse.

14. J. *bivalvis. Linn. Sp. Pl. ed. Willd.* 1. 82.

Shrubby. *Leaves* lanceolate. *Peduncles* axillary, and terminal, three to six cleft; lateral pedicels often trifid; ultimate divisions bearing a small head of sessile flowers, between two, unequal, linear bractes. *Capsule* clavate.

A native of the *Moluccas.*

Folium tinctorium of Rumphius, quotted under the preceding species, does not so well agree with J. bivalvis. I think, *Adel-Odagam, Rheed. mal.* ix. 81. *t.* 43. belongs to neither of them.

15. J. *chinensis. Vahl. En. Pl.* 1. 110.

Perennial, retroflex. *Leaves* cordate. *Umbellets* axillary, from three to five, and with the pedicels (from three to five), distichous, involucels obovate, unequal. *Capsules* round and thin.

J. *chinensis. Linn. Sp. Pl. ed. Willd.* 1. 80. *Burm. ind.* 8. *t.* 4. *f.* 1. is bad if intended for this plant.

A middle-sized, straggling, suffruticose species introduced from China into the Botanic Garden, where it flowers from August, till December.

Stem scarcely any; *branches* several, scarcely ligneous, straggling and ascending, slender, somewhat six-angular, pretty smooth; striking root where they rest on the ground; from one to two feet long.—*Leaves* opposite, petioled, ovate-cordate, entire, somewhat acuminate, smooth, from one to two inches long.—*Peduncles* axillary, from three to five in each axil, distichous; of various length, each supporting a little, distichous umbellet of from three to five, one or two-flowered, short pedicels.—*Involucres* of the pedicels two, subulate, opposite, of the flowers two, sessile, unequal, obovate, or oblong-ventricose, embracing between them one or two, complete flowers, and one or two rudiment.—*Bractes* (or inner calyx), ensiform, two to each flower, and rudiment.—*Flowers* small, pale-pink.—*Calyx* five-parted, segments ensiform.—*Corol; tube* length of the involucres, a little twisted.—*Border* bilabiate, *upper lip* linear, three-toothed; *under lip* oval, and slightly emarginate.—*Anthers* two on each filament, purple.—*Capsule* pedicelled, small, flat, round, villous, two-celled, with two thin round seeds in each cell.

Obs. I have two varieties (if not distinct species) from the Coast of Coromandel. In the first, the leaves are much narrower, and run more down on the petioles, and, with the bractes of the flowers, longer, more cuspidate, and ciliate. In the second variety, the leaves are cuneiform-lanceolate, and the bractes of the flower broad-ovate, short cuspidate, and hairy.

16. J. *latebrosa. Kön. MSS.*

Herbaceous, branches straggling, round, smooth. *Leaves* petioled, cordate-oblong, pointed. *Tube* of the corol twisted, lips almost equal; the upper one three-dentate, the lower one bidentate. *Seeds* solitary.

Grows always among bushes, or under their shade, in a dry barren soil. Flowering time, the wet and cold seasons.

Root perennial.—*Stem* scarcely any; *branches* many, straggling, weak, round, smooth, jointed, ramous.—*Leaves* opposite, petioled, cordate-oblong, pointed, entire, smooth; size very various.—*Flowers* terminal, large, of a bright red colour.—*Bractes* numerous, lanceolate.—*Calyx* single.—*Corol; tube* twisted at the middle; *lips* equal, linear; the *upper* one three-toothed and the *under* one emarginate.—*Stamens* as long as the lips of the corol, projecting along the under lip. *Anthers* double.—*Stigma* two-cleft.—*Capsules* two-seeded, one in each cell.

17. J. *bicalyculata. Linn. Sp. Pl. ed. Willd.* 1. 81.

Annual, erect, six-sided, scabrous. *Leaves* short-petioled, ovate-cordate, oblong, pointed, calyx double; tube of the corol twisted, border bilabiate, seeds solitary.

Dianthera malabarica. Suppl. 85.

J. *malabarica. Ait. Hort. Kew.* 1. 27.

J. ligulata. *Lam. illustr.* 1. 40. *t.* 12. *f.* 2. *Cavan. icon.* 1. 52. *t.* 71.

Beng. Nasa-bhaga.

Teling. Chebiera.

A poor looking plant, a native of rubbish, road sides, and hedges. Flowering time the wet, and cold seasons.

Stem erect, annual, ramous, six-angled, jointed, scabrous, and hairy, from three to five feet high.—*Leaves* opposite, petioled, ovate-cordate, hairy, somewhat scabrous, size very various.—*Flowers* terminal, and axillary, numerous, red, generally three-fold, or rather sub-panicled, and divided by threes.—*Bractes* small.—*Calyx** double, or rather some *bractes* embracing the proper perianth.—*Tube* of the corol with a half twist about the middle; *upper lip* narrowest, linear, three-toothed; *lower lip* broad, oblong, emarginate.—*Filaments* shorter than the corol, projecting along the under lip. *Anthers* red, two on each filament.—*Capsules* four-seeded, two in each cell.

Sect. 3. Corols ringent. Anthers double.

18. J. *Adhatoda. Willd.* 1. 96.

Sub-arboreous. *Leaves* broad-lanceolar. *Spikes* short, long-peduncled; exterior bractes ovate, permanent. *Corol* ringent, upper lip vaulted, and emarginate.

Sans. वैद्यमातृ, Vidyumatri, सिंही, Singhee, वासिका, Vasika, वृषः, Vrishu, अटरूषः, Uturoosha, सिंहास्यः, Singhashya, वासुकः, Vasooka, वाजिदन्तकः, Vajiduntuka.

Hind. and *Beng.* Bakus, Vasooka.

* The outer calyx may be called common, for there are always rudiments or buds of more than one flower in it, which often expand in succession.

This small tree or large shrub, is common over most parts of *India*. Flowering time the cold season.

Trunk straight, bark pretty smooth, ash-coloured. *Branches* sub-erect with bark like that of the trunk, but smoother.—*Leaves* opposite, short-petioled, broad-lanceolar, long, taper-pointed, smooth on both sides, about five or six inches long, and one and a half broad.—*Spikes* from the exterior axills, solitary, long-peduncled, the whole end of the branchlet forming a leafy panicle, flower-bearing portion short, and covered with large bractes.—*Flowers* opposite, large, with small ferruginous dots; the lower part of both lips streaked with purple.—*Bractes* three-fold, opposite, one-flowered. *Exterior* one of the three, large, ovate, obscurely five-nerved. *Interior* pair, much smaller, and sub-lanceolate; all are permanent.—*Calyx* five-parted to the base; divisions nearly equal.—*Corol* ringent. *Tube* short; *throat* ample; *upper lip* vaulted, emarginate; *lower lip* broad, and deeply three-parted; both streaked with purple.—*Filaments* long, resting under the vault of the upper lip. *Anthers* twin.

Obs. The wood is soft, and esteemed very fit for making charcoal for gun-powder.

19. J. *decussata*. *R*.

Shrubby, erect; branches twiggy, villous, and decussated. *Leaves* from ovate to oblong. *Flowers* in opposite fasciculi on the leafless branchlets. *Anthers* double, all calcarate.

A tall, stout shrubby plant, a native of the country above *Rangoon;* and from thence introduced into the Botanic Garden, where it is in flower during the mouth of February and March.

Stem straight to the top of the plant, ligneous; in two years, old plants as thick as a walking cane. *Branches* regularly decussate, straight, all the younger shoots villous, and as they shoot into flower nearly leafless.—*Leaves* short-petioled, ovate-oblong, entire, pretty smooth; from three to twelve inches long.—*Flowers* in numerous, a little, opposite fascicles, of three to six, appearing subverticelled, over the slender, leafless twigs, subsessile, pale rose colour.—

Bractes numerous, slender, and very downy.—*Calyx* five-parted; segments ensiform, villous.—*Corol* ringent; *tube* gibbous near the base; *upper lip* sub-lanceolate, concave, erect; *under lip* broad and three-lobed.—*Filaments* nearly as long as the upper lip; apex much enlarged. *Anthers* two on each filament, all with a calcarate base.—*Germ* oblong, sitting in a large, glandular, yellow ring. *Style* the length of the stamens. *Stigma* simple.

20. J. *Gendarussa.** *Linn. Sp. Pl. ed. Willd.* 1. 87.

Shrubby, diffuse, smooth. *Leaves* narrow-lanceolar. *Spikes* terminal, sub-verticelled. *Lower* anthers calcarate.

Vada-Kodi. *Rheed. mal.* 9. *p.* 79. *t.* 42.

Gendarussa. *Rumph. Amb.* 4. *p.* 70. *t.* 28.

Beng. Jug*ut*-m*u*d*u*n.

A handsome shrub. I have never met with it wild, but in gardens it is common, and grows readily from slips and cuttings; it is in flower during the wet season chiefly. I am told it is indigenous on the *Malay* Islands.

Stem scarcely any. *Branches* numerous, long, straggling, except when kept cut, they are then more erect. *Bark* of the young parts generally dark purple, and very smooth; in some varieties green.—*Leaves* opposite, short-petioled, lanceolar, obtuse, frequently a little scolloped, smooth; nerve and veins dark purple, from three to six inches long, and from half an inch to an inch broad.—*Spikes* terminal, erect, verticelled.—*Anthers* double, the lower ones have a spur-like process projecting downwards and outwards.

21. J. *Betonica. Linn. Sp. Pl. ed Willd.* 1. 96.

Shrubby. *Leaves* sub-ovate. *Spikes* terminal. *Bractes* white, with

* Willdenow has spoiled this name. Linneus wrote it Gandarussa, which is nearer the truth. The word is গন্ধরুস, G*u*ndh*a*r*u*s*a*.

green veins. *Corol* ringent. *Anthers* two-lobed, the lower calcarate. *Seeds* turbercled, two in each cell.

Bem-Curini. Rheed. mal. 2. *p.* 33. *t.* 21.

A perennial species; a native of *Coromandel.* Flowering time the dry season chiefly, though it is more or less in blossom all the year, particularly when in a cultivated state.

Stems several, while young, erect by age, leaning in various directions, round, smooth, swelled above the joints, the height of the whole plant from two to four feet.—*Leaves* opposite, petioled, nearly ovate, entire, or slightly waved, smooth on both sides; length from one to six inches.—*Spikes* terminal, secund, sub-cylindric, erect.—*Bractes* of the rachis opposite, ovate-oblong, villous, white with green veins, the posterior one always abortive; those of the flowers opposite, and resembling those of the rachis, but narrower.—*Flowers* pretty large, whitish, with rose-coloured veins.—*Calyx* single, five-cleft; divisions ensiform.—*Corol* ringent; upper lip vaulted, and two-toothed; lower three-parted.—*Anthers* twin, the lower one with a spur from the lower end.—*Capsule* longer than the calyx, villous.—*Seeds* two in each cell, tending to be round, rough with numerous sharp points.

22. J. *ramosissima. R.*

Shrubby, creeping, very ramous. *Leaves* sub-ovate. *Spikes* terminal, secund. *Bractes* lanceolate. *Lower* lobe of the anthers calcarate.

A native of Coromandel. Flowering time, in the Botanic Garden, the cold season.

Stem none, *branches* numerous, rising in every direction, as well as spreading on the ground, and striking root at the swelled joints; the tender parts striated, and slightly villous.—*Leaves* opposite, petioled, sub-ovate, rather obtuse, tapering down the petioles nearly to their base.—*Spikes* terminal, secund, erect, many times longer than the leaves.—*Bractes* in opposite pairs, lanceolate, coloured green and white; the posterior one of the rachis always abortive, the other two are calycine.—*Flowers* scarcely longer than the bractes, dull white, with a few rosy veins on the centre of the lower lip.—*Calyx* of five linear-lanceolate, acute segments.—*Corol* ringent; *upper lip* vaulted, with a bidentate apex; *under lip* broad, and three-cleft.—*Anthers* double; lower lobe with a long curved, sharp spur.—*Capsule* oblong, four-seeded, very protuberant at the seeds.—*Seeds* nearly sphærical, rugose.

Obs. In the foliage, spikes, flowers, and capsules, this species is exceedingly like *J. Betonica*, which however, is always a tall, erect plant, with coloured bractes, while this is very bushy, spreads on the ground, and strikes root abundantly at the joints; the bractes also are narrower, and greener.

23. J. *tranquebariensis. Linn. Sp. Pl. ed. Willd.* 1. 99.

Herbaceous. *Leaves* round, or broad-cordate. *Spikes* terminal, four-sided, occasioned by four rows of roundish ciliate bractes. *Flowers* solitary, in two rows on the fore-part of the spikes. *Anthers* calcarate.

Common amongst bushes on a dry barren soil, over the coast of Coromandel. Flowering time the wet, and cold season.

Roots perennial, somewhat granulated.—*Stems* in young plants erect, in old ones uncertain; *branches* annual, small, straggling, round, a little downy, jointed, and swelled above the joints, from one to two feet long.—*Leaves* opposite, short-petioled, oval, or round, entire, pretty smooth.—*Spikes* terminal, peduncled, long, erect, generally four-sided, occasioned by four rows of orbicular, short-petioled, opposite, ciliate bractes; besides these there are a pair of small pointed ones which press the calyx laterally.—*Flowers* a mixture of pale yellow, and red; standing in the axills of the two rows of bractes that occupy the fore part of the spikes; on the back part they are generally wanting.—*Calyx* single.—*Corol; upper lip* emarginate, erect; *under lip* broad, three-cleft. *Anthers* double, the lowermost calcarate.—*Seeds* two in each cell, bristly.

24. J. *glabra. Kön. MSS.*

Herbaceous, erect, branchy, round, smooth. *Leaves* petioled, ovate, pointed. *Peduncles* trichotomous. *Upper lip* of the corol vaulted. *Lower anthers* calcarate. *Capsules* four-seeded.

A native of similar places with the last described. Flowering time the same.

Root perennial.—*Stems* several, erect, branchy; annual, round, smooth, jointed.—*Leaves* opposite, oblong-cordate, pointed, entire, smooth.—*Panicles* axillary, small, three-forked.—*Flowers* rose-coloured, yellow, and whitish green mixed.—*Bractes* small, pointed; generally two below every division of the panicle, and two or three pressing upon the calyx.—*Calyx* single.—*Corol; tube* short, widening; *upper lip* erect, vaulted; *under lip* broad, and three-parted.—*Filaments* concealed under the upper lip. *Anthers* double, the lower ones calcarate.—*Stigma* slightly bifid.—*Capsule* four-sided, two seeds in each cell.

25. J. *tomentosa. R.*

Herbaceous, sub-erect, hairy. *Leaves* lanceolar, tomentose. *Spikes* terminal, strobiliform. *Anthers* distinct. *Capsules* conic, four-seeded.

A native of wet, rich, pasture ground. Flowering time the wet and cold seasons.

Root nearly spindle-formed, perennial.—*Stems* several, sub-erect, round, hairy, from one to two feet high.—*Leaves* opposite, petioled, broad-lanceolar, entire, very downy; the lower many times larger than the upper ones, the former being often twelve inches long, and the latter (or floral *leaves*) not more than one.—*Spikes* terminal, oval, strobiliform, sessile on little terminal branchlets.—*Flowers* small, bright, bluish purple.—*Bractes* numerous, imbricated, oval, pointed, concave, hairy, one-flowered.—*Calyx* four-cleft; the lower leaflet two-cleft.—*Corol; tube* gibbous on the under side, contracted at the mouth; *upper lip* ovate, half two-cleft, divisions pointed; *under lip* very broad, three-parted.—*Anthers* twin, equal, oval.

—*Stigma* two-cleft; segments broad.—*Capsules* pointed, eight-seeded.

26. J. *orixensis*. *Kön. MSS.*

Herbaceous. *Branches* straggling, foursided. *Leaves* cordate, repand. *Spikes* terminal. *Bractes* orbicular, ciliate. *Capsules* length of the bractes, containing about four echinate seeds.

Obs. It is a native of high, dry, gravelly ground, in the vicinity of *Samulcota.* Is very like *J. tranquebariensis.*

27. J. *repens*. *Linn. Sp. Pl. ed. Willd.* 96. *Corom. Pl.* 2. *No.* 152.

Herbaceous, diffuse. *Leaves* sub-sessile, lanceolate. *Spikes* axillary and terminal, compressed. *Bractes* ovate, white, membrane-bordered, posterior rows neuter. *Lower anthers* crescent-shaped.

Adhadota. spicata flosculos ex foliolis membranaceis producens. *Burm. Zeyl.* 7. *t.* 3. *f.* 2.

A native of pasture ground. Flowering time the wet and cold season.

Root perennial.—*Stems* many, diffuse, round, jointed, smooth, sometimes rooting at the joints; one to two feet long.—*Leaves* opposite, short-petioled, broad-lanceolate, entire, sharp-pointed, smooth.—*Spikes* axillary peduncled, erect, somewhat compressed, secund.—*Bractes; exterior,* four rows, oval, with a broad, white, membranaceous margin; the two rows on the back of the spike neuter; the two on the fore part embracing the flowers; the calyx of each flower is also embraced laterally by two smaller lanceolate bractes.—*Flowers* alternate, pale rose-coloured.—*Corol, stamens, pistil* and *capsule* as in *J. procumbens,* which it much resembles.

28. J. *procumbens*. *Linn. Sp. Pl. ed. Willd.* 1. 87.

Root perennial; *branches* diffuse, herbaceous. *Leaves* sessile, from orbicular to linear-lanceolate. *Spikes* terminal, four-sided. *Bractes* lanceolate, rigid. *Lower* anthers calcarate.

Euphrasia, Alsines angustiore folio, &c. *Pluk. Alm.* 142. *t.* 56. *f.* 3.

Very common on pasture ground over the coast of *Coromandel*, *flowering* most part of the year.

Root often perennial.—*Stems* many, spreading, jointed, five-striated, often rooting at the joints, extremities sub-erect; size very various, but generally from six to twelve inches long.—*Leaves* opposite, subsessile, from round to linear-lanceolate, entire, a little downy; size various.—*Spikes* terminal, erect, four-sided.—*Flowers* opposite, decussate, small, rose colour.—*Bractes* three-fold, lanceolate.—*Calyx;* upper divisions very minute.—*Corol; tube* short; *upper lip* erect, two-cleft; *under lip* broad, three-parted.—*Nectary* a yellow ring surrounding the base of the germ.—*Anthers* double; the lower calcarate or sharp-pointed.—*Stigma* slightly bifid.—*Capsules* four-seeded, two in each cell.

Obs. When this plant grows in wet places, it seems altogether another species, being many times larger, nearly erect, with linear-lanceolate leaves. I believe Dr. König described this variety as a distinct species, under the name of *J. juncea*, or *graminea*.

29. J. *quinqueangularis*. *Kön. MSS.*

Perennial in a dry soil; branches five-sided. *Leaves* sessile, lanceolate. *Spikes* terminal. *Bractes* linear. *Capsules* rather larger than the bractes, four-seeded.

A native of the borders of rice-fields about *Samulcota*. It is but little removed from *procumbens*, when growing in the same soil.

30. J. *pectinata*. *Corom. Pl.* 2. *N.* 153.

Herbaceous, diffuse. *Leaves* short-petioled, lanceolar. *Spikes* sessile, secund. *Bractes* five-fold, broad, white, membrane-bordered: lower anthers calcarate.

Compare with *J. parviflora*. *Retz. Obs.* 5. *p.* 9.

A very beautiful, delicate, diffuse species, growing in shady places. Flowering time the cold season.

Stems many, diffuse, filiform, round, smooth, jointed. *Branches* alternate.—*Leaves* opposite, short-petioled, depending, lanceolate, entire, smooth; the pair are generally unequal in size.—*Spikes* axillary or terminal, sessile, much shorter than the leaves, secund, generally three together, the middle one largest.—*Flowers* very small, bright blue.—*Bractes*, to each flower five, of different sizes, oval or lanceolate, with beautiful white, membranaceous margins.—*Corol*; upper lip entire, acute.—*Anthers*, the lower one of each filament has a long obtuse horn from the under end.—*Capsules* four-seeded, two in each cell.

31. J. *lamifolia. Kön. MSS.*

Whole plant very downy, procumbent, annual. *Leaves* oval. *Spikes* terminal, imbricated. *Calyx* four-leaved.

Found in the vicinity of Calcutta, in uncultivated shady places.

Root fibrous.—*Stems* and branches very numerous, round, spreading close on the earth, clothed with long soft hair, and down, somewhat glutinous, length from one to three feet.—*Leaves* opposite, sub-sessile, oval, entire, clothed and glutinous like the branches; about an inch long.—*Spikes* terminal, oblong, closely imbricated with the bractes, and flowers.—*Flowers* numerous, small, variegated with deep and light purple.—*Bractes* ovate, one-flowered, hairy, clammy.—*Calyx* four-leaved, the under leaflet with its apex bifid.—*Corol* ringent. *Tube* gibbous; throat compressed; *upper lip* two-parted, and streaked with deep purple; *under lip* broad, and three-parted.—*Filaments* short, inserted into the mouth of the tube. *Anthers* twin, singly oval, and all equal.—*Stigma* of two, broadish, spreading lobes.—*Capsule* conical.—*Seeds* eight or ten, roundish.

Obs. It has the calyx of a *Barleria*, and the stamina of a *Dianthera*. Mr. R. Brown informs me that he thought it would belong to a new genus, which he calls *Nelsonia*.

32. J. *verticillata. R.*

Erect, hairy, spines verticelled with their apices bifid, or trifid.

Leaves petioled, ovate. *Flower* amongst the verticils of spines, sub-sessile.

Hind. Custoola.

Found by Dr. Hunter in the vicinity of *Ougein*, in seed and flower in March.

GRATIOLA. Schreb. Gen. N. 37.

Corol one-petalled, irregular. *Sterile* filaments two, affixed to the lower lip of the corol. *Anthers* double, and connected. *Capsules* superior, two-celled, two-valved. *Seeds* numerous.

1. G. *amara. R.*

Annual (in Bengal), creeping. *Leaves* petioled, cordate, serrate. *Peduncles* axillary, four to six flowered. *Calyx* four-leaved, exterior pair cordate, inner subulate, apices of the sterile filaments free. *Capsule* ovate, hid in the calyx.

Caranga amara. *Vahl. Enum. vol.* 1. 100.

Serratula amara. *Rumph. amb.* 5. *p.* 459. *t.* 170. *f.* 1. which has been quoted by mistake, I presume, for *Scutellaria indica.*

This pretty plant has been accidentally introduced into the Botanic Garden, among spice and other plants from the *Moluccas.* Flowering time the cold season.

Root fibrous.—*Stems* and *branches* creeping, with their extremities sub-erect, four-sided, coloured, smooth.—*Leaves* opposite petioled, cordate, deeply serrate, somewhat rugose, of a soft texture, but free of pubescence, general length about two inches.—*Peduncles* axillary, solitary, rather longer than the petioles, supporting at their extremities, two or three pairs, of opposite pedicelled flowers —*Bractes* small, oblong, one under the insertion of each pedicel. —*Calyx* of two pair of opposite leaflets; *exterior* pair large, cordate, often notched; the upper one larger, and incumbent over the rest of the calyx, and all the corol except the under lip, *inner* or lateral pair, linear-lanceolate, and much smaller than the exterior pair.—*Corol* bilabiate; *tube* contracted at the middle; *upper lip* straight, narrow, with a two-lobed apex; *under lip* broad, three-lobed, drooping.—*Stamina*, the fertile pair of filaments lodged under the upper lip of the corol with twin anthers, adhering by pairs, the sterile pair from the under lip, with large yellow, free, clavate apices.—*Stigma* two-lobed.—*Capsules* ovate, two-celled, hid between the exterior leaflets of the calyx.—*Seed* numerous, very small.*

Obs. As Rumphius observes, the leaves of this plant are exceedingly bitter, and might, no doubt answer valuable purposes in medicine.

2. G. *grandiflora. Linn. Sp. Pl. ed. Willd.* 1. 105. *Corom. Pl.* 2. *N.* 179.

Annual, creeping. *Leaves* sessile, broad-lanceolate, pointed, serrate. *Flowers* axillary, solitary. *Capsules* linear.

A native of moist places. Flowers during the wet and cold seasons.

Stems several, annual, creeping, four-sided, smooth; from six to twelve inches long.—*Leaves* opposite, sessile, oblong, sharply-serrate, pointed, smooth.—*Peduncles* axillary, about half the length of the leaves, erect, one-flowered.—*Flowers* light blue.—*Corol; tube* cylindric, upper lip entire; *under lip* broad, three-parted.—*Ctamens*, the two fertile filaments lodged under the upper lip, approaching arch ways. *Anthers* twin, united; the sterile filaments resting on the under lip and adhering to it; yellow, curved.—*Capsules* lanceolate, longer than the calyx.

3. G. *alata. R.*

Annual. *Leaves* petioled, ovate, serrate-dentate. *Racemes* terminal. *Calyx* five-toothed, and five-winged. *Sterile* filament free. *Capsule* linear, hid in the calyx.

A native of the *Moluccas*, is one of the largest of our *Indian Gratiolas*, and readily known by the five large membranaceous wings which run the whole length of the calyx.

* In November 1815, I received specimens of this plant from Silhet, under the *Khasee* name *Khai-Nim*. It is used medicinally there.—N. W.

4. G. *involucrata*. *R.*

Leaves unequally petioled, round-cordate, with large rounded lobes, many-nerved, serrulate, villous. *Flowers* collected into long peduncled, bracted heads. *Sterile* filaments free. *Capsule* hid in the villous, gibbous, five-toothed calyx.

This species is also a native of the *Moluccas.*

5. G. *rotundifolia*. *Linn. Sp. Pl. ed. Willd.* 1. 103. *Corom. Pl.* 3. *N.* 204.

Annual, erect. *Leaves* sessile, round-cordate, serrate-dentate. *Peduncles* axillary, solitary, longer than the leaves. *Capsules* globular.

A native of moist places, appearing and flowering during the wet and cold seasons.

Stems erect, four-sided, from four to six inches high.—*Leaves* opposite, sub-sessile, round-cordate, serrate, or notched, a little hairy.—*Flowers* axillary, or between the leaves, white. *Peduncles* longer than the leaves.—*Corol; stamens;* and *stigma* as in the foregoing species—*Capsules* round, as high as the calyx.

6. G. *integrifolia*. *R.*

Annual, while young straight, afterwards straggling, four-sided. *Leaves* sessile, ovate, three-nerved, entire. *Peduncles* solitary, axillary, one-flowered. *Capsules* the length of the calyx.

Tsjanga-pupum. *Rheed. mal.* 9. *t.* 57.

A native of *Bengal,* where it appears as a weed in gardens, during the rains.

7. G. *lucida*. *Linn. Sp. Pl. ed. Willd.* 1. 103. *Ind. Pl.* 3. *N.* 202.

Annual, spreading, four-sided. *Leaves* petioled, ovate-cordate, serrate. *Flowers*. long-peduncled, axillary, and terminal. *Capsules* ovate, hid in the calyx.

Crusta ollæ minor. *Rumph. amb.* 5. *p.* 460. *t.* 170. *f.* 3.

Annual, and like the rest, appearing in moist places during the rainy season.

Stem scarcely any. *Branches* generally opposite, diffuse, four-sided, from six to twelve inches long.—*Leaves* short-petioled, cordate, serrate, smooth.—*Flowers* terminal, and also from the exterior axills, long-peduncled, of a bright deep blue colour.—*Capsules* ovate, rather shorter than the calyx.

8. G. *veronicifolia*. *Linn. Sp. Pl. ed. Willd.* 1. 103. *Corom. Pl.* 2. *N.* 154.

Annual, while young erect, afterwards creeping. *Leaves* sessile, lanceolate, serrate. *Spikes* terminal. *Flowers* opposite. *Capsules* linear.

Crusta ollæ major. *Rumph. amb.* 5. *p.* 460. *t.* 170. *f.* 2.

A native of similar places with the last described ; flowering time the same.

Stem several, creeping. *Branches* erect, four-sided, smooth ; from six to eight inches long.—*Leaves* opposite, sessile, decussate, oblong, glandular, smooth, fleshy, finely and most sharply serrate ; serratures ending in a minute bristle.—*Racemes* terminal.—*Flowers* opposite, decussated, of a violet colour.—*Corol*, &c. as in the last species.

9. G. *racemosa*. *R.*

Annual, erect. *Leaves* sub-sessile, linear-lanceolate, entire. *Racemes* terminal, brachiate. *Flowers* long-pedicelled. *Capsules* linear, three or four times longer than the calyx.

A native of cultivated land over various parts of *India*, appearing in the rainy season.

10. G. *oppositifolia*. *Linn. Sp. Pl. ed. Willd.* 1. 105. *Corom. Pl.* 2. *N.* 155.

Annual, sub-erect. *Lower* leaves opposite to each other or to a one-flowered peduncle, lanceolate, serrate. *Capsules* linear.

G. *oppositifolia*. *Retz. obs. t.* 4. *N.* 5.

A native of moist pasture ground. *Flowering* in the wet season.

Stems several, nearly erect, ramous, four-sided, smooth; from three to six inches high.—*Leaves* sessile, opposite to each other, or a flower, linear-lanceolate, serrate, smooth.—*Peduncles* leaf-opposed, or opposite to each other, short, horizontal, clubbed.—*Bractes* solitary, embracing the insertion of the peduncle.—*Flowers* small, blue.—*Anthers* two pairs united. *Sterile filaments* small, white.—*Capsules* linear, twice the length of the calyx.

11. G. *serrata*. *R.*

Annual, erect, four-sided, smooth. *Leaves* opposite, stem-clasping, oblong, smooth, most acutely serrate. *Racemes* in the division of the branches, and terminal, long-peduncled. *Capsules* linear, diverging.

Beng. Bhoomi-nim.

Pee-tsjanga-puspum. *Rheed. mal.* 9. *t.* 59. is exceedingly like this plant, though quoted by Linnæus for *Ruellia antipoda*, which I have never met with, if this be not it.

A native of *Bengal*, and in flower during the rains.

Root annual.—*Stem* erect, four-sided, ramous, smooth, from two to eight inches high.—*Leaves* opposite, sessile, or stem-clasping, oblong, very obtuse, most acutely serrate; smooth on both sides.—*Racemes* solitary, terminal, and also from the divisions of the plant, on long, four-sided peduncles.—*Flowers* opposite, short-pedicelled, verging, pink-coloured.—*Bractes* lanceolate, the lower pair of the racemes dentate.—*Calyx* parted half way.—*Corol*, upper lip small, incumbent.—*Capsules* lanceolate, twice the length of the calyx.

12. G. *reptans*. *R.*

Annual, creeping. *Leaves* opposite, from oblong to round, serrate. *Racemes* terminal. *Sterile* filaments free. *Capsule* linear.

A native of the *Moluccas*; from thence introduced into the Botanic Garden, where it blossoms during the rains.

Stems and branches round, smooth, and creeping close to the earth, with short, (about an inch long) floriferous, erect shoots.—*Leaves* opposite, sub-sessile, on the creeping shoots round, on the erect flower-bearing ones oblong, with rounded apices; all serrate, and smooth on both sides.—*Peduncles* or racemes terminal, on the short, erect shoots, bearing several, approximate pairs of short-pedicelled, opposite, large, pale bluish pink flowers.—*Bractes* opposite, one-flowered, ovate.—*Calyx* five-leaved.—*Corol* with an enlarged throat, and ringent border; *upper lip* of one pale coloured segment; *under lip* broad, deeper coloured, and three-parted.—*Sterile* filaments large, yellow, clavate, with the apices a little recurved, perfectly free.—*Stigma* of two, thin, broad, round lobes.—*Capsule* linear, as long as the calyx, two-celled, two-valved.

13. G. *parviflora*. *R. Cor. Pl. N.* 203.

Annual, erect, ramous. *Leaves* sessile, ovate-oblong, three-nerved, serrate. *Flowers* long-peduncled, solitary. *Sterile* filaments two-cleft. *Capsules* oblong.

Stems erect, ramous, four-sided; with their angles acute, about six inches high.—*Leaves* opposite, sessile, ovate-oblong, slightly serrate, acute, when large three-nerved.—*Flowers* axillary, solitary, long-peduncled, small; I have seen some plants with white flowers, and others with blue ones.—*Anthers* twin. *Sterile* filament two-cleft, each division headed.—*Capsules* rather oblong, pointed, longer than the calyx.

14. G. *Monnieria*. *Linn. Sp. Pl. ed. Willd.* 1. 102.

Annual, creeping. *Leaves* sessile, long, obovate, entire. *Peduncles* axillary, solitary, one-flowered. *Calyx* seven-leaved. *Corol* campanulate; *stamina* four.

Beng. Adh*a*-birni.

Teling. Sambran*i*-chittoo.

A native of moist places, borders of brooks, &c. Flowering time the wet and cold seasons.

Stems several, annual, creeping, round, jointed, very ramous, smooth, succulent.—*Leaves* opposite, sessile, obovate, wedge-shaped, or oblong, smooth, entire, obtuse, fleshy, dotted with minute spots.—*Peduncles* axillary, alternate, solitary, round, smooth, shorter than the leaves, one-flowered.—*Flowers* blue.—*Bractes* two-awled, pressing on the calyx laterally.—*Calyx* five-leaved, the exterior three leaflets large, oblong, the two interior small, linear, all are concave, smooth, pointed and permanent.—*Corol* companulate ; border five-parted, nearly equal.—*Stamens* as in Didynamia.—*Anthers* two-cleft, at the base, blue.—*Stigma* large, somewhat two-lobed.—*Capsule* ovate, two-celled, two-valved.—*Seeds* very numerous.

Obs. This plant certainly ought to be placed in the 14th class, it has no one claim, that I can observe, to a place in this genus.*

The natives use the expressed juice mixed with Petroleum, to rub on parts affected with rheumatic pains.

15. G. *hyssopioides*. *Linn. Sp. Pl. ed. Willd.* 1. 104. *Corom. Pl.* 2. *N.* 128.

Annual, erect. *Leaves* ensiform, stem-clasping, much shorter than the stem joints. *Peduncles* axillary. *Capsules* linear-oblong.

G. *hyssopioides*. *Kön. in Retz. obs.* 4. *p.* 8. *Linn. Spec. Pl. ed. Reichh.* 1. 48. Appears with the former species during the rains.

16. G. *Juncea*. *Corom. Pl.* 2. *N.* 129.

Annual, erect, columnar, nearly naked. *Lower* leaves lanceolate, stem-clasping ; above minute. The two pair of anthers not united. *Capsules* globular.

Like the rest, this species delights in a wet situation, where it appears and flowers during the rains.

Stem erect, almost simple, round, jointed, smooth, from six to twelve inches high.—*Leaves* opposite, stem-clasping, lanceolate, obtuse, entire. *Floral leaves* minute, triangular.—*Peduncles* axillary, solitary, one-flowered.—*Flowers* rose-coloured, small.—*Anthers* twin, not united. The sterile filaments in this species stand between the fertile ones, and are very small.—*Capsules* globular.

17. G. *cuneifolia*. *R.*

Annual, erect, simple, nearly naked. *Leaves* one pair at the base,

* *Monniera*, Browne. *Herpestis*, Rob. Brown.—N. W.

cuneiform, obtuse; above one or two minute pairs. *Flowers* solitary, long peduncled. *Capsules* round.

A native of wet places on the coast of *Coromandel,* and nearly allied to G. *lobelioides,* and *juncea.*

18. G. *lobelioides. Linn. Sp. Pl. ed. Willd.* 1. 104. *Retz. obs.* 4. 7. *Vahl. Enumer.* 1. *p.* 92.

Annual, on the lower part of the simple stem grew from two to four pair of approximate, lanceolate, entire, sessile leaves. *Flowers* toward the apex, remote, solitary, long-peduncled. *Capsules* round.

A native of *Coromandel,* growing in wet places during the rainy season.

UTRICULARIA. Schreb. Gen. N. 41.

Calyx two-leaved.—*Corol* ringent, and generally calcarate. *Capsule* superior, one-celled. *Seeds* numerous.

1. U. *stellaris. Linn. Sp. Pl. ed. Willd.* 1. 113.

Floating, leafless. *Scape* with a verticil of bladders. *Corol* without a nectary.

Teling. Natsoo.

Grows in sweet waters, appears and flowers during the rains. *Root* conferva-like, swimming in, and not on the water; I have always found small utriculi adhering to its filaments.—*Scape* sub-erect, simple; raceme above the verticil of bladders, which rest on the surface of the water.

2. U. *fasciculata. R.*

Floating. *Leafless,* scapes naked; racemes four to eight flowered. *Utriculi* sub-ovate, two-horned, scattered amongst the fibres of the root. *Nectary* horn-shaped, ascending to the edge of the under lip. *Capsules* beaked, drooping.

Teling. Natsoo.

Beng. Janjee.

Found swimming in stagnant water in the vicinity of *Calcutta*, at the end of the cold, and beginning of the hot seasons.

Stem jointed, length various, but often extending some feet.—*Radicles* most numerous, in alternate fascicles, multifid, filiform, armed with minute, acute, bristles pointing forward.—*Utriculi* scattered, semi-ovate, dotted, becoming black by age, with two horns from the mouth.—*Scape* erect, from four to eight inches high, bearing generally from four to eight, alternate, large, yellow, pedicelled flowers.—*Bractes* ovate, obtuse, one-flowered.—*Calyx* of two large, permanent leaflets.—*Corol* as in the genus. *Nectary* horn-shaped, blunt, nearly as long as the under lip, and bending up to its margins.—*Capsules* conical, pointed, reflected.

3. U. *biflora*. *R.*

Floating, leafless. *Scape* naked, two-flowered. *Utriculi* ovate, with filaments from their apices. *Nectary* horn-shaped, ascending, as long as the under lip. *Capsules* globular, erect.

Beng. Chota-janjee.

Found with the last described in stagnant water near *Calcutta*, during the same seasons. A much smaller plant.

Root of a few, scarcely compound, filiform fibres, and some scattered, subulate nigrescent utriculi, with two or more filaments from their apices, or mouths.—*Scape* about three inches high, filiform, naked, generally two-flowered.—*Flowers* yellow, small.—*Calyx* and *corol* as in the genus, with the faux closed, and a horned, ascending nectary as long as the under lip.—*Capsules* globular, erect.

Addition by N. W.

4. U. *nivea*. *Vahl. Enum.* i. 203.

Naked, suberect, filiform. *Scape* with a few gibbous *scales*, ending in a few-flowered *raceme*. *Bractes* three-parted, the middle segment adnate and sharp at both ends, the lateral ones minute. *Nectary* ascending longer than the lower lip. *Capsules* globular nodding.

I found this delicate plant growing among rice in the vicinity of Serampore during the rainy season.

Root consisting of a few capillary simple and short fibres.—*Scape* about six inches long, sometimes divided in two equal branches, when fruit-bearing elongated and twining.—*Flowers* from four to eight, white with a yellowish entire palate, very small on short alternate pedicels. *Calyx* permanent, of two orbicular segments, the uppermost larger and concave.—*Upper lip* very short, linear erect, ending in two toothlets; *lower lip* ovate, larger. *Spur* almost twice as long as the lower lip, conical.—*Capsule* half hid in the calyx.

LYCOPUS. Schreb. Gen. N. 44.

Corol four-cleft, with one of the divisions emarginate. *Stamina* distinct. *Seeds* four, retuse.

1. L. *dianthera. Buch.*

Annual, erect, four-sided. *Leaves* petioled, rhomb-ovate, serrate. *Calyx* bilabiate.

A native of Nepala; from thence Dr. Buchanan sent the seeds to this garden, (in 1802,) where the plants thrive during the cold season, and blossom in March and April.

Stems annual, straight, four-sided; *sides* slightly grooved; angles somewhat hairy. *Branches* axillary, decussate, four-sided, &c. like the stem; height of the whole plant from one to two feet.—*Leaves* opposite, petioled, declined, rhomb-ovate; anteriour margins serrate, dotted with minute, darker green glands underneath, general length about one inch, or less.—*Racemes* terminal.—*Bractes* opposite, lanceolate, one-flowered.—*Flowers* pedicelled, small, very pale pink, almost white.—*Calyx* bilabiate; *upper lip* ascending, tridentate; *lower lip* bipartite, and, defended with hair on the inside.—*Corol; tube* short, somewhat four-sided; *upper* division of the border broad and emarginate; *lateral* two of the same length, but narrower and semicircular; *under* one large, and nearly round.—*Filaments* four, short, in the angles of the tube of the corol.—*Anthers* twin on the long filaments, and tending to be so on the shorter ones, but very small, and abortive.—Is nearly allied to *Thymus*.

SALVIA. *Schreb. Gen. N.* 50.

Corol irregular. *Filaments* two-forked; *anthers* on the superior ones. *Seeds* naked.

1. S. *bengalensis*. *Kön. Mss.*

Shrubby. *Leaves* linear-ovate, lanceolate, entire. *Racemes* verticelled, verticils distinct, many-flowered. An *anther* to each division of the filaments.

The natives on the Coromandel Coast have no name for it. From Bengal it was introduced into the gardens on the Coast a few years ago, by Mr. Parsons; and is only found in gardens, where it grows to be a large, straggling shrub.

Trunk seldom erect, woody, sometimes as thick as a man's arm. *Bark* cracked, and peelling off in irregular pieces. *Young shoots* downy, round.—*Leaves* as in *salvia officinalis*.—*Racemes* terminal, often compound, verticelled. *Verticils* approximate, globular, many-flowered.—*Flowers* white.—*Calyx* gibbous, downy, three or four toothed.—*Corol*, both lips recurved, or spreading.—*Stamens*, there are sometimes three or even four filaments, with their extremities bifid, each division bearing an oval proper *anther*.

Obs. The leaves of this plant smell and taste considerably stronger, I think, than those of S. *officinalis*, and are applied to the same uses.

2. S. *brachiata*. *R.*

Annual, erect, brachiate. *Leaves* oblong, crenate, tomentose. *Racemes* verticelled; *verticils* six-flowered; *seeds* elevated on a receptacle.

Is a native of moist places, over various parts of India. Flowering time the cold season.

Stem annual, erect, ramous, four-sided, four-grooved, downy; from one to two feet high. *Branches* opposite, cross-armed, ascending.—*Leaves* opposite, petioled, running down on the petioles, oblong, irregularly crenulate, waved, rugose, a little downy; two or three inches long.—*Racemes* terminal, long, verticelled. *Verticils* six-flowered, six-bracted.—*Flowers* small, pale purple. *Stamens* as in the genus, with the rudiments of two additional sterile filaments between the large pair.—*Germ* elevated on a large fleshy receptacle.—This plant is slightly aromatic.

3. S. *lanata. R.*

Herbaceous, four-sided, villous. *Leaves* sessile, ovate-oblong, entire, woolly. *Flowers* verticillate.

Found by Colonel Hardwicke on the most elevated mountain near Adwaanee, on the road from Hurdwar to Sirinagur. It is his S. *integrifolia*. See *Asiatic Researches. vol.* 6. *p.* 349.

Obs. Salvia rosea of Vahl, is the same as S. *coccinea*, an American plant, which though in a manner naturalized here has no right to a place in this Flora.

VERONICA.

Corol four-cleft, the lower segment smaller. *Capsules* two-celled.

1. V. *undulata. Wall.*

Smooth, erect. *Leaves* linear-lanceolate, waved, unequally serrated. *Racemes* terminal and axillary, elongated. *Peduncles* covered with short glandular hairs.

This little plant was discovered in the Turraye by Mr. W. Jack, an assistant surgeon on the Honourable Company's Bengal establishment, who most obligingly communicated the following description.

Root creeping.—*Stem* herbaceous, erect, fistulous, round, smooth, ramous.—*Leaves* opposite, sessile, stem-clasping with their broad base, linear-lanceolate, acute, gradually narrower towards their apex, with waved, unequally serrated margins, smooth.—*Racemes* long, axillary and terminal.—*Peduncles* and *pedicels* hispid with short, glandular hairs.—*Bractes* linear, acute, smooth, entire, one under each pedicel.—*Calyx* deeply four-parted, smooth, persistent, with straight, lanceolate, acute *segments*.—*Corol* white, a little shorter than the calyx; tube very small; three segments of the limb ovate, acute, equal; the fourth and lowermost shorter, lanceolate.—*Stamina* two, upright, short. *Anthers* oblong.—*Pistil* as long as the stamina. *Stigma* capitate.—*Capsule* roundish, compressed, with a furrow on each side, emarginate at the top, beset with glandular hairs, two-celled. *Seeds* numerous, minute, round.—N. W.

BOERHAAVIA. Schreb. Gen. N. 13.

Calyx inferior, gibbous, entire, permanent, and becoming an envelope for the seed. *Germ* one-celled; ovula single, erect. *Corol* campanulate, inserted on the calyx. *Seed* solitary. *Embryo* conduplicate, with inferior *radicle*, and central *perisperm*.

1. B. *procumbens. Herb. Banks.*

Root fusiform, perennial. *Branches* procumbent, smooth. *Leaves* variously cordate, covered with a silver coloured pellicle underneath. *Flowers* terminal, in long-peduncled heads. *Stamina* three.

Sans. पुनर्नवा, Poon*u*rn*u*v*a*, शोथघ्नी, Shot*u*ghnee.

Beng. Gadha-poorna, the red variety; and Shwet*a*-poorna, the white one.

Talu dama. *Rheed mal.* 7. *p.* 105. *t.* 56. good for a young luxuriant plant.

B. diandra, and erecta. *Burm. Flor. Ind. p.* 3. *t.* 1. seem to be our plant, taken at different ages.

B. *erecta. Gært. Carp.* ii. 209. *t.* 127.

This species, whatever it may be, whether *diffusa, erecta, repanda* or *diandra*, or all of them, is the only one I have yet found in India. It is not only common every where, but is one of the most troublesome weeds we have. The long, fusiform, perennial roots, strike so deep, as to render it no easy task to dig them up. It produces blossoms and ripe seed the whole year.

Root perpendicular, fusiform, slender, perennial. *Stem* none; *branches* many, herbaceous, with alternate, bifarious, round, smooth, jointed, often coloured branchlets spreading close on the ground, to an extent of many feet in a good soil but never striking root.—*Leaves* opposite, unequal in size, one of the leaves being alternately smaller in each pair; petioled, variously cordate, margins more or less scalloped, waved, and often coloured; sometimes acute, sometimes obtuse; all are smooth above, and covered with a silver coloured pellicle underneath; size very various.—*Petioles* shorter than the leaves, channelled.—*Peduncles* solitary, from the naked swelled joints between the leaves, but nearest to the small leaf. At the ends of the *branches*, where the joints approximate, they are so numerous as to form a panicle, each supporting, from one to five, or six small heads of sessile, red or white, minute flowers.—*Calyx* (apparently the germ), beneath, five-sided, covered with headed glatinous glands; *mouth* entire, and much contracted.—*Corol* campanulate, plaited, inserted on the crown of the calyx.—*Nectary*, a small fleshy, three-toothed cup, surrounding the base of the germ, in the divisions of which the filaments are inserted. It requires a powerful lens, and much patience to understand the structure of this part of the flower.—*Filaments* generally three, as long as the style. *Anthers* of two round lobes.—*Germ* oblong, hid in the belly of the calyx, one-celled; *ovulum* single, attached to the bottom of the cell. *Style* slender. *Stigma* peltate.—*Pericarp* turbinate, five-sided, covered with clammy pedicelled glands, &c. exactly as in Gærtner's figures of his *Boerhaavia* erecta.

Obs. There are two varieties of this plant in Bengal; one with red, the other with white flowers; in other respects they are the same. In the St. Helena plant, B. *repanda* of the *hortus bengalensis*, the leaves are more angular, the inflorescence terminal, racemes verticelled, with one bracte only to each pedicel.

FRAXINUS.

Polygamous. *Calyx* none, or four parted. *Corol* none, or four-petalled. *Samara* one-seeded, with lanceolate wings.

1. F. *chinensis. R.*

Arboreous, with erect branches. *Leaflets* from three to seven, lanceolate, serrate. *Panicles* axillary, and terminal. *Flowers* without petals.

A native of China, from thence introduced into the Botanic Garden by the late Col. Robert Kyd, prior to 1793; where they blossom in April, but have not produced seed, or even full grown seed vessels.

Our small trees have little or no *trunk;* but several, erect, straight *branches* and *branchlets;* giving an entire height of from ten to twelve feet.—*Leaves* opposite, pinnate, and ternate. *Leaflets* from three to seven, opposite, lanceolar, serrate, smooth. The pairs short-petioled, much smaller and shorter in proportion to their breadth than the terminal one, which is protruded on a longer winged petiolet.—*Petioles* channelled, sometimes with a marginal wing.—*Panicles* axillary, and terminal, small, drooping, and composed of most slender, simple, and compound, sub-verticelled ramifications.—*Calyx,* in all, four-toothed.—*Corol* none.—*Filaments* in the completely *hermaphrodite flower* much longer than the calyx. In the *female* wanting, or abortive, and hid within the calyx. *Male flowers* not found.—*Germ* ovate. *Style* long, and curved. *Stigma* emarginate.

2. F. *floribunda. Wall.*

Leaflets oblong, tapering, acuminate, acutely and unequally serrated; male flowers with a corolla.

A native of Nepala, from whence I received specimens with male flowers and others with seeds, in April, 1818, under the name *Kangu* and *Tahasee.*

Grows to be a large tree with smooth, ash-coloured, dotted

branches; branchlets compressed.—*Leaves* opposite, with three pairs of smooth leaflets, which encrease somewhat in size, the terminal or odd one being the largest, ovate and equal; they vary in their figure from oblong to ovate-oblong, measuring from three to four inches, about one inch distant from each other, acute and more or less unequal at the base, tapering into a long narrow serrated acumen, perfectly smooth above, reticulated and nerved below.—*Petioles* slender, grooved, opposite, *terminal* one an inch long.—*Panicles* terminal, showy and large, compound, smooth, loose, interspersed with a few floral leaves, with opposite branches and branchlets and long slender racemes consisting of innumerable short peduncled or sessile fascicles of white scentless flowers on capillary pedicels.—*Calyx* truncated, four-toothed.—*Petals* very long, linear-clavate.—*Filaments* capillary, almost as long as the corolla; *anthers* linear, erect.—*Rudiment* of the female organs none.—*Capsule* linear, about an inch long, one-celled; wing scarcely two lines broad, veined and marked with elevated dots, slightly retuse at the apex. *Seed* solitary, possessing a bitterish somewhat aromatic taste; in other respects like *Fr. ex elsior, Gært. Carp.* i. 222. *tab.* 49.

Obs. I have not seen the female flowers of this ornamental tree which unquestionably comes near to *F. Ornus* (the Manna or Flowering Ash), differing chiefly in having taper-pointed, acutely serrated leaves.—N. W.

LIGUSTRUM.

Corol four-cleft. *Berry* superior of two cells with two seeds in each cell.

1. L. *nepalense. Wall.*

Leaves ovate, or ovate-oblong, acuminate, villous below. *Panicle* terminal, villous, consisting of sub-erect crowded racemes.

A native of the mountains of Nepala where it is called *Goomgacha* and grows to be a considerable tree, producing profuse clusters of white, sweet smelling flowers from April to June; *berries* ripen in four to six months.

Branches round, opposite, ash-coloured, dotted with callous spots; all the younger parts covered with abundance of soft hairs.—*Leaves* opposite, spreading, on very short channelled petioles, varying considerably in size and figure, mostly ovate with a rounded base, sometimes oblong or lanceolar, generally tapering into a long apex, terminated by a short cylindric point; from one to two, or even three inches long, of a firm texture, shining above, densely villous underneath, when old entirely smooth, with very fine remote nerves communicating with each other in reticulated arches near the margin.—*Panicle* somewhat contracted, ovate, from four to six inches long, sessile, consisting of pretty crowded racemes, the uppermost very short and alternate.—*Peduncles* round, villous, with a deciduous small lanceolate bract under each division.—*Flowers* short-pedicelled, smooth, disposed in small bunches.—*Calyx* truncate, with four hardly discernible teeth.—*Corol; tube* equalling the calyx; segments of the limb spreading, lanceolate, acute, twice as long as the tube.—*Stamina* almost as long as the corol; *filaments* inserted within its mouth between two opposite fissures; *anthers* oblong, bursting lengthways on both sides.—*Pistil* shorter than the stamina, smooth; *ovarium* roundish, two-celled, with two pendulous ovula in each cell; *style* short; *stigma* fleshy, oblong, acute.—*Berries* oval, dark blue with a beautiful bloom on them, rather smaller than those of the common *Privet;* pulp brown.—*Seeds* one or two; *cotyledons* broad-ovate. In other respects like those of Ligustrum. *Gært. Carp.* ii. 72. *tab.* 92.

Obs. This species differs from L. *japonicum,* Thunb. and *lucidum,* Ait. in the figure and villosity of its leaves and the contracted shape of its panicle. It may perhaps be found the same as L. *sinense,* Lour. especially if the racemes, which in the *flora cochinchinensis* are attributed to that tree, are considered as panicles, which indeed has been done by the illustrious author of that article in Rees's New Cyclopædia.

I possess specimens taken from old branches, with more expanded panicles, which together with the leaves are perfectly smooth.—N. W.

VERBENA. See *Didynamia Angiospermia.*

TRIGYNIA.

PIPER. Schreb. Gen. N. 59.

Ament filiform, imbricated with peltate scales. *Corol* none. *Germ* one-celled with a single, erect *ovulum*. *Berry* one-seeded. *Embryo* inverse, and furnished with an ample perisperm.

1. P. *nigrum. Linn. Sp. Pl. ed. Willd.* 1. 159.

Leaves bifarious, obliquely ovate-cordate, acuminate, polished, from five to seven-nerved.

Pepper-vine. Marsden's History of Sumatra, *p.* 105.

Molago-codi. *Rheed. mal.* 7. 23. *t.* 12.

Sans. वेल्लोजं, Vellojung, मरीचं, Mureechung, मरिचं, Murichung, कोलकं, Kolukung, कृष्णमूषणं, Krishnumooshunung, धर्म्मपत्तनं, Dhurmuputtunung.

Hind. and *Beng.* Murich, or Gol-murich.

Tam. Moloovoo-kodi.

Pers. Pilpil.

Arab. Filfil.

Malay. Ladda.

Cultivated in various parts of India, and its Islands.

2. P. *trioicum. R.*

Triœcous, shrubby, rooting. *Leaves* bifarious, obliquely-ovate, acuminate, from five to seven-nerved, glaucous. *Aments* leaf-opposed, cylindric, pendulous.

Teling. Murial-tiga.

Roots long, striking deep into the earth.—*Stem* jointed, winding, when old woody, and scabrous, running along the ground to a great extent, or up trees, &c. when trained to them; from each joint issue roots which take firm hold of whatever they meet with. *Branches* numerous, alternate; the young ones smooth, the old ones woody, and scabrous like the stem.—*Leaves* alternate, petioled, cordate, broad-oval, ovate, or oblong, pointed, from five to seven nerved, (generally five,) above, smooth and glossy, below, lighter coloured, often somewhat bubbled, from four to six inches long, and from two to four broad.—*Petioles* channelled, smooth, an inch or an inch and half long.—*Stipules* solitary, spathiform falling off when the leaf begins to be unfolded.

MALE PLANT. *Calyx* an *ament*, leaf-opposed, peduncled, filiform, pendulous, closely imbricated with five spiral rows, of fleshy, oval, one-flowered scales.—*Scales* oblong, peltate, sessile, one-flowered. —*Corol* none.—*Filaments* three, very thick, and very short, scarcely elevating the *anthers* above the margins of the scales of the ament. *Anthers* four-lobed.—*Pistil*, in some, a minute cylindric gland in the centre, in others, not the smallest rudiment of one.

FEMALE PLANT. *Calyx* an *ament*, leaf-opposed, shorter, thicker, and more rigid than in the male, imbricated with three spiral rows of scales.—*Scales* as in the male.—*Corol* none.—*Stamens* none.—*Germ* sessile, globose, immersed in the substance of the ament. *Style* none. *Stigma* three-lobed, white, glandular.—*Pericarp*, a small, round, red, somewhat fleshy berry.—*Seed* one, globose.

Obs. When I described the three vines included in this one species, I had not seen *Piper nigrum*, and took it for granted that this was it; but as soon as I had an opportunity of seeing that famous plant, I was immediately convinced that they were distinct species. In *trioicum*, the leaves have a glaucous appearance, which readily distinguishes it from P. *nigrum*, which has shining dark green leaves.

This vine I have found wild amongst the chain of mountains directly north from Coringa, in the Raja-mundri Circar. It delights in a moist, rich soil, well shaded with trees; to them it adheres most firmly, by means of the roots which issue from the joints. Flowering time, in their wild state, during the latter part of the wet season, September and October. The pepper ripens in March. With me, in a cultivated state, they flower almost all the year round, but chiefly during the forementioned period.

I have not met with any author, or any sort of information, that could lead me to think black pepper was the produce of a dioecous plant, (male and female on distinct vines,) till upon examination, I found it was so. Attention to this circumstance will, I think, render the culture of pepper much more certain, and successful, viz. by planting to each prop tree a male and female plant, the male on the sides from whence the most prevailing wind blows; and the female on the other.

The vegetable world is full of proofs that the *sexual system* of the immortal Linneus is founded on the soundest principles, and not a single plant have I ever found in India, that does not corroborate this fact. The Arabians from time immemorial knew, that to render the *female* date-tree prolific, it was necessary to bring it in contact with the *male*; which they do by making a slit in the spathe of the female flower, just before it is ready to burst and thrusting therein a branch of the *male* spadix. I have therefore the utmost reason to conclude the pepper vine will be much more productive, if the above mentioned circumstance be attended to, by the cultivators. I think, if the Malays on Sumatra had known it, the accurate Mr. Marsden, would not have neglected mentioning so material a circumstance, when describing this plant, and the method of cultivating it there.

Soon after the above description was made, I found a third vine bearing aments with *hermaphrodite* flowers; or *hermaphrodite* and *female* flowers mixed on the same aments. At the same time I found that the pepper of the female vine did not ripen properly, but dropped while green, and immature from the plant, and that when dried it had not so much pungency as common pepper, whereas the pepper of this third sort ripens perfectly, when dry is exceedingly pungent, and has been, by pepper merchants at Madras, reckoned equal, if not superior to the best pepper of the Malabar Coast, or Ceylon; consequently this must be the sort that is found cultivated; the other two being, I conjecture, entirely neglected.

This *hermaphrodite* plant grows wild, with the male and female, in the moist, uncultivated, over-run shady vallies, up amongst the mountains; and also upon the mountains, where springs keep them moist. Such places are common in the cliffs of the rocks and there the vines thrive with the greatest luxuriance. In its stem, branches, leaves, and stipules, it agrees perfectly with the other two. The aments are also the same in every respect, except that here are four spiral rows of flowers; the *scales* of the *ament* are as in the other two, viz. the male and female plants.—*Stamens*, generally two, fleshy, clubbed filaments, shorter than the germ, and placed laterally, so as to press upon it. It frequently happens that they are entirely wanting, or only one is present. *Anthers* two oval pits in the apex of each filament.—*Germ* globular, immersed in the substance of the ament. *Style* none. *Stigmas* three, spreading.—*Berry* globular, size of a small pea, red, smooth, one-celled.—*Seeds* solitary.

3. P. *longum*. *Willd*. 1. 161.

Dioecous, shrubby, creeping. *Lower* leaves petioled, broad-cordate, floral leaves sessile, stem-clasping, oblong-cordate; all are from five to nine-nerved.

Cattu-tirpali. *Rheed. mal.* 7. *p.* 27. *t.* 14.

Sans. कृष्णा, Krishna, उपकुल्या, Oopukoolya, वैदेही, Videhee, मागधी, Magudhee, चपला, Chupula, कणा, Kuna, उषणा, Ooshuna, पिप्पली, 'Pippulee, शौण्डी, Shoundee, कोला, Kola.

Beng. Pippul, the root Pippula-moola.

English. Long-pepper.

Pippul-chittoo is the Telinga name of the plant, and Pippuloo the pepper.

The plant I have found wild amongst bushes, on the banks of water courses, up towards the Circar mountains. It flowers and bears fruit during the wet, and cold seasons.

Root woody, perennial.—*Stems* many, creeping, jointed, round; *joints* swelled; *young* shoots downy. *Branchlets* bearing the fruit are erect, with the leaves sessile, or nearly so.—*Leaves* on the creeping branches largest, petioled, broad-cordate, seven-nerved; on the erect, fruit-bearing branchlets (*floral leaves*) stem-clasping, oblong-cordate, five-nerved; all are smooth, somewhat wrinkled; below pale green; size various.—*Stipules* of the petioled leaves two, joining lengthways to the petioles, lanceolate; of the sessile leaves within the leaf, single, spathiform.

FEMALE FLOWERS. *Ament* sessile, leaf-opposed, peduncled, erect, cylindric, imbricated with five, or more, spiral rows of small, orbicular, permanent, peltated, one-flowered scales.—*Calyx* none.—*Corol* none.—*Stamens* none.—*Germs* numerous, aggregate, sessile, sub-orbicular. *Style* none or exceedingly short. *Stigmas* three or four-lobed.—*Pericarp* aggregate, sub-cylindrical, composed of firmly united, one-seeded drupes.—*Seed* ovate, smooth.

Obs. It is in Bengal only, so far as I have been able to learn, that this plant is cultivated for its pepper. When the pepper (ament) is full grown, it is gathered and daily exposed to the sun, till perfectly dry; after which it is packed up in bags for sale.

The roots, and thickest parts of the creeping stems, when cut into small pieces and dried, form a considerable article of commerce all over India, under the name of *Pippula moola;* for which purpose it is particularly cultivated in many of the vallies amongst the Circar mountains. This sort is more esteemed, and bears a higher price than that of Bengal; where by far the largest proportion is cultivated. It is, as well as the *pepper,* chiefly employed medicinally, and the consumption of both these drugs is very great.

CULTIVATION IN BENGAL. The long pepper is not propagated by seed, but by suckers, and requires to be cultivated upon a rich, high, and dry soil. The suckers are transplanted soon after the setting in of the periodical rains, and the pepper (which is preserved merely by drying it in the sun), is gathered in the month of January, after which the stalk, and branches of the plant wither, and the roots only remain alive. A bigha of land (the third of an English acre) will yield in the first year about a maund (eighty-four pounds) of the pepper, in the second year four maunds; and in the third six; after which, as the plant becomes annually less and less productive, the roots are grubbed up, dried, and sold; and fresh roots, or young shoots are set in their stead, the earth requiring merely a slight covering of manure. The plants are never to be watered, and at the commencement of the hot season the roots are to be carefully covered with straw to preserve them against the heat of the sun. The plants should be set about five feet asunder. Large quantities of this pepper and also of the roots are exported to Bombay, and Surat; where both are in great demand, the first for culinary, the latter for medicinal purposes. The Ryots in this part of the country, usually sow radishes, or barley, or plant *brinjals* (*Solanum melongena*), in the intermediate space between the plants.

4. P. *Chaba*. W. Hunter in *Asiat. Res.* ix. 391.

Shrubby, creeping. *Leaves* short-petioled, ovate-lanceolate, base unequal, scarcely triple-nerved : aments leaf-opposed, erect, cylindrico-conical, firm and fleshy.

Sans. चव्यं, Chuvyung, चविका, Chuvika, चवी, Chuvee, चविकं, Chuvikung.

Beng. Choee.

Piper-longum. *Rumph. amb.* 5. *p.* 333. *t.* 116. *f.* 1.

Obs. Cattu-tirpali. *Rheed. mal.* 7. *p.* 27. *t.* 14. represents a very different species, the fruit of which is also used over the continent of India, and particularly in the western part thereof, as *long pepper*, and is much cultivated in Bengal, chiefly for its root, which the natives call *Pippula*. Woodville's figure, in his *Medical Botany*, is very bad, for it answers neither to this, nor P. *longum*, *Linn. Sp. Pl. ed. Willd.* 161. which may be called the long pepper of the continent of India. Blackwell's figure is still worse.

5. P. *sylvaticum*. R.

Leaves all petioled, broad-cordate, from five to seven-nerved, obtuse; lobes of the base large, equal, circular. *Aments* erect, short-peduncled, columnar, male flowers tetrandrous.

A native of the mountains on the North-west border of Bengal, where the natives call it Pahari peepul, or *mountain* long-pepper and use it, both green and ripe, in their dishes. In the Botanic Garden it blossoms, and the berries ripen during the rains.

Root perennial.—*Stem* or rather branches creeping on the ground, or rooting on trees like Ivy, and most of the East India species of pepper; all the young parts polished.—*Leaves* alternate, petioled, equally-cordate, obtuse, from five to seven nerved, smooth, from three to five inches long, by from two to four broad.—*Petioles* from one to two inches long, grooved.—*Stipules* interfoliaceous, &c. as in the genus.—*Male aments* leaf-opposed, short-peduncled, or columnar, slender. *Scales* one-flowered.—*Corol* none.—*Filaments* generally four, oval, fleshy, very short. *Anthers* one celled.—*Germ* none.—*Female* aments on a different plant, leaf opposed, short-peduncled, cylindric. —*Germ* oval, one-celled, ovula single, erect, that is, attached directly to the bottom of the cell.

6. P. *peepuloides. R.*

Leaves equally-ovate, and ovate-lanceolate, short-petioled, from three to five-nerved, smooth, acuminate. *Aments* sub-sessile, cylindric.

Pippul the vernacular name in Silhet, and on the adjacent range of mountains where it is indigenous, and considered the *Pippul*, or *long-pepper* of that country, and as such is used in medicine, &c. It differs specifically, no doubt, from the long-pepper plant of lower Bengal, as well as from that of Rumphius, called P. *Chaba* by Dr. W. Hunter, who found it common on the Island of Pulo-Pinang. It is immediately known from P. *longum* by the shape of the leaves, as well as by their being all equally petioled; and from P. *Chaba*, by both sides of the base being equal, or nearly so; whereas in *Chaba* the lower side is much larger and longer.

Root perennial.—*Stems* and alternate branches creeping, young shoots smooth.—*Leaves* alternate, short-petioled, from ovate-oblong, to lanceolate, entire, acuminate; generally five-nerved, of which the exterior two are minute, and when absent the leaf is perfectly three-nerved, from three to five inches long, by from one to three broad. —*Stipules* stem-clasping, and petiolary.—The *Female aments* have only been found, which induces me to think this species is dioecous, as most of them are; they are short-peduncled, solitary, and directly opposite to the leaves.—*Berries* smaller than in common black-pepper (P. *nigrum*) obliquely oval; pulp in small quantity, one-celled.—*Seeds* solitary, conform to the berry.—*Perisperm* conform to the seed.—*Embryo* minute, lodged in a little hollow in the apex of the perisperm.

7. P. *Betle*. *Linn. Sp. Pl. ed. Willd.* 1. 159.

Perennial, dioecous, creeping. *Leaves* alternate, bifarious, cordate, from five to seven-nerved, smooth, entire. Female aments sub-cylindric, drooping.

P. qui saururus, &c. *Burm. Zeyl. p.* 193. *t.* 83. *f.* 2.

Beetla-codi. *Rheed. mal.* 7. 29. *t.* 15.

Sans. तम्बूलवल्ली, Tamboo*luvu*llee, ताम्बूली, Tamboolee, नागवल्ली, Nag*uvu*llee.

Hind. and *Beng.* Pan.

I have never met with this plant in its wild state. It is much cultivated over India and the Malay countries.

Root woody, ramous.—*Stems* woody, creeping on the earth, or rising up trees, poles, &c. when carefully trained to them; smooth in every part; innumerable roots issue from the joints, by which the plant is well supported after they have taken hold.—*Leaves* alternate, petioled, bifarious, from broad-cordate to obliquely ovate-oblong, somewhat pointed, from five to seven-nerved, entire, smooth; from four to six inches long, and from two to four broad.—*Petioles* of various lengths, and channelled.—*Stipules* as in P. *nigrum*.—*Female aments* leaf-opposed, peduncled, filiform, sub-pendulous, closely imbricated with several spiral rows of scales, and corollets, exactly as in P. *trioicum*.—The *male flowers* I have never found.

Obs. The plants are constantly raised from the slips, and cuttings, these are carefully planted in a rich moist soil, well inclosed, and shaded, so that they are in great measure, protected from both sun and wind. In some places, small plantations of *Æschynomene grandiflora*, are made to train them to, and to keep off the sun; in others, poles are employed for the first, and a thin shed of mats over them for the latter purpose, they require to be frequently watered during the dry weather. Every history of India describes, and every body knows, the use the Hindoos of all ranks make of the leaves of this plant.

8. P. *Cubeba. Gært. Carp.* ii. 67. *Linn. Sp. Pl. ed. Willd.* 159.

Shrubby, creeping. *Leaves* petioled, ovate-lanceolate, scarcely triple-nerved. *Aments* leaf-opposed, on peduncles longer than the petioles. *Berries* pedicelled.

Piper canninum. *Rumph. amb.* 5. *t.* 28. *f.* 2.

A native of Prince of Wales Island and the countries to the eastward of the Bay of Bengal.

9. P. *lanatum. R.*

Creeping, every part very woolly. *Leaves* petioled, ovate-lanceolate, scarcely three-nerved. *Aments* leaf-opposed, one-peduncled, as long as the petioles. *Berries* pedicelled, clavate.

A native of the Malay Archipelago.

10. P. *methysticum. Linn. Sp. Pl. ed. Willd.* 161.

Leaves short-petioled, from salient to obliquely re-entering, cordate, from seven to nine-nerved, entire, smooth.

A native of the Malay Archipelago. There is no doubt but this is the Sirium *decumanun. Rumph. amb.* 5. *t.* 27. but whether it be the South sea Island intoxicating species, I cannot take upon me to say. The leaves are often a foot long, by six or eight inches broad, on sheathing petioles not more than an inch long. I have not seen the fruit.

11. P. *arborescens. R.*

Arboreous, creeping. *Leaves* petioled, ovate-elliptic, acute, five-nerved. *Aments* leaf-opposed, long, slender, cylindric, on peduncles about as long as the petioles.

Sirium arborescens tertium. *Rumph. amb.* 5. *t.* 28. *f.* 1.

A native of the Molucca Islands.

12. P. *lanceolatum. R.*

Shrubby, creeping. *Leaves* short-petioled, broad-lanceolate, prominently five-nerved, smooth. *Aments* leaf-opposed, long-peduncled, filiform.

A native of the Molucca Islands.

13. P. *Malamiris. Linn. Sp. Pl. ed. Willd.* 160.

Shrubby, creeping. *Leaves* rather long-petioled, from salient to re-entering, cordate, five, or seven-nerved. *Aments* leaf-opposed, filiform, on peduncles that are generally shorter than the petioles.

Amalago. *Rheed. mal.* 7. *t.* 16.

Sirium. *Rumph. amb.* 5. *t.* 116. *f.* 2. is no doubt P. *Betle,* consequently must be excluded from the list of synonyms belonging to this species.

14. P. *rostratum. R.*

Creeping, tender parts hairy. *Leaves* short-petioled, obovate, cuneate, obtuse-pointed, villous, five or seven-nerved, downy, meeting above the base. *Aments* leaf-opposed, short-peduncled. *Berries* beaked.

A native of the Moluccas.

15. P. *diffusum. Vahl. Enumer. Pl.* 1. 333.

Leaves long-petioled, round-cordate, sharp-pointed, five or seven-nerved, smooth. *Aments* cylindric, short-peduncled.

Sirium *frigidum. Rumph. amb.* 5. 345. *t.* 119. *f.* 2.

A native of the Moluccas.

16. P. *sarmentosum. R.*

Erect, ramous at the top, sarmentose. *Leaves* from broad-cordate to oblong, smooth, five-nerved, the superior ones sub-sessile. *Aments* leaf-opposed, short-peduncled.

P. *latifolium.* Hunter in *Asiat. Res.* ix. 392.

Gadu. Cando, or Gadukh, of the Malays.

A native of the Malay Islands, and brought by Mr. Smith into the Botanic Garden, where it thrives luxuriantly, and is in fruit most part of the year.

Stems erect, from six to eighteen inches high, ramous at top, jointed, and smooth. *Sarmentose* shoots of a great length, issuing in abundance from the top of the root, and the lower part of the stem, by which the plant is quickly propagated to any extent.—*Leaves* alternate, the lower ones petioled, broad, re-entering, cordate ; the upper ones sub-sessile, obliquely-oblong ; all are smooth on both sides, and shining on the upper one.—*Aments* solitary, opposite to the leaves, short-peduncled, oblong, very small, being rarely more than a quarter of an inch long.—*Germs* oval, one-celled, *ovulum* one attached by the base to the bottom of the cell.—*Fruit* compound, sub-cylindric, about the size of an infant's finger, when ripe softish, and of a dark green or livid colour. *Berries* numerous, most of the germs prove abortive, one-celled. *Seed* solitary, oval, attached to the bottom of the cell. *Integuments* single, thin, light brown.—*Perisperm* conform to the seed, friable.—*Embryo* shape of a small broad inverted cone lodged in the apex of the perisperm. *Radicle* superior.

17. P. *angustifolium. R.*

Scandent, or creeping, smooth. *Leaves* sessile, obliquely-lanceolate, very smooth. *Aments* leaf-opposed, short-peduncled.

A native of the Malay Archipelago.

18. P. *saxatile. Wall.*

Stems furrowed, creeping, with villous joints and fascicled sub-erect branches. *Leaves* sub-sessile quatern, obovate-cuneate, fleshy, convex and villous beneath, with three-coloured nerves, veinless. *Spike* terminal, filiform, elongated, peduncled.

A native of the neighbourhood of Katmandu in Napal*a*, where it grows on rocks and blossoms during the first four months of the year.

A small succulent, probably annual species, which creeps to a considerable extent, sending forth fascicled, sometimes single, opposite or alternate branches from each joint, which scarcely rise four inches above the surface on which the plant grows.—*Stems* thread-like, pubescent, with four or five furrows; *branches* once or twice subdivided into small opposite branchlets.—*Leaves* generally quatern, rarely tern, four or five lines in length, obtuse, cuneate at the base, shining and somewhat concave above, with copious short hairs below, slightly ciliated, without veins or ribs, and losing even the three pale-coloured nerves when dry; the lowest verticils many times, the others two or three times shorter than the interstices between the joints.—*Leaves* of the young shoots linear-oblong, measuring sometimes an inch in length.—*Petiols* very short, villous, erect, with a gland-like body in their axils, villous.—*Spike* terminal, very slender, from an inch to an inch and a half long, round, villous, excavated on its surface with innumerable small pores in which the flowers are lodged, on a peduncle about half its length.

Obs. All the parts of this elegant little species have a faint pungent taste. I have not been able hitherto to examine its organs of fructification in a satisfactory manner.

Addition to Jasminum heterophyllum, p. 99.—*By.* N. W.

Since the above page was printed I have through the kindness of the Honourable Mr. Gardner been favored with flowering specimens of that interesting species which have enabled me to add the following description.

J. *heterophyllum. R.*

Arboreous. *Leaves* alternate, petioled, simple or ternate, from oblong to broad ovate, acuminate, waved, lucid, firm.—*Panicles* terminal, spreading, trichotomous, fastigitate.—*Calyx* urceolate with subulate teeth.—*Segments* of the *corol* oblong, equalling the tube.

Nepala names, *Gooje* and *Javana.*

It grows to be a midling-sized tree with long branches which have a tendency to become rambling. *Leaves* varying so much in size, as well as figure, as to appear to belong to different species, while in reality they are the produce of one and the same tree, only taken from differently aged branches; oblong and tapering, some times ovate-oblong, or even very broad-ovate almost cordate, generally alternate, in younger shoots opposite, from three to six inches long, two to three or even four inches broad, terminating into a long acumen, rounded, sometimes unequal at the base, perfectly smooth, shining above, pale below, with a very elevated rib, and long, sub-opposite, oblique, arched nerves which unite into sub-marginal reticulated veins. The leaves are promiscuously single and ternate, or even binate on one and the same branch; but in general they have a tendency to become ternate on younger branches.—*Petiols* slender, grooved, as well as the peduncles covered with short hairs, which occasionally continue along the principal vessels on the under side of the leaves; from one to two inches long; partial ones, short; they are all articulated at both extremities.—*Panicles* more or less elevated above the leaves, large, spreading, many times branched, the first divisions being long, the subsequent ones gradually shorter.—*Peduncles* slender, lax, together with the calyx hairy, with subulate deciduous bractes under each ramification.—*Flowers* very numerous and dense, ternate, yellow, delightfully fragrant, elevated on longish pedicels to nearly the same level; the intermediate ones sessile.—*Tube* of the corol about half an inch long, cylindric, striated.—*Stamens* slightly elevated above the mouth.

Obs. This ornamental Jessamine is probably the largest of the genus, growing as I am informed, to a considerable tree.—N. W.

Class 3d.

TRIANDRIA

MONOGYNIA.

VALERIANA. Schreb. Gen. N. 60.

Calyx none. *Corol* superior, one-petalled, gibbous on one side of the base. *Seed* one.

1. V. *Jatamansi.* Sir W. Jones in *Asiat. Res.* vol. ii. p. 405, and vol. iv. p. 109. Roxb. *ibid.* 433.

Flowers triandrous. *Leaves* entire, four-fold, the inner radical pair petioled, and cordate; *cauline* sessile, lanceolate. *Seeds* crowned with a pappus.

V. *spica. Vahl. En. Pl.* 2. 13.

2. V. *Hardwickii. Wall.*

Triandrous villous. *Radical leaves* numerous, fleshy, ovate-cordate, acute, unequally sinuated, on long groved petiols; *cauline* sessile small, opposite, remote, pinnate or laciniate at the base. *Corymb* terminal gradually elongated ; pedicels dichotomous. *Seeds* crowned with a downy pappus.

Valeriana. *Hardw. in Asiat. Res.* vi. 350.

A native of the mountains of Sirinagur and Nepal*a* ; where it blossoms in the beginning of the year.

Newarree Name, *Chammaha Soa.*

Root fleshy, as thick as the little finger, sending out a great number of pale, cylindric, slightly pubescent fibres.—*Leaves* numerous, spreading in all direction, from one to three inches long, with slightly waved and unequally repand, sometimes deeply gashed, margins, covered on both sides with very short soft hairs, dark green and somewhat shining above; very pale with fleshy prominent rib and nerves and slightly reticulated veins below; the lobes at the base more or less marked, sometimes decurrent or entirely wanting. As the stem grows in length they are elevated with it a few inches above the ground and are then perfectly opposite in very approximate pairs.—*Stem leaves,* two to three or even four very remote pairs, pinnate or pinnatifid, the terminal *leaflets* lanceolate, sinuate-dentate about an inch long, the others in one or two pairs linear-lanceolate, very small.—*Petiols* cylindric, pubescent, with a deep channel above, tapering from a broad and thick base, twice or thrice longer than the leaf.—*Stems* several, erect, striated, fistulous, slightly villous, from one to four feet high, as thick as a good goose-quill.—*Corymb* terminal, crowded at first, but afterwards much lengthened, with opposite slender pubescent peduncles, the lowest one or two pairs of which are generally distant and appear like smaller axillary inflorescences; *pedicels* three or four times forked, capillary, ending into short, few-flowered or crowded racemes.—*Bractes* linear opposite under each subdivision of the corymb; lowermost almost as long as the peduncles.—*Flowers* very small, pale, with scarcely any other calyx than an elevated margin at the top of the ovarium. *Corol* funnel-shaped with a five-cleft regular border; its tube naked at the base, villous within.—*Seed* compressed, with a series of white hairs in the furrows of the ribbed side; in other respects like that of V. *rubra. Gært. Carp.* ii. *p.* 35. *t.* 86.

Obs. I have received abundance of specimens with flowers and fruit from the Honourable Mr. Gardner. He also sent me roots of this interesting plant, which at present (May 1818), are shooting forth numerous large fleshy leaves. Its fibrous root acquires on being dried a strong smell like that of the common valerian and appearing to be permament. It is used medicinally by the natives of Nepal*a*.

The flowers and seeds seem like those of V. *Jatamansi,* but in regard to the root and leaves the two plants differ widely.—N. W.

OLAX. *Schreb. Gen. N.* 61.

Calyx entire. *Corol* three-petalled. *Nectary* of a few abortive filaments inserted on the petals. *Germ* one-celled; *ovulum* one, erect. *Drupe* half hid in the enlarged calyx, one-seeded. *Embryo* inverse, and amply furnished with a perisperm.

1. O. *scandens. Corom. Pl.* 2. *N.* 102.

Shrubby, scandent. *Leaves* ovate-oblong. *Petals* and nectarial filament bifid.

Fissilia psittacorum. Linn. Sp. Pl. ed. Willd. 1. 194.

Roxburgia baccata. König's Mss.

Teling. Turka-vepa.

A large, woody, climbing shrub, a native of forests. Flowers all the year round.

Trunk leaning, often as thick as a man's thigh. *Branches* numerous, climbing.—*Thorns* one here and there over the old woody parts only, very large and strong, shaped like the horn of a Rhinoceros.—*Leaves* alternate, bifarious, petioled, oval, entire, smooth, about two inches long, and one broad.—*Racemes* axillary, solitary, half the length of the leaves, few-flowered.—*Flowers* small, white.—*Calyx* cup-form, entire, permanent, and enlarging with the fruit.—*Petals* three; one, two, or all three half two-cleft; *divisions* linear, spreading.—*Nectary*, a bifid, filiform, yellow body, rising from the inside of the base of each division of the petals.—*Filaments* three, short, standing between the nectaries, inserted into the petals below their fissures. *Anthers* oblong.—*Germ* superior, ovate, one-celled, containing one *ovulum* attached to the bottom of the cell. *Style* length of the corol.—*Stigma* simple.—*Drupe* or berry globular, three-fourths covered with the enlarged calyx, yellow, pulpy, one-celled. Birds are fond of them.—*Seed* single, globular. *Integuments,* two; *exterior* nuciform; *inner* one white and spongy. *Perisperm* conform to the seed, amygdaline. *Embryo* inverse, minute, lodged in the apex of the perisperm. *Cotyledons* two, equal, minute, semi-linear. *Radicle* larger than the *cotyledons,* oval, superior.

2. O. *imbricata. R.*

Shrubby, scandent. *Leaves* bifarious, from ovate-lanceolate to oblong, entire, glossy. *Racemes* axillary, before the flower expands bifariously imbricated, (like the spiculæ in *Briza.*) *Drupes* ovate.

A very large climbing shrub, or tree, if any scandent plant can be so called; a native of Chittagong, where it flowers in the hot season.

LOEFLINGIA. Schreb. Gen. N. 71.

Calyx five-leaved. *Corol* five-petalled. *Capsules* superior, one-celled, three-valved.

1. L. *indica. Linn. Sp. Pl. ed. Willd.* 1. 191. *Retz. Obs.* 48.

Annual, diffuse, branchy. *Leaves* sub-verticelled, cuneiform *Flowers* terminal. *Style* three-cleft. *Calyx* fleshy.

Pharnaceum depressum. *Mant.* 564.

A small, diffuse, annual plant; appearing and flowering in the cold season.

Stems many, pressing on the ground, branchy.—*Leaves* opposite sessile, with many smaller ones from their axills, which makes them appear verticilled, oval, smooth, from a quarter to half an inch long. —*Stipules* four-fold, membranaceous, with many small similar scales about the flowers.—*Flowers* sessile in the divisions of the *branches,* and terminal. *Petals* linear, with three or four-toothed apices.

HIPPOCRATEA. Schreb. Gen. N. 73.

Calyx five-parted. *Petals* five. *Germ* three-celled; *ovula* a few, attached to the axis. *Capsules* three, one-celled, two-valved. *Seeds* membrane-winged. *Embryo* erect, without perisperm.

1. H. *indica. Linn. Sp. Pl. ed. Willd.* 1. 193. *Corom. Pl.* 2. *N.* 130.

Shrubby, climbing by tendrils. *Leaves* oblong, ventricose, serrate. *Corymbs* axillary. *Capsules* oblong, two-seeded.

A large, twining shrub, native of forests, hills, and wild uninhabited places, over various parts of India. Flowers about the beginning of the hot season.

Tendrils simple, woody.—*Leaves* opposite, short-petioled, oval, pointed, serrate, smooth, shining, about two inches long.—*Panicles* (*Umbells*) axillary, opposite, peduncled, about as long as the leaves, dichotomous, expanding.—*Flowers* numerous, very small, of a rusty yellow colour.—*Bractes* minute.—*Petals* linear-oblong.—*Germ* superior, three-lobed, three-celled, with two *ovula* in each, attached to the axis.—*Style* shorter than the stamens.—*Capsules* from one to three, oblong, one-celled, one-valved.—*Seeds* two, broad-winged, affixed to the bottom of the capsules by the apex of the wing.

2. H. *obtusifolia. R.*

Shrubby, climbing by the help of cirrhose branchlets. *Leaves* elliptically-oblong, entire, obtuse. *Corymbs* terminal and axillary. *Capsules* obovate, four-seeded.

A native of the coast of Coromandel. Flowering time in the Botanic Garden, March and April; the seeds take one year to ripen.

Stems and larger *branches* ligneous, and covered with cracked, spongy grey *bark*, climbing to a great extent by the help of the smaller, round, smooth, diverging, opposite branchlets, which take a turn, or two round their supporters.—*Leaves* opposite, short-petioled, oblong, entire, (large luxuriant leaves are sometimes obscurely serrate,) obtuse, firm, polished, length from three to four inches. *Stipules* of the leaves four-fold, minute, subulate; those of the branchlets larger. —*Corymbs* terminal, or from the exterior axills, often compoundly dichotomous, and shorter than the leaves.—*Bractes* minute, ovate. —*Flowers* small, pale greenish yellow.—*Calyx* five-leaved.—*Petals* five, ovate-lanceolate.—*Nectary*, a large, fleshy receptacle receiving the base of the germ; its margin becomes thin, and is divided into three, tapering, antheriferous segments, which are at first erect, but soon become revolute.—*Filaments* no other than the three segments of the nectary. *Anthers* of two, round, one-celled lobes.—

Germ three-lobed, immersed in the thick, fleshy nectary, three-celled, each cell containing several seeds, attached to the axis.—*Style* the length of the filaments; *stigma* three-toothed.—*Capsules* three, obovate, thin, smooth, slightly veined, about two inches long, by one and half broad, one-celled, two valved.—*Seeds* generally four in each capsule, two on each side, lodged about its middle, of a narrow-oblong shape, with a large, oblong, membranaceous wing proceeding from the base, and attached by its extremity to the very base of the capsule. *Integuments* two; *exterior* a very thin, smooth membrane; *interior* rather thick and somewhat spongy.—*Perisperm* none. —*Embryo* erect. *Cotyledons* oblong, pale green while recent. *Plumula* two-leaved. *Radicle* short, inferior.

Obs. In the following species there are only two seeds in the capsule, in every other respect it resembles exactly this species.

3. H. *arborea*. *R.*

Sub-arboreous, with climbing branches. *Leaves* opposite, oblong, serrulate, cuspidate. *Corymbs* axillary. *Capsules* linear-oblong, two-seeded.

B. Kath*a*-paharia.

A native of the interior parts of India. It blossoms in July, and the seeds ripen in March.

Trunk tolerably straight, but short, about as thick as a man's thigh, covered with smooth, dark brownish ash-coloured bark. *Branches* and *branchlets* spreading and climbing to a great extent, young shoots round and smooth, indeed highly polished; many of them form the claspers.—*Leaves* opposite, short-petioled, drooping, oblong, taper-pointed, serrate, very smooth on both sides; from six to seven inches long, and about three broad.—*Stipules* none, but two or three small, dark brown, scaly gems are found in each axil.—*Corymbs* axillary, dichotomous, with a pedicelled flower in the divisions. —*Bractes* opposite, small, permanent.—*Flowers* very numerous and small, pale greenish yellow.—*Calyx* beneath, generally four, though sometimes five or even six-parted; divisions rounded, and only about half the length of the corol.—*Corol* four, five, or six-petalled, (six is the most common, and I suppose the natural number,) in a double series, equal, ovate-oblong, smooth. *Nectary* a three-sided, fleshy cup surrounding the insertion of the stamens and germ.—*Filaments* three, membranaceous, shorter than the corol. *Anthers* four-lobed. —*Germ* ovate. *Style* short. *Stigma* simple.—*Capsules* three, cuneiform, striated, three inches long, and about one and a half broad, one-celled, two-valved; *valves* exactly boat-shaped.—*Seeds* two, ovate, compressed, enlarged with an oblong, soft, thin, spongy wing, the attachment is between its apex and the base of the capsules. *Integument* single, ochraceous.—*Perisperm* none.—*Embryo* oblique, compressed. *Cotyledons* conform to the seed, yellow. *Plumula* two-lobed. *Radicle* oval, sub-inferior.

JOHNIA. R.

Calyx inferior, five-leaved, or five-parted. *Corol* five-petalled. *Nectary* or receptacle of the stamina and pestillum sub-globular. *Germ* three-celled; *ovula* one or two in each cell, peltate. *Berry* one or more-seeded. *Embryo* without perisperm, direction various.

In honour of the Rev. Dr. John of Tranquebar. It is nearly allied to *Salacia. Gen. Pl. ed. Schreb.* 1380. Its proper place is probably the third order of this class.

1. J. *salacioides.* R.

Leaves opposite, broad-lanceolate, entire, firm and polished. *Calyx* five-leaved. *Petals* sessile. *Anthers* sessile on the three divisions of the nectary.

A native of Tipperah, Chittagong, and other parts of the eastern frontier of Bengal. It flowers in the Botanic Garden during the hot season; and the fruit ripens about the close of the rains.

A large, very ramous shrub, with its *branches* spreading much, and often drooping elegantly. *Bark* of the ligneous parts brownish and somewhat scabrous, that of the young twigs smooth.

Leaves opposite, short-petioled, recurved, broad-lanceolate, entire, firm, polished, obtuse-pointed; about four inches long, and half as much broad.—*Stipules* minute, caducous.—*Peduncles* axillary, several, longer than the petioles, one-flowered.—*Flowers* small, deep greenish orange colour.—*Calyx* of five round, villous margined, minute leaflets.—*Petals* round, sessile, with the margins recurved.—*Filaments* no other than the three, short, triangular segments of the fleshy urceolate nectary, elevating the three, large, twin *anthers* even with the stigma.—*Germ* above, conical; three-celled, with two *ovula* in each, attached to the axis.—*Style* scarcely any; *stigma* acute.—*Berry* size of a crab-apple, round, smooth, dull-red. *Pulp* in considerable quantity, white, of a soft, tough, gelatinous consistence, and eaten by the natives.—*Seeds* two, or three, separated by a thick partition of pulp; transversely oblong; flat on the inside, convex on the exterior.—*Integuments*, single, spongy, brown.—*Perisperm* none. *Embryo* consisting of two equal *cotyledons*, conform to the seed, and until vegetation begins, so firmly, and closely united, that they seem to be of one entire piece.—*Radicle* small, lodged at the umbilicus, relative, centripetal.

2. J. *coromandeliana.* R.

Leaves opposite, serrulate, polished. *Calyx* five-toothed. *Petals* unguiculate.

A small, straggling tree, or rather a large climbing shrub, a native of forests and mountains of Coromandel.

Leaves opposite, short-petioled, oval, smooth, shining, sometimes most slightly serrulate; about three inches long, and one and a half broad.—*Peduncles* axillary, several, bowing, one-flowered.—*Calyx* inferior, one-leaved, five-cleft.—*Petals* five, clawed, round, or oval, spreading.—*Nectary* a large globular, fleshy receptacle, elevating the stamens and pistil.—*Filaments* three, spreading, shorter than the petals, inserted into the top of the nectary round the germ. *Anthers* small, twin.—*Germ* half immersed in the nectary, three-celled, with one or two *ovula* in each, attached to the axis. *Style* very short. *Stigma* simple.—*Berry* superior, globular, upon the remaining nectary, or receptacle, size and colour of a small red cherry; one-celled, one-seeded.

IRIS. *Schreb. Gen.* N 97.

Corol six-petalled, petals unequal, alternate, jointed and spreading. *Stigmas* petals, from cowled to bilabiate.

1. I. *chinensis. Curtis's Bot. Mag. N.* 373.

Root creeping. *Leaves* ensiform, drooping, *scape* ramous, many-flowered.

This elegant species has been introduced from China, into the Botanic Garden, where it thrives best in cool, moist, shady situations. Flowering time the beginning of the hot season, viz. March and April.

MOREA. Schreb. Gen. N. 80.

Corol six-petalled; the three inner spreading and narrower. *Stigma* three-cleft.

1. M. *chinensis. Linn. Sp. Pl. ed. Willd.* 245. *Thunb. Diss. N.* 19.

Herbaceous, stems compressed. *Leaves* bifarious, ensiform. *umbels* terminal, proliferous, all the six-petals expanding.

Belameanda-shular mini. *Rheed. mal.* 11. *t.* 37.

Ixia chinensis. *Curt. Magaz.* 171.

Pardanthus chinensis. *Annals of Bot.* 1. 416.

Beng. Dusbaha, or *Dusbichundi.*

A native of India, China, &c. *Flowers* in the rainy season in the Botanic Garden near Calcutta.

Is common in gardens over India, where it blossoms during the rains, and ripens its seeds in the cold season.

COMMELINA. Schreb. Gen. N. 86.

Calyx inferior, three-leaved. *Corol* three-petalled, often dissimilar. *Nectaries* or *sterile* filaments three, with a cruciform head. *Capsules* two or three-celled. *Seeds* one or more. *Embryo* simple, and furnished with a perisperm.

1. C. *communis. Linn. Sp. Pl. ed. Willd.* 1. 249.

Polygamous, creeping. *Leaves* ovate-lanceolate, smooth; *spathes* many flowered. *Petals* unequal. *Anthers* dissimilar. *Capsules* tricoccous.

Beng. J*u*ta-kansh*i*ra.

Ephemerum dipetalum. *Kempf. amœn. p.* 888. *t.* 889.

This species is common over the low moist parts of *India;* flowering time the rainy season chiefly.

Root fibrous.—*Stems* numerous, creeping, dichotomous, jointed, round, smooth; length very various.—*Leaves* alternate, sessile on tubular, vaginated sheaths, ovate-lanceolate, acute; margins waved; smooth on both sides, with delicate, parallel veins runing length ways on the under side; from two to four inches long; mouths of the sheathing petioles a little ciliate.—*Spathes* terminal, or sub-fasicled, semi-cordate.

2. C. *bengalensis. Linn. Sp. Pl. ed. Willd.* 1. 250.

Polygamous, creeping. *Leaves* cordate. *Spathes* three-flowered. *Petals* unequal. *Anthers* dissimilar. *Capsules* three-celled, six-seeded.

Ephemerum bengalense, &c. *Pluck.* 1. 27. *f.* 3.

Beng. Kansh*i*ra, also Kan*u*raka.

Like the last described, this species is a native of wet places over Bengal, but much scarcer and smaller; flowering time the same.

Stems creeping, dichotomous, jointed, hairy.—*Leaves* alternate, petioled, cordate, nervous, a little hairy, about an inch long and three-fourth of an inch broad.—*Petioles* sheathing, hairy, and ciliate.—*Spathes* terminal, and axillary, of a shape between that of a short, inverted cone and falcate; open on the upper side only; each containing a common peduncle of two hermophrodite flowers, and a long-peduncled male one.—*Flowers* small, bright blue colour.—*Calyx* as in the former.—*Corol* three-petalled; two of them larger, and long-clawed, as in the former species, the third small, and sessile, but coloured.—*Nectary, stamens* and *pistil* as in C. *communis,* only the style is here beautifully spiral.—*Pericarp* three-celled, three-valved, with generally two seeds in each cell.

3. C. *molucca na. R.*

Annual, creeping. *Leaves* sessile, oblong, about five-nerved. *Flowers* terminal, solitary, sub-sessile. *Petals* equal.

A native of the Moluccas. Compare with Vahl's. C. *attenuata.*

4. C. *salicifolia. R.*

Polygamous, creeping. *Leaves* narrow-lanceolate. *Petals* three, equal. *Spathes* from four to six-flowered. *Anthers* dissimilar. *Capsules* pointed, three-celled, six-seeded.

Sans. लाङ्गली, Lang*u*lee.

Beng. Panee-kansh*i*ra.

Found with the former in wet places near Calcutta, it is also a native of the Coromandel Coast; flowering time the wet season.

Stems creeping to a great extent, ramous, round, smooth, jointed, very succulent.—*Leaves* alternate, sessile on their cylindric sheaths, narrow-lanceolar, a little hispid on the upper side and margins, when the finger is drawn backward; from four to six inches long, and not quite one inch broad.—*Sheaths* cylindric, striated; mouth ciliate.—*Spathes* leaf-opposed, long-peduncled, semi-cordate, long-pointed.—*Flowers* hermaphrodite, and male; large, of a beautiful deep azure colour, disposed as in the former species, only there are fewer to the common shorter peduncle, and often two on the longer one.—*Calyx* as in the former.—*Petals* three, equal, except that the lower one having a shorter claw is nearly sessile.—*Nectaries, stamens,* and *pistilum,* as in the former.—*Capsules* oblong, pointed, three-celled, three-valved, with two seeds in each. Cattle are fond of this species.

5. C. *nana. R.*

Creeping. *Leaves* cordate-lanceolate, stem-clasping. *Flowers* terminal, sub-panicled. *Petals* equal. *Capsules* three-celled, many-seeded.

A native of wet places over India; it appears during the rainy and cold season.

Root annual.—*Stems* or branches creeping, enveloped in the sheaths of the leaves; from three to six inches long.—*Leaves* stem-clasping, sheathing linear-cordate, somewhat ciliate, and waved; *sheaths* a little downy.—*Flowers* terminal, sub-panicled, small, bright blue.—*Calyx* equal, permanent.—*Corol* of three, roundish, equal petals. *Nectarial* filaments as in the other *Indian* species, but naked.—*Filaments* three, from the middle downwards clothed with hairs, converging over the stigma. *Anthers* blue.—*Style* shorter than the stamens, straight and tapering.—*Capsules* three-celled, with from three to six seeds in each.

Obs. Were the style longer, or even spirally bent, I should have concluded this plant to be *Commelina spirata. Mant.* 2. *p.* 176. Burman's figure of his *C. diffusa. Flor. Ind. tab.* 7. *f.* 2. is so bad that no comparison can be made; they may be the same.

6. C. *nudiflora. Linn. Sp. Pl. ed. Willd.* 1. 252.

Frequently creeping. *Leaves* linear, sessile, sheathing. *Flowers* terminal, racemed, diandrous. *Petals* equal. *Capsules* three-celled; *cells* two-seeded.

Beng. Kand*u*li.

Tal*i*-pull*i*. *Rheed. mal.* 9. *t.* 63.

Tradescantia malabarica. *Sp. Pl.* 412.

A native of most of the warmer humid parts of *India.* Flowering time the rainy season.

Stems none, but it has many short, creeping, smooth, jointed *branches,* with their flower-bearing extremities sub-erect.—*Leaves* alternate, sessile, linear, sheathing, smooth, excepting a few fine hairs about the mouths of the sheaths.—*Racemes* terminal, long-peduncled, corymbed, bearing on their upper side many small blue flowers in succession.—*Bractes* solitary, one-flowered, concave, caducous.—*Calyx* three-leaved, equal.—*Corol* of three, obovate, equal, blue petals. *Nectarial filaments* with a very few, moniliform, blue hairs; corpuscles two, clavate.—*Filaments* three, two of them ascending, larger, and thicker, well clothed with beautiful jointed blue hairs. *Anthers* blue, oval, the third filament sterile. *Style* short, ascending.—*Capsules* three-celled, with two scabrous seeds in each.

7. C. *cespitosa*. *R.*

Polygamous, creeping. *Leaves* lanceolate. *Spathe* cordate, peduncled, few-flowered; *petals* nearly equal. *Capsules* three-celled, with one or two seeds in each.

A native of Chittagong, from thence the plants were sent to the Botanic Garden by Mr. W. Roxburgh. It is in blossom most part of the year.

Stems and *branches* numerous, slender, smooth, jointed, creeping close over a space of some feet and striking deep root at the joints. *Leaves* alternate; sessile on their smooth cylindric sheaths, ovate-lanceolate, the upper ones narrower and longer; all are entire, and smooth on both sides.—*Sheaths*, besides the usual leaf-bearing one, there is a second or inner annular one wherever there is a peduncle.—*Peduncles* solitary, leaf-opposed, about an inch long, round, and smooth. *Spathe* cordate, or rather semi-cordate in their natural folded state, each embracing two common pedicels; the superior one bearing two, three, or four, pedicelled male flowers; and the lower about as many hermaphrodite ones; all blossoming in succession.—*Calyx* of three, equal, obovate, concave, pale blue, pellucid leaflets.—*Petals* three; the upper two reniform and pedicelled; the lower one cordate, and sub-sessile, all of a most beautiful light blue colour. *Nectaries* three, erect, on the upper side opposite to the stamens.—*Filaments* three, equal, ascending. *Anthers*, the middle one larger, and sagittate.—*Germ* ovate, (in the male flowers wanting.) *Style* ascending, as long as the stamens. *Stigma* incurved, simple.—*Capsules* oblong, three-celled, with one or two seeds in each.

8. C. *scapiflora*. *R.*

Root perennial. *Racemes*, radical, erect, with alternate, many-flowered spathes. *Petals* equal. *Leaves* in a tuft, ensiform.

Beng. Kurelli.

A native of Hindoostan and introduced into the Botanic Garden by Colonel Hardwicke, where it blossoms in March and April, at which time the plant is destitute of leaves, nothing being seen but the racemes ornamented with large lively blue flowers round the apex.

Root perennial, composed of several, smooth, elongated tubers, with a few filiform radicles.—*Stems* none, except the sheathing bases of the leaves which appear a month or two after the flowers; these are ensiform, waved, acute, smooth, and marked with several straight veins; length from four to eight inches, breadth about one.—*Racemes* radical, erect, straight, smooth, invested at each of the remote two or three joints, with a small, solitary sheath; these sheaths toward the top are more approximated, each of them there embracing the insertion of a branchlet which bears several pedicelled flowers.—*Calyx* the three leaflets thereof broad-lanceolate, and considerably shorter than the corol.—*Petals* equal, round, concave.—*Nectaries* alternate with the stamens, hairy.—*Filaments* three, alternate with the petals, longer than the filaments of the nectaries, and like them hairy about the middle. *Anthers* blue.—*Pistil* declining. *Stigma* minutely three-toothed.

9. C. *herbacea*. *R.*

Herbaceous. *Stems* jointed, ascending. *Leaves* lanceolate. *Panicles* terminal. *Petals* equal. .*Capsules* three-celled, many-seeded.

A native of Sumatra, from thence introduced into the Botanic Garden by Dr. Charles Campbell, where it blossom and ripens its seeds during the whole year.

Root perennial, consisting of long, firm, thick fleshy fibres. *Stems* several from the same root, herbaceous, jointed, ascending, round, smooth, much swelled at the joints, from two to three feet in height, nearly the whole enveloped in the sheath of the leaves.—*Leaves* alternate, sessile on their sheaths, lanceolate, smooth, slightly marked with five parallel nerves, margins entire, waved, somewhat membranaceous, and coloured. *Sheaths* shorter than the joints, striated, and smooth, except a few white hairs round their mouths.—*Panicles* terminal; there are also smaller ones from the superior axils, which are sub-globular, and smaller. *Ramifications* alternate, at first divaricate, afterwards recurved.—*Flowers* numerous, small, short-pedicelled.—*Bractes* triangular, acute.—*Calyx*, leaflets three, equal, oblong, concave, spreading, fleshy, furrowed on the outside.—*Petals* three, equal, oval, much larger than the calyx, concave, entire, spreading and of a delicate translucent pale bluish white.—*Nectarial* filaments alternate with the longer stamina, bearded (like the latter) towards the base, with delicate, slight blue hairs, and crowned with large incumbent, three-lobed, yellow glands.—*Germ* somewhat three-sided. *Style* oblique, length of the filaments. *Stigma* small, crowned with a tuft of hairs.—*Capsules* three-celled, three-valved, with generally from four to six, or seven, angular, rough seeds in each cell. *Embryo* lodged in a deep pit, in the back of the seed, small, oval.

SONERILA. R.

Calyx superior, three-toothed. *Petals* three on the mouth of the calyx, alternate with the stamina. *Germ* three-celled, cells many-seeded, attachment central. *Capsule* three-celled. *Seeds* numerous, minute.

1. S. *maculata. Roxb.*

Leaves opposite, unequally-cordate, bristle-spotted, serrulate, ciliate.

Soneri-ila. *Rheed. mal.* 9. *p.* 127. *t.* 65.

Sootli, the vernacular name in the Khassee language.

It is indigenous amongst the mountains on the North-East border of Bengal, and used by the natives as a potherb.* Flowering time the beginning of the rains. In its *natural character* it agrees very exactly with *Burmannia.*

Root fibrous.—*Stem* short, being only a few inches high, and when old leaning much, *branches* opposite, longer than the stem, the lower reclinate, with their apices resting on the ground and striking

* I have had specimens from Nepala under the name of Lushee-Sowa.—N. W.

root, all are nearly round and bristly; height of the whole plant from six to twelve inches.—*Leaves* opposite, long-petioled, more or less unequally broad ovate-cordate, nerved, bristle-serrulate, pointed, surface bullate, and marked with numerous small white specks, from the centre of each rises a slender, subulate, bristle; veins and nerves bristly, length from two to six inches, by from one to three broad.—*Racemes* axillary, spiral, their peduncles as long as the petioles, deeply coloured and hairy.—*Flowers* springing from the upper convex side of the raceme, short-pedicelled, red, bristly.—*Bractes* subulate, minute.—*Calyx* superior, three-sided, and tubular; mouth three-toothed, permanent.—*Petals* three, lanceolar, inserted into the mouth of the calyx, alternate with its segments, smooth on the inside, on the outside are frequently found some slender bristles. *Nectary* of three, laciniated scales round the base of the style.—*Filaments* three, alternate with the petals, and about their length. *Anthers* sagittate, one-third as long as the filaments, until the flower has been expanded some time they are inverted within the tube of the calyx, afterwards erect. —*Germ* clavate, three-sided, three-celled; *ovula* numerous, attached to their respective receptacles, rising from the axis, exactly as in Osbeckia chinensis. *Gært. Carp.* 2. *t.* 126. *Style* length of the stamens. *Stigma* simple.—*Capsules* erect, three-sided, clavate; pedicel included, about half an inch long, three-celled, opening at the apex.—*Seeds* very numerous, and too minute to admit of my ascertaining their internal structure.

2. S. *emaculata. Roxb.*

Leaves opposite, unequally narrow-cordate, acute, bristly, but void of spots, serrulate, ciliate.

A native of the Khassee mountains on the North-East border of Bengal, where it flowers in July, and differs from the preceding species, in the leaves being destitute of the little round white spots, into which the bristles are inserted, (viz. one in the centre of each spot.) It is also less robust than that plant, with longer peduncles to the racemes, which are sometimes leaflet-bracted at the middle.

The flowers, and seed vessels are alike.

3. S. *angustifolia. Roxb.*

Leaves opposite, unequally oblong-lanceolar, serrulate, hairy.

A native of the mountains on the North-East border of Bengal, and like the two former species in habit, but the shape of the leaves widely different, and only a little hairy.

4. S. *moluccana. R.*

Sub-caulescent. *Leaves* in pairs, entire, hairy, one minute and nearly round, the other unequally cordate. *Racemes* umbelled.

A native of the Molucca Islands.

5. S. *squarrosa. Wall.*

Stems short, cylindric, marked with numerous stipulary inoffensive *prickles* which are deciduous, together with the sub-sessile crowded, sub-verticilled, cuneate-lanceolate, serrulate, smooth *leaves.*

Growing among ferns on the Khassee mountains, where it is called *Tearanga.* It blossoms during the rains.

This little perennial plant rises to the height of from four to six inches.—*Root* consisting of capillary fibres.—*Stem* ash-coloured, about the thickness of a crow-quill, erect, marked with large elevated tubercles to which the leaves were attached, and innumerable small, subulate, axillary sub-erect *prickles* which together with the leaves are deciduous leaving the lower portion of the plant naked. They are about two lines long and interspersed with small elevated dots. —*Branches* few, simple.—*Leaves* of a dark green colour, sometimes verging towards brown, crowded near the top of the plant, spreading, somewhat fleshy, slightly dotted, veinless, scarcely above half an inch long, acute, with adpressed, bristly, purplish serratures, entire toward their base and tapering into a very short flat *petiol* which articulates with the corresponding tubercle of the stem or branch. —*Stipules,* two axillary, opposite, prickles.—*Raceme* terminal or axillary on a filiform peduncle, two or three times longer than the leaves.—*Flowers* unilateral on short pedicels, large in comparison with the size of the plant, rose-coloured, imbricated during their æstivation.—*Calyx* slightly coloured, with pubescent angles and the segments obscurely keeled.—*Petals* oval, acute, contracted at the base and inserted into the mouth of the calyx so as to occupy its whole circumference.—*Filaments* naked.—*Anthers* beaked, subulate, with a cordate base, bilocular, opening on both sides along their whole length.

Obs. I received this pretty little plant from my indefatigable and zealous assistant, Mr. Smith at Silhet, in 1817. It seems to thrive very well on an elevated bed, in a soil mixed with pebbles, and while I write this, in July the following year, it commences shooting out numerous fresh leaves from the top of the stem and branches. The structure of its flowers agrees entirely with that of the first suecies; but they are about one third smaller in size.—N. W.

XYRIS. Schreb. Gen. N. 89.

Head with roundish one-flowered scales. *Perianth* beneath, three-leaved. *Corol* three-petalled, equal, waved. *Nectaries* three, bifid. *Capsule* one-celled, three-valved. *Seeds* numerous, on parietal receptacles.

1. X. *indica. Linn. Sp. Pl. ed. Willd.* 1. 254.

Leaves ensiform. *Heads* globular; *scales* round.

Kotsjiletti-pullu. *Rheed. mal.* 9. *p.* 139. *t.* 71.

Beng. Cheena ghauza. Dab*i* doob*a*.

Grows on a low clayey soil, over many parts of Coromandel, and Bengal. Flowering time November and December. Seeds ripen in January and February.

Root fibrous, annual.—*Leaves* radical, bifarious, straight, sword-shaped, on one edge slit into a sheath for the scape, pointed, smooth; from six to twelve inches long.—*Scape* naked, round, striated, erect, length of the leaves, each supporting a round, flower-bearing head.—*Flowers* a beautiful bright yellow.—*Bractes*, or scales one-flowered, orbicular, concave, hard, smooth.—*Calyx* three-leaved, hid within the scale, membranous.—*Petals* three, each supported on an unguis just long enough to raise their expanding, oval, crenate borders above the scales.—*Nectary,* three filaments inserted, alternately with the petals, round the base of the germ; apex two-cleft, each division ending in a pencil of fine yellow hairs; adhering firmly at the cleft to the edges of the petals, near the apex of the claws.—*Filaments* three, short, broad, erect, inserted on the inside of the apex of the claws of the petals. *Anthers* twin, erect, united by a continuation of the filament.—*Germ* superior, three-sided. *Style* length of the claws of the petals, from thence three-cleft. *Stigmas* torn.—*Capsule* three-valved, one-celled. *Seeds* numerous, attached to a heel down the inside of each valve.

Obs. The following account of the virtues of this plant I have been favoured with, by the Honourable John Hyde, who informs me that "the natives of Bengal esteem it a plant of great value because they think it an easy, speedy, and certain cure for the troublesome irruption called ring worms." This accords with what Van Rheede says of it, at page 139 of the 9th volume of the Hortus Malabaricus, viz. *Foliorum succus cum aceto mixtus impetigini resistit. Folia cum radice oleo incocta contra lepram sumuntur.*

FUIRENA. Schreb. Gen. N. 90.

Ament imbricated on all sides with tailed scales. *Corol* three-petalled. *Seed* naked.

1. F. *ciliaris. R.*

Culms erect, about twelve inches, columnar, jointed, leafy. *Leaves* hairy. *Heads* axillary and terminal. *Scales* very hairy. *Stigmas* three. *Seeds* three-sided.

Scirpus *ciliaris. Linn. Sp. Pl. ed. Willd.* 1. 309. *Rottb. gram.* 5. *t.* 17. *f.* 1.

Teling. Elagabora gaddi.

A native of moist places on the Coast of Coromandel, with the *Scirpi,* &c.

Root fibrous.—*Culms* nearly erect, about a foot high, round, striated, leafy, jointed, but not piped.—*Leaves* sheathing, shorter than the culms, hairy, ciliated on the margin, concave on the inside, not keeled.—*Heads* sub-globular, compound, axillary, and terminal, the terminal one is considerably longer, they are composed of several oval, imbricated, many-flowered, sessile spikes. *Involucre* of the terminal head one or at most two small leaves; of the others, nothing more than the sheath of the leaf from whence they issue.—*Scales* of the aments, or spikes, *corol*, &c. as in the genus.

KYLLINGIA. Schreb. Gen. N. 9.

Ament imbricated. *Flowers* with *calyx* and *corol* of two chaffy valves each. *Seed* one.

1. K. *monocephala. Linn. Sp. Pl. ed. Willd.* 1. 256. *Vahl. in Pl.* 2. 379.

Rottb. gram. 13. *t.* 4. *f.* 4.

Culms six to eighteen inches high, triangular; *head* terminal, single; *involucres* three or four. *Seeds* obcordate, without angles.

Pee-Mottenga. *Rheed mal.* 12. *p.* 99. *t.* 53.

Gramen capitatum. *Rumph. amb.* 6. *p.* 8. *t.* 8. *f.* 2.

Beng. Sweta gothoobi; Nirbishee, its fragrant aromatic root being accounted an antidote to poisons.

Common over *India* on low shady pasture ground, when totally shaded the heads are generally more green, which is rather uncommon.

Root fibrous, and stoloniferous.—*Culms* erect, from six to eighteen inches high, generally about eight; three-sided, smooth, naked except near the base; *angles* sharp, and sides hollowed.—*Leaves* sheathing, as long as the culm, smooth, sharp keeled.—*Head* terminal, sessile oval, generally white.—*Involucre* three-leaved, unequal, the largest leaf as long as the culm. Back of the glumes of the corol hispid.—*Stamens* three. *Stigma* three-cleft.—*Seed* obliquely obcordate, much compressed, smooth, light brown.

2. K. *triceps. Linn. Sp. Pl. ed. Willd.* 1. 256.

Culms from six to twelve inches high, three-cornered. *Heads* terminal, sessile, from three to six; *involucres* three or four; *florets* diandrous; *Seed* oblong, without angles.

Mottenga. *Rheed. Hort. mal.* 12. *p.* 97. *t.* 52.

A native of low shady pasture ground.

Root somewhat tuberous, with capillary fibres.—*Culms* straight, from six to twelve inches high, smooth, naked, except at the base, obtusely three-sided; *angles*, rounded.—*Leaves* sheathing, nearly as long as the culm, smooth.—*Head* terminal, composed of from three to six sessile, oval, white spikes, the largest one is in the centre, the rest in the circumference.—*Involucre* as in K. *monocephala.*—*Stamens* seldom more than two.—*Stigma* two-cleft.—*Seed* oblong, much compressed, white, no angles.

Obs. It differs from K. *monocephala* not only in having a compound head but in being diandrous, with a bifid style, and oblong lanceolate seed. I have found luxuriant plants with more than one flower in the calyx.

3. K. *umbellata. Linn. Sp. Pl. ed. Willd.* 1. 257. *Rottb. gram.* 15. *t.* 4. *f.* 2.

Culms from one to two-feet high, three-sided, umbel of sessile and peduncled, cylindrical, ragged spikes, involucres many; involucels none. *Seeds* three-sided, linear-oblong.

Kol pullu. *Rheed. Hort. mal.* 12. *t.* 63.

Mariscus *umbellatus. Vahl. Enum. Pl.* 2. 376.

Beng. Bura-gothoobi.

A native of wet or marshy places.

Root tuberous, rust-coloured, with rust-coloured fibres.—*Culms* erect, from one to two feet high, three-fourths naked, three-sided, smooth.—*Leaves* sheathing nearly as long as the culm.—*Umbel* terminal, sessile, composed of sessile and peduncled, cylindrical spikes of numerous diverging flowers.—*Involucre; universal,* many leaved, unequal; the largest leaf often as long as the culm; *partial,* none. *Calyx* and *corol,* &c. as in the genus. *Seeds* three-sided.

Obs. As this plant wants the partial involucre it may be K. *sumatrensis* of Retzius.

4. K. *cyperoides. R.*

Culms from six to eight inches high, three-sided. *Umbel* compound, consisting of globular heads of numerous spikelets, *involucre* three-leaved. *Seed* oblong, three-sided.

An elegant cyperus-looking small species, a native of moist pasture ground.

Root fibrous.—*Culms* erect, two-thirds naked, from six to eight inches high, three-sided, smooth.—*Leaves* sheathing, as long as the culm. *Involucre; universal,* three-leaved, very unequal, the largest leaf being as long as the culm, and the smallest one inch, *partial,* minute.—*Calyx, corol,* &c. as in the genus, excepting that there are sometimes two flowers in the same calyx.

TUNGA. R.

Ament ovate, imbricated on all sides. *Calyx* one-valved, one-flowered. *Corol* two-valved. *Seed* naked.

The plants which come under this definition I cannot by any means reconcile to the character of *Schœnus,* nor indeed to any genus known to me. I have therefore constituted a new one of them, under the Telinga name *Tunga,* which is applied to the whole family of *Calamariæ.* Linn. *Cyperoideæ.* Juss. This genus is probably the same as Vahl's *Hypaelyptum.*

1. T. *triceps. R.*

Culms from four to eight inches high, a little compressed; *spikes* two or three, terminal; *involucre* two-leaved.

A native of Coromandel, growing with other *cyperi.*

Root fibrous.—*Culms* erect, mostly naked, from four to eight inches high, a little compressed, smooth.—*Leaves* sheathing, broad.

nearly as long as the culm.—*Spikes* terminal, generally three, sessile, oval, imbricated with innumerable small rust-coloured flowers.—*Involucre* two-leaved, the largest is often erect, and looks like a continuation of the culm, it is from two to three inches long; besides these there are two or more oval, lateral, scales.—*Calyx,* a scale, single, nearly as in the *scirpi,* one-flowered, wedge-shaped, striated.—*Corol* two-valved, membranaceous, shorter than the scale.—*Stamens* three.—*Stigma* three-cleft.—*Seed* oblong, three-sided, without bristles.

2. T. *lævigata. R.*

Culms from one to two feet high, three-sided. *Spikes* several, terminal. *Involucre* three-leaved.

A native of moist vallies.

Root fibrous. *Culms* erect, from one to two feet high, nearly naked, three-sided, smooth.—*Leaves* numerous, sheathing, half as long as the culm —*Head* terminal, generally composed of about seven oval, sessile, closely imbricated spikes.—*Involucre* two or three-leaved, very unequal, the largest from five to eight inches, the shortest one, or one and a half.—*Calyx* as in T. *triceps,* only sometimes three-lobed at the apex.—The rest as in T. *triceps.*

3. T. *diandra. R.*

Culms leafy, two or more feet high, three-sided; *corymbs* terminal; *involucres* alternate. *Flowers* diandrous. *Style* two-cleft.

A stout, erect, smooth, long-leaved species; a native of Amboyna.

Culm jointed, three-sided, smooth.—*Leaves* one at each joint, sheathing, rising much higher than the culm, linear, smooth, acute, three-nerved, about three feet long.—*Corymbs* terminal, decompound or more.—*Involucres* several, the largest at the first division of the corymbs, and like the leaves of the culm, but smaller; *spikelets* oval, minute, composed of many small brown imbricated flowers as in the *scirpi.*—*Calyx,* a glume, roundish, smooth —*Corol;* the two valves, small, placed transversely with respect to the calyx, and smaller than it.—*Filaments* two, one rising laterally from the bosom of each valve of the *corol; anthers* linear. *Germ* oblong, *style* deeply two-cleft. *Seed* oval, rather longer than the calyx.

SCHŒNUS. Schreb. Gen. N. 92.

Calyx, or *Corol, glumes* several; *seed* one, naked.

1. S. *articulatus. R.*

Culms three-sided, erect, from four to five feet high, jointed, leafy. *Corymbs* terminal and axillary, compound and decompound. *Style* undivided. *Seed* obovate, beset with bristles, and crowned with the conic base of the *Style.*

Teling. Konda-*tunga.*

A native of marshy places up amongst the Circar mountains. It may be referred to Vahl.'s *Rynchospora Enum. Pl.* 2. 229.

Culms erect, three-sided, leafy, jointed, smooth, from four to five feet high, angles sharp, between the joints fistulous.—*Leaves* of the root as long as the culm, smooth, except that the margins are a little hispid, slender considering their great length, those of the culm from five to eight, sheathing, shorter than those of the root.—*Umbels* the terminal one decompound, those only from the exterior axils compound.—*Involucre* of the terminal umbel, from three to four-leaved, about as long as the umbel, leaf like; the axillary umbels have no involucre, or, at most, a very trifling one.—*Flowers* sub-solitary, chesnut-coloured.—*Calyx* none. *Corol; petals* four. *Style* undivided, with a permanent conical base.—*Stigma* entire, acute. *Bristles,* six permanent ones surrounding the germ.—*Seed* a little compressed, crowned with a large brown, spongy, conical body, the base of the style.

Obs. There is frequently a very slender male flower within the large or inner glume of the corol; it consists of one glume or scale, and three stamens. The plant is so very coarse that cattle will not eat it.

CYPERUS.

Glumes chaffy, bifariously imbricated. *Corol* none. *Seed* one, naked.

Sect. 1st, with culumnar Culms.

1. C. *setaceus. Linn. Sp. Pl ed. Willd.* 1. 269. *Retz. Obs.* 5. *p.* 10.

Culms setaceous, from two to three inches high; *spike* terminal, solitary, naked.

A native of Coromandel, appears and flowers during the rainy season chiefly.

2. C. *mucronatus. Linn. Sp Pl. ed. Willd.* 1. 273. *Retz. Obs.* 5. 10.

Culms from six to eight inches high, round, incurved, half naked. *Spikes* lateral, sessile. *Involucre* one-leaved, besides the continuation of the culm. *Seed* oval, rounded.

A native of moist sandy places, over various parts of India.

Root creeping, invested in brown sheaths.—*Culms* half naked, issuing singly from the root at some small distance from each other, nearly round, ascending, from six to eight inches long; the lower half is involved in the sheath of a single, short leaf.—*Leaf* sheathing, generally one to each culm, which is only about half its length.—*Head* about an inch and a half below the extremity of the culm, it consists of five or six small, oval, ten or twelve-flowered spikes. —*Involucre* one-leaved, about as long as the spikes; besides these there are some small, chaffy, calyx-like scales.—*Seed* naked, oval, compressed, white.

3. C. *lateralis. Forst. descript.* 13. *N.* 28.

Culms erect, columnar, from eight to twelve inches long, naked; *Head* of five or six sessile spikes beneath the straight apex of the culm; *Stigma* bifid; *Seed* oval, compressed.

A native of Coromandel.

Root creeping, dark rusty brown.—*Culms* straight and perfectly erect, from eight to twelve inches high, round, and naked.—*Leaves* no other than a short sheath or two, embracing the base of each culm.

—*Spikes* sessile, five or six, forming a small head on one side of the culm about an inch and half below its apex. *Involucre*, a minute, subulate leaflet immediately under the spikes, and of the same length with them.—*Scales* of the spikes numerous, ovate-oblong, white, smooth and rather obtuse.—*Stamina* three.—*Stigma* bifid.—*Seed* oval, flat on the inside, and convex on the exterior, lead-colour or grey, and smooth.

Obs. This differs from the plant which I consider to be C. *mucronatus of Retz. fasc.* 5. *p.* 10. in being perfectly erect and destitute of foliage; whereas that species has leaves, is smaller, and always ascends in a curved direction; in that the scales of the spike are acute, in this obtuse, in other respects they are very much alike. At present I am inclined to think C. *mucronatus* of Vahl. and Rottboel may be referred to this, consequently Willdenow's C. *mucronatus. Sp. Pl.* 1. 273. includes these two species.

4. C. *nudus. R.*

Culm from three to four feet high, round, naked, inwardly intercepted. *Leaves* none; *Umbel* compound; *involucre* chaffy. *Stigma* three-cleft. *Seed* three-sided.

A native of stagnating, or slowly running sweet water over India.

Root perennial, fibrous.—*Culms* erect, round, naked, articulated, smooth, from three to four feet high, above as thick as a common large quill.—*Leaves* no other than a rusty coloured sheath or two embracing the base of each culm.—*Umbel* terminal, generally compound, though sometimes decompound.—*Involucre* several small, chaffy scales.—*Umbellets* long, slender, peduncled, bowing, composed of from eight to twelve linear, sessile, rust-coloured spikes.—*Involucels* like the involucres.—*Seed* sub-linear, three-sided, smooth, light brown.

5. C. *semi-nudus. R.*

Culms from two to six feet high, round, tending to be angular at the top, naked. *Leaves* none. *Umbels* decompound, with an involucre half its height. *Stigma* three-cleft; *Seed* three-sided.

Beng. Gola-methee.

Teling. Godoo tunga kooda.

This species is with the former a native of wet places.

Root perennial, creeping, fibrous.—*Culms* erect, from two to six feet high, round, smooth, naked, except at the base, which is embraced by three or four sheaths.—*Leaves* no other than the sheaths just mentioned.—*Umbels* generally decompound, erect.—*Involucre* three-leaved, equal, shorter than the umbel, sides and keel hispid.—*Umbellets* and partial *umlellets* pedunc'ed, composed of linear-lanceolate, yellowish spikes. *Seed* three-sided, smooth.

Obs. Cattle are not fond of it, and it is only eaten occasionally by buffaloes.

Sect. 2d, with a three-sided Culm.

6. C. *dubins. Rottb gram.* 20. *t* 4. *f.* 5. *Linn. Sp. Pl. ed. Willd.* 1. 273.

Culm from six to twelve inches high. *Head* terminal, roundish, sessile. *Involucre* of four or five long leaves, some of them longer than the culm.

Cyperus longus. *Rumph. amb.* 6. *p.* 5. *t.* 2. *f.* 1.

C. *kyllingœoides. Vahl. Enum. Pl.* 2. 312.

Teling. Tagada tunga.

Beng. Chhota-gothoobi.

This species is much like *Kyllingia monocephala,* and is with it, a native of shady, moist pasture lands.

Root, a somewhat tuberous head, with many rust-coloured capillary fibres.—*Culms* erect, from six to twelve inches high, three-sided, smooth, the lower third or fourth part involved in the sheaths of the leaves.—*Leaves* sheathing, radical, embracing the base of the culm, very smooth, many of them longer than the culm.—*Head* globular, size of a large cherry, composed of numerous, lanceolate, sessile, white-pointed spikes.—*Involucre* four or five-leaved, unequal, some of them longer than the culm.—*Seed* three-sided, pointed, dark rust-coloured, smooth.

Obs. Cattle eat it.

7. C. *monocephalus. R.*

Culms from twelve to eighteen inches high, straight; *Head* terminal, ovate, compound; *Involucre*, from three to four-leaved; *spikelets* ovate. *Style* undivided. *Seed* three-sided, elevated on a spongy receptacle.

Beng. Gothoobi.

A native of low wet places over Bengal.

Root fibrous.—*Culms* straight, three-sided, smooth, general height from twelve to eighteen inches.—*Leaves* three, four, or five, investing with their sheaths, the lower part of the culms straight and smooth, length various, the longest about as long as the culm. *Involucre, leaves* several, diverging, of very unequal lengths, the longest as long as the culm, and the shortest about an inch.—*Heads* terminal, composed of a larger one in the centre, and four or five smaller ones in the circumference, all rigidly sessile.—*Spikelets* ovate, incurved, ten or twelve-scaled, each spikelet embraced on the under side by an ovate-cordate, smooth, dark brown, bracte.—*Scales* boat-shaped, pointed; *keel* greenish; *sides* deep, polished, chesnut colour.—*Stamens* three. *Style* simple and undivided to the extremity.—*Seed* three-sided, elevated on a spongy whitish receptacle.

Obs. The undivided style and spongy receptacle of the seed distinguish this from all the other species I have yet found in India.

8. C. *diffusus. R.*

Diffuse, from two to four inches high. *Heads* terminal, glomerate, sessile; *spikelets* incurved. *Scales* daggered. *Involucre* many-leaved.

Mulen-pullu. *Rheed. Hort. mal.* 12. *p.* 101. *t.* 54.

A native of moist pasture ground.

Root fibrous.—*Culms* three-sided, several, diffuse, from two to four inches long, smooth, towards the base leafy, and of a dark chesnut colour.—*Leaves* sheathing, two or three embracing the base of each culm and about the same length. *Head* terminal, sessile, composed generally of one large capitulum in the centre, and four or five in the circumference; these are again composed of small, incurved, from eight to twelve-flowered spikes. *Involucre* many-leaved; the exterior three larger, being as long as the culm or longer.—*Stigmas* two, revolute. *Seed* three-sided.

9. C. *squarrosus. Linn. Sp. Pl. ed. Willd.* 1. 274.

Culms erect, from four to six inches high, leaves as long as the culms. *Head* terminal, glomerate, round. *Involucre* many-leaved. *Scales* with acute, recurved points.

C. *pygmæus. Rottb. gram.* 20. *t.* 14. *f.* 4.

A native of Coromandel.

10. C. *aristatus. Linn. Sp. Pl. ed. Willd.* 1. 275. *Rottb. gram.* 23. *t.* 6. *f.* 1.

Culm from three to six inches high, with sharp angles. *Heads* generally single, and sessile; *scales* with a long recurved point. *Involucre* three-leaved.

Scirpus *intricatus. Mant* 182.

A native of moist pasture ground. It is in general from three to six inches high, though sometimes I have met with luxuriant specimens ten inches in height.

Root fibrous.—*Culms* erect, from three to six inches high, three-sided, two-thirds naked, smooth; angles sharp.—*Leaves* sheathing, erect, short, smooth; sheaths purple —*Head* generally simple, and sessile, though sometimes, there is a second, which is peduncled; and in the larger plant above mentioned there are two or three sessile, and two or three peduncled, they are composed of small, sessile, ragged, from six to eight-flowered spikelets.—*Involucre* three or four-leaved, broad, unequal, the largest being about the length of the culm.—*Calyx, scales* tapering to a long, fine recurved point, which gives the spikes the ragged appearance they have.—*Seed* three-sided.

11. C. *pectinatus*. *R.*

Culms three-sided, leafy, from eight to twelve inches high, straight. *Umbel* terminal, sessile. *Involucre* length of the umbel. *Spikes* all sessile, linear, pectinate, *scales* mucronate. *Seeds* three-sided.

A native of Coromandel.

Culms erect, straight, three-sided, leafy toward the base; from eight to twelve inches high.—*Umbel* terminal, simple, sessile, globular, composed of about twelve sessile, linear, diverging spikes.—*Involucre* from two to three-leaved, the longest leaflet about as long as the umbel.—*Scales* with long, acute, recurved apices.—*Stamina* three.—*Style* with a three-cleft apex.—*Seed* obovate, three-sided, smooth, dark ash-coloured.

12. C. *niveus*. *Linn. Sp. Pl. ed. Willd.* 1. 273. *Retz. Obs.* 5. *p.* 12.

Culms from five to eighteen inches high. *Umbels* or rather heads, sessile. *Involucre* two-leaved. *Style* three-cleft. *Seed* three-sided.

A native of shady moist pasture land.

Root creeping, fibrous.—*Culms* erect, from five to ten inches high, mostly naked, three-sided, smooth.—*Leaves* surrounding the base of the culm, sheathing, those that are nearest the root are little more than sheaths, those within are nearly as long as the culm, and waved.—*Umbel* terminal, simple, sessile, about an inch each way, composed of linear, many flowered, sessile spikes.—*Involucre* two-leaved, length of the spikes.—*Seed* three-sided.

13. C. *jemenicus*. *Linn. Sp. Pl. ed. Willd.* 1. 281.

Culms from two to four inches high, obscurely three-cornered; base *bulbiferous*. *Umbels* simple. *Involucres* and *spikelets* alternate. *Scales* acute. *Seed* roundish, three-sided.

Puri-gaddi is the Telinga name of the plant, and Puri-dumpa that of the roots.

A native of dry sandy pasture ground near the sea.

Root bulbiferous, stoloniferous. When the small oval bulbs are taken up entire they appear covered and entirely surrounded with two, three, or four black and brown, longitudinally striated sheaths, which burst and fall off in succession. There are many capillary fibres issuing from the base of the culm. The stolones generally push their way up through the sheaths, and out at their mouths, run horizontally, producing from their extremities the new plant. On removing the sheath a small truncated, pointed ball is observed, resting in the upper side of the base of the culm, which (onion-like) is formed by the united sheaths of the leaves. This little bulb, which is slightly connected to the base of the culm on which it rests, is the eatable part; it is composed of two or three coats, and about the size of a pea; when its proper exterior brown membranous sheath is removed, it is very white, inviting to the eye, and pleasant to the taste.—*Culm* erect, two-thirds naked, between round, and three-sided, two, three, or four inches long.—*Leaves* a few, filiform on the upper side, grooved, four or five inches long, below the surface of the earth, their sheaths form the little firm culm above-mentioned; from its head the capillary roots and stolones issue.—*Umbel* simple, composed of from six to ten sessile, linear, diverging spikes.—*Involucre* generally three-leaved, unequal, the largest twice as long as the spikes; they issue not as usual from the same point, but alternately one above another.—*Scales* oblong, striated.—*Stigma* two-cleft.—*Seed* roundish, three-sided, grey.

Obs. The little bulbs are gently roasted or boiled, then rubbed between the hands or the folds of a cloth, to take off the sheaths, which is all the preparation the natives give them, to make them a pleasant wholesome part of their diet, which they have frequent resource to, particularly in times of scarcity. Some dry them in the sun, grind them into meal, and make bread of them, while others stew them in their curries and other dishes; they are palatable, tasting like a roasted potatoe. I wish they could be enlarged by culture, when they might soon prove a valuable acquisition, for on account of their smallness, it is troublesome and tedious gathering and cleaning any quantity of them.

14. C. *polystachyos. Linn. Sp. Pl. ed. Willd.* 1. 275. *Rottb. gram.* 39. *t.* 11. *f.* 1. *Vahl. Enum. pl.* 2. 315.

Culm twelve inches long, *leaves* short, *umbel* compound, sessile, crowded. *Involucre* many-leaved. *Style* two-cleft. *Seeds* linear, obcordate, without angles.

Gramen cyperoides, &c. *Pluck. mant.* 97. *tab.* 416. *fig.* 6. is not unlike this plant.

A native of moist pasture land, of ditches, &c.

Root fibrous.—*Culms* erect, about a foot high, simple, three-sided, smooth, lower part involved in the sheaths of the leaves.—*Leaves* shorter than the culm, smooth.—*Umbel* terminal, compound, crowded, sessile, about an inch each way. *Umbellets* sub-sessile, crowded, composed of numerous, sessile, linear-lanceolate spikes. —*Involucre* from four to six-leaved, unequal, the longest being from six to eight inches long, and the shortest only one.—*Stigma* two-cleft.—*Seed* compressed, oblong-obcordate.

15. C. *punctatus. R.*

Culms from two to six inches high, shorter than the leaves. *Umbel* compound. *Involucres* longer than the leaves. *Seeds* oval, compressed, dotted.

Gramen cyperoides elegans, &c. *Pluck. t.* 192. *f.* 3.

A small elegant species; a native of moist pasture ground.

Root fibrous.—*Culms* erect, from two to six inches high, half naked, three-sided, smooth.—*Leaves* sheathing, the lower ones little more than the sheaths, the superior ones longer than the culm.—*Umbel* terminal, compound, about an inch each way. *Umbellets* some sessile, some peduncled, composed of linear, many-flowered spikes.—*Involucre* from three to four-leaved, unequal, longer than the culm; *scales* daggered.—*Seed* compressed, obovate, beautifully marked with white dots, on a dark-coloured ground, hence the specific name.

Obs. In Königs catalogue he called this C. *pygmæus,* but Retzius says the culm of that species is round, whereas in this plant it is three-sided.

16. C. *capillaris. Kön. Mss.*

Culm filiform, from one to two feet high, umbel of one sessile, and two pedicelled umbellets. *Involucre* three-leaved, filiform, florets diandrous. *Style* two-cleft. *Seed* obovate, without angles.

A native of moist pasture ground.

Root fibrous.—*Culms* erect, from twelve to eighteen inches high, three-sided, filiform, three-fourths or more naked, smooth.—*Leaves* sheathing, generally two to each culm; the lower one short, the superior one as long as the culm, very slender and smooth.—*Umbel* small, compound, generally composed of one sessile, and two peduncled umbellets, these are again composed of linear, many-flowered spikes.—*Involucre* three leaved, filiform, the largest is longer than the culm; the shortest about an inch long, and the other between the two.—*Scales* membranaceous, obtuse.—*Stamens* two.—*Style* two-cleft.—*Seed* obovate, compressed, brown, pretty smooth, half hid in the rachis.

17. C. *compressus. Linn. Sp. Pl. ed. Willd.* 1. 282. *Vahl. Enum. Pl.* 2, 324.

Culms from six to ten inches high, shorter than the leaves, angles blunt. *Umbel* simple or compound. *Involucre* three-leaved. *Scales* daggered. *Seed* obcordate, triangular.

Gramen cyperoides rarius, &c. *Pluck. t.* 192. *f.* 1.

Beng. Choonch*a*.

Teling. Sal*i*tung*a*.

Delights in a moist soil.

Root fibrous.—*Culms* erect, from six to ten inches high, two-thirds naked, three-sided, smooth; angles blunt.—*Leaves* numerous sheathing, many of them longer than the culms, smooth and glossy.—*Umbel* terminal, simple, or compound; when compound the umbellets are peduncled; from one to three; and consist of from three to six, green, linear, sessile, many-flowered spikes.—*Involucre* three-leaved, smooth, the largest of them is nearly as long as the culm.—*Scales* acute.—*Seed* turbinate, three-sided, smooth, dark-brown colour.

Obs. The upper flowers of the spike are incomplete.

18. C. *castaneus. Linn. Sp. Pl. ed. Willd.* 1. 278.

Culms from two to six inches high, length of the leaves. *Umbels* generally compound; length of the involucre. *Scales* with long recurved points, monandrous. *Seed* rugose, obtusely triangular.

A most beautiful delicate, small species; delights in moist, sandy pasture land.

Root fibrous.—*Culms* erect, from two to six inches high, naked, three-sided, smooth.—*Leaves* sheathing, filiform, scarcely so long as the culm.—*Umbel* terminal, one or one and half inches, simple, or compound, but in general composed of from one to five small peduncled umbellets; and one or two sessile ones.—*Spikelets* linear, many-flowered.—*Scales* three-keeled, with a recurved point, green keel and chesnut coloured sides.—*Stamen* single, (at least I have not found more.)—*Style* two-cleft.—*Seed* obtusely three-sided, oblong, rugose.

19. C. *difformis. Linn. Sp. Pl. ed. Willd.* 1. 280. *Vahl. Enum.* 2. 337.

Culm from nine to eighteen inches high, angles sharp. *Umbels* compound, *umbellets* globular, *spikelets* crowded. *Involucre* from two to three-leaved. *Scales* imbricated. *Seed* obovate, three-sided.

Gramen cyperoides, &c. *Pluck. alm.* 192. *f.* 3.

Beng. Behooa.

Grows equally in either a dry, or wet sandy soil.

Root fibrous.—*Culms* two-thirds naked, erect, from nine to eighteen inches high, three-sided, smooth, angles sharp.—*Leaves* sheathing, as long as the culm, keeled, smooth.—*Umbel* compound, small, being only about an inch and a half each way. *Umbellets* from two to four, sub-sessile, and from two to four longer or shorter peduncled, globular, composed of innumerable, very small, oval spikes.—*Involucre* from two to three-leaved, very unequal, the longer being half as long as the culm; and the third (when present) less than an inch.—*Scales* imbricated, margins membranaceous.—*Seed* three-sided.

20. C. *pumilus. Linn. Sp. Pl. ed. Willd.* 1. 282. *Rottb. gram.* 29. *t.* 9. *f.* 4. *Vahl. Enum. Pl.* 2. 336.

Culms about six inches high, angles blunt. *Umbels* compound; umbellets sub-globular. *Spikelets, Scales* daggered.

Delights in a moist soil.

Root fibrous.—*Culms* erect, six inches high, half naked, obsoletely three-sided, smooth, angles rounded.—*Leaves* sheathing, shorter than the culm.—*Umbel* compound, *umbellets* from three to six, shorter or longer peduncled, globular, composed of ten or twelve linear, many-flowered spikes.—*Involucre* three or four-leaved, unequal, the largest as long as the culm. *Scales* acute.

21. C. *cruentus. Linn. Sp. Pl. ed. Willd.* 1. 275. *Rottb. gram.* 21. *t.* 5. *f.* 1.

Culms three-sided, from six to twelve inches long. *Leaves* sub-radical, *umbel* compound, *involucre* three or four-leaved. *Scales* obtuse, *flowers* diandrous. *Style* two-cleft. *Seed* round and smooth.

A native of low pasture land all over Bengal, &c. where it is found growing in tufts during the rainy season. Like most of this natural order.

Obs. Cattle rarely eat it, except when pressed by hunger.

22. C. *incurvatus. R.*

Culms triangular, base leafy, incurvate. *Umbel* compound. *Involucres* two or three, incurvate. *Spikes* needle-shaped, incurved. *Scales* oblong, obtuse. *Stigma* three-cleft.

A native of the moist banks of the Ganges. Flowers during the cool season.

Roots creeping, stoloniferous, with dark-coloured fibres.—*Leaves* ensiform, polished, shorter than the culms.—*Culms* slightly incurved, leafy round the base, triangular, with sharp angles, concave sides, and from one to two feet high.—*Involucre* two; or three-leaved, incurved, one of them two or three times longer than the umbel—*Umbel* compound, *spikes* needle-shaped, incurved.—*Scales* oblong, obtuse, with a green rib.—*Stamina* three.—*Germ* oblong. *Stigma* three-cleft.—*Seed* linear, oblong, three-sided, smooth, brown.

23. C. *tortuosus*. *Kön. Mss.*

Culms obtusely three-sided, from six to eight inches high. *Umbel* compound, length of the involucre. *Spikelet* linear, incurved, flowers diandrous. *Seed* obovate, compressed, without angles.

A native of moist pasture ground.

Root somewhat tuberous, with filiform fibres.—*Culms* erect, from six to eight inches high, naked, three-sided, smooth.—*Leaves* radical, shorter than the culm.—*Umbel* compound. *Umbellets* from four to eight, shorter and longer peduncled, composed of large, linear, many-flowered, chesnut-coloured, curved spikes.—*Involucre* three-leaved, longer than the umbel.—*Stamina* two.—*Style* two-cleft.—*Seed* obcordate, compressed, smooth.

24. C. *rotundus*. *Linn. Sp. Pl. ed. Willd*, 1. 283. *Vahl. Enum. Pl.* 343.

Root tuberous. *Culms* from one to two feet high, obtusely three-sided. *Umbel* compound, *spikelets* alternate; *involucre* three-leaved, length of the umbel. *Seed* angled.

C. rotundus. *Rumph. amb.* 6. *p.* 1. *t.* 1. *fig.* 1. 2.

C. *hexastachyos*. *Rottb.* 28. *t.* 14. *f.* 2.

Sans. मुस्ता, Moosta, मुस्तक, Moostuka.

Beng. Mootha.

Teling. Shaka-tunga.

This is by far the most common species we have in India; it delights in a moist sandy soil, though it grows abundantly every where.

Root tuberous, with filiform fibres, *tubers* about the size of filberts or acorns, of irregular form, outwardly a dark brownish rust colour; inwardly firm, tender, white, and very fragrant.—*Culms* erect, from one to two feet high, naked, three-sided, smooth, with the angles rounded. —*Leaves* radical, or only investing the base of the culm, sheathing, shorter than the culm, smooth, glossy.—*Umbels* terminal, compound in very luxuriant plants, sometimes decompound, from two to three inches each way *Umbellets* from two to eight, all having peduncles of from one-fourth of an inch to two inches long.—*Involucre* generally three-leaved, unequal, smooth, the largest is only a little longer than the umbel.—*Spikes* linear, sub-sessile.—*Seed* obsoletely three-sided; brown, a little rugose.

Obs. Cattle eat it. Hogs are remarkably fond of the roots. Dried and powdered they are used as a prefume at the weddings of the natives. It is by far the most troublesome weed we have in our gardens, there is no extirpating it as every little bit of the root grows readily.

25. C. *pertenuis. R.*

Culms from three to four feet high, sub-rotund at the base, three-cornered above. *Leaves* few. *Umbel* compound and decompound. *Spikelets* filiform, many-flowered. *Seeds* three-sided.

Beng. Nagur-Mootha.

This most delicate, tall, slender species, I have only found in low wet places, in the vicinity of Calcutta.

Root somewhat tuberous, with many dark-coloured villous fibres.—*Culms* naked, except at the base, straight, generally three or four feet high, slender, tapering much, toward the base nearly round, becoming more and more three-sided, till they are acutely so at the umbel.—*Leaves* one or two at the base of each culm, slender, about one third the length of the culm —*Umbel* compound and decompound, very small for the height of the plant, generally composed of one sub-sessile umbellet, and three or four, on naked, smooth, filiform peduncles, of different length, and these again composed of from four to twelve, most slender, long, filiform, many-flowered spikelets.—*Involucre* three-leaved, the longest about the length of the umbel, the other two shorter, all slender, and tapering to most acute points.—*Scales* membranaceous.—*Stamens* three.—*Style* three-cleft.—*Seed* three-sided.

Obs. Its naked delicate form, small and compound umbel, short slender leaves, and scanty involucre, immediately distinguish it from all the other Indian species I have yet met with.

26. C. *tuberosus*. *Linn. Sp. Pl. ed. Willd.* 1. 281. *Rottb. gram.* 28. *t.* 7. *f.* 1. *Vahl. Enum. Pl.* 2. 340.

Root tuberous. *Leaves* linear, length of the culms. *Umbels* compound, or decompound. *Involucre* three-leaved, longer than the umbel. *Style* three-cleft. *Seed* oblong, three-sided.

This seems to me to be only a luxuriant variety, if so much, of C. *rotundus*. It grows in the same places, the roots are tuberous, &c. in short, I can scarcely perceive any difference.

27. C. *tenuiflorus*. *Linn. Sp. Pl. ed. Willd.* 1. 284. *Rottb. gram.* 30. *t.* 14. *f.* 1. *Vahl. Enum. Pl.* 2. 374.

Culm from two to four feet high, sharp angled. *Involucre* three or four-leaved, much longer than the decompound *umbel*. *Spikelets* alternate. *Scales* obtuse; *seeds* obcordate, three-sided.

Teling. Gelleba-*tunga*.

Grows in standing sweet water.

Root creeping, with ramous fibres.—*Culm* erect, from four to six feet high, naked, three-sided, smooth; *angles* sharp.—*Leaves* mostly radical, shorter than the culm, deeply channelled, smooth.—*Umbel* terminal, decompound, from four to eight inches each way.—All the *umbellets* peduncled, oblong, spreading, composed of alternate, sessile, lanceolate, from twenty to thirty-flowered spikes.—*Involucre* three or four-leaved, very unequal, the largest being from one to two feet long, and the shortest about one or two inches.—*Scales* obtuse, apex often emarginate and membranaceous.—*Seeds* three-sided, obovate.

28. C. *strictus*. *R.*

Culms about twelve inches high, stiff and straight, shorter than the leaves; *angles* sharp. *Umbel* decompound, straight, *flowers* diandrous. *Seeds* obovate, compressed, without angles.

Delights in a moist uncultivated soil.

Root fibrous.—*Culm* straight, rigid, about a foot high, obtusely three-angled, three-fourths naked, smooth.—*Leaves* mostly radical, sheathing, many of them longer than the culm, erect, rigid, much keeled, smooth.—*Umbel* erect, linear, thin, sometimes decompound, though in general only compound. *Umbellets*, one or two sessile; and from three to eight with peduncles of unequal lengths; the largest peduncles are compound.—*Involucre* from three to five-leaved, very unequal; the largest nearly as long as the culm, the smallest not more than an inch long.—*Spikes* sessile, lanceolate in small plants, in large ones linear.—*Stamens* two.—*Style* two-cleft.—*Seed* compressed, obcordate.

29. C. *flavidus*. *Linn. Sp. Pl. ed. Willd.* 1. 279. *Retz. Obs.* 5. 13 *Vahl. Enum. Pl.* 2. 334.

Culms from six to eight inches high, generally shorter than the leaves. *Umbel* decompound, longer than he involucre; *umbellets* globular, *spikelets* crowded; *flowers* monandrous. *Seeds* obcordate, three-sided.

This is a small delicate species, a native of moist places on the coast of Coromandel.

Root fibrous, dark purple.—*Culm* erect, from six to eight inches long, naked, three-sided, smooth.—*Leaves* sheathing, very short, embracing the base of the culm.—*Umbel* compound, or decompound, from one to two inches high, *umbellets* one or two sessile, and from two to eight on peduncles of various lengths, the highest sometimes bears a small partial umbellet. *Involucre* two-leaved, shorter than the umbel. *Spikes* lanceolate, chesnut-coloured, *flowers* monandrous. *Seeds* white, short, three-sided, a little rough.

30. C. *Iria*. *Linn. Sp. Pl. ed. Willd.* 1. 286 *Vahl. Enum.* 2. 360.

Culms from one to two feet high, *angles* sharp, *umbels* decompound, shorter than the involucres; *spikelets* alternate; *flowers* rather distinct, *scales* imbricated. *Seeds* obovate, angular.

Ira. Rheed. mal. 12. *p.* 105. *t.* 56. does not very well agree with my plant.

Gramen cyperoides, &c. *Pluk. Almag t.* 191. *f.* 7. is much more like this plant than the above-quoted plant of Van Rheede.

Beng. Bura-choocha.

Is a native of moist, cultivated lands.

Root fibrous.—*Culms* erect, from one to two feet high, four-fifths naked, three-sided, sharp-angled.—*Leaves* sheathing, nearly the length of the culm, keeled, smooth.—*Umbel* decompound, from two to four inches high, *umbellets* from four to eight, one or two sessile, the rest unequally peduncled; *partial* umbellets lanceolate, raceme-like, being composed of alternate, linear, from six to twelve-flowered spikes.—*Involucre* from three to four-leaved, the largest two or three times as long as the umbel.—*Scales* imbricated, with membranaceous sides.—*Style* two-cleft.—*Seeds* three-sided, length of the scales.

31. C. *inundatus. R.*

Culms from two to four feet high, exactly triangular. *Leaves* as long as the culms. *Involucre* from four to five-leaved, one very long. *Umbel* decompound. *Spikelets* lanceolate. *Stamens* two. *Style* two-cleft. *Seed* obcordate, a little compressed.

Beng. and *Hind.* Patee.

Found in great abundance on the low banks of the Ganges and rivulets near Calcutta; where the tide rises high over it, it thrives most luxuriantly and helps much to bind, and protect the banks from the rapidity of the water.

Root jointed, creeping, stoloniferous, perennial.—*Culms* erect, from three to four feet high, exactly three-sided, about as thick as the little finger.—*Leaves* numerous, radical, or surrounding the base of the culms, smooth, deeply channelled on the inside, and keeled on the back.—*Universal involucre* composed of four or five leaves of very unequal lengths, the largest being two feet or more long, and the shortest as many inches; *partial involucre* subulate.—*Umbel* decompound, erect, about a span long.—*Spikelets* sessile, diverging, lanceolate, rigid, obtuse.—*Style* two-cleft.—*Seeds* obcordate, a little compressed, smooth.

32. C. *Pangorei*. *Rottb. gram.* 31. *t.* 7. *f.* 3. *Linn. Sp. Pl. ed. Willd.* 1. 284.

Culms from three to four feet high, naked. *Leaves* scarcely any. *Umbels* decompound; *involucre* three or four-leaved; *involucels* setaceous. *Spikelets* filiform. *Seeds* oblong, compressed.

Beng. Chumati Patee.

Is common on the banks of the Ganges, and serves, with C. *inundatus*, the same useful purposes, though in an inferior degree.

Root jointed, creeping, stoloniferous, perennial.—*Culms* erect, naked, smooth, three-sided towards the apex, sides concave, from two to four feet high, and about as thick as a goose quill.—*Leaves* one or two, sword shaped, very short, involving with their sheaths, the base of the culm. *Universal involucre* three or four-leaved, very smooth, the longest twice the length of the umbel; *partial* minute, setaceous.—*Umbel* decompound, composed of one sessile umbellet in the centre, and from four to eight spreading, compound ones in the circumference, supported on fine, compressed peduncles of unequal lengths.—*Spikelets* filiform, from ten to twenty-flowered. —*Scales* obtuse.—*Stamens* three.—*Seed* oblong, compressed.

Obs. Its naked, exactly three-sided culms, and scarcity of short leaves readily distinguish it from all the other species I have yet described. C. *tegetum* has the angles rounded, and no leaves.

33. C. *procerus*. *Rottb. gram.* 29. *t.* 5. *f.* 3.

Culms from three to six feet high, angles sharp. *Umbel* decompound. *Involucres* none; *spikelets* alternate, linear, *flowers* diandrous. *Seeds* obcordate, compressed, without angles.

Pota-pullu. Rheed. mal. 12. *p.* 93. *t.* 50.

Is a native of moist vallies, rivulets, &c. amongst the mountains.

Obs. In Bengal this plant, or one exceedingly like it, (for it differs from it only in having three stamens, the culm, leaves, umbels, scales, and seeds being the same) grows on the banks of the Ganges and so low as frequently to be entirely under water during the high tides; yet it thrives and helps to bind the banks of the rivers where it grows and is one of those plants that prevent their giving way so much to the rapidity of the stream as they otherwise would do.

34. C. *spinulosus. R.*

Culms from three to five feet high, rigid. *Umbel* scarcely decompound; *spikes* long, and short-peduncled, globular, *spikelets* rigid, from three to four-flowered; *involucre* and *involucels* many-leaved.

Teling. Kurra-tunga.

A large strong coarse species, grows in ditches and water courses.

Root fibrous.—*Culms* erect, from two to five feet high, very rigid, obsoletely three-sided, smooth, four-fifths naked.—*Leaves* sheathing, length of the culm, rigid, striated, margins hispid.—*Umbel* terminal, decompound, from six to eight inches each way. *Umbellets* with peduncles of various lengths, composed of peduncled, globular heads, and these again of numerous small, sessile, linear, lanceolate, rigid, spinulous-pointed, from three to four-flowered spikelets.—*Involucre* from four to six-leaved, unequal, the largest being from two to three feet long, and the shortest about the same number of inches, margins hispid, like those of the leaves. *Involucel* from three to six-leaved, length of the umbellets.—*Scales* linear.

Obs. I have not seen the pistil nor seed. It is a remarkably coarse species; no animal eats it.

35. C. *elatus. Linn. Sp. Pl. ed. Willd.* 1. 287. *Vahl. Enum.* 2. 363.

Culms from six to twenty feet long, sharp-angled. *Umbel* scarcely decompound. *Spikes* all sub-sessile, linear, erect. *Spikelets* most numerous, alternate, from six to eighteen-flowered, anthers ending in a thread. *Seeds* oblong, three-sided.

Beng. Gol-mulunga.

Teling. Rakisha.

This is by far the largest species on the coast of Coromandel. It grows in sweet standing water.

Root fibrous.—*Culms* erect, from six to twenty feet high, and thick in proportion, naked, except near the base, three-sided, angles sharp, rigid and scabrous, so as to cut the hand readily, other parts pretty smooth.—*Leaves* sheathing, in small plants, as long as the culm, about an inch in diameter at the broadest part, triangularly concave, with the keel and margins sharply serrate.—*Umbel* terminal, decompound, erect, small for the size of the plant, being only about a foot high in middle-sized plants. *Umbellets* one, or two, sub-sessile and from eight to ten peduncled, composed of erect, linear, short pedicelled, compound spikes, of simple, sessile, lanceolate, from six to ten-flowered spikelets which surround every part of the compound spikes.—*Involucre* three-leaved, with some smaller within these three, margins and keel as in the leaves.—*Involucels* small, one-third the length of the umbellets.—*Scales* membrane-bordered, acute.—*Filaments* three, membranaceous. *Anthers* ending in a thread.—*Seeds* three-sided.

36. C. *umbellatus. Vahl. Enum. Pl.* 2. 362.

Culms from four to six feet, blunt-angled. *Umbel* decompound; *spikes* all sub-sessile, linear, spreading; *spikelets* numerous, alternate, from three to four-flowered. *Seed* elliptic, three-sided.

Teling. Pedda shaka.

A large species, growing in standing fresh water.

Root fibrous.—*Culms* erect, from four to six feet high, pretty smooth, naked, except at the base, obsoletely three-sided, angles rounded, sides hollowed.—*Leaves* mostly radical, sheathing, the longest of them about as long as the culm.—*Umbel* decompound, spreading, about six inches each way. *Umbellets,* two or three sub-sessile, and from five to ten on pretty long peduncles of unequal lengths; they are composed of small partial, short-pedicelled umbellets of three to six linear compound spikes, composed of numerous, small, sessile, from three to four-flowered, simple spikelets. *Involucel* from three to four-leaved, of the length of the umbellets.—*Involucre* from three to six-leaved, unequal; the longest from two to three feet, the shortest as many inches.—*Seeds* three-sided, smooth.

37. C. *digitatus. R.*

Culms from one to six feet high. *Umbel* decompound, fastigiate. *Involucre* and *involucels* many-leaved; *spikes* digitate, linear, *spikelets* diverging, many-flowered, *scales* with long recurved points.

A middle sized species; grows in marshy places, ditches, &c.

Root, a somewhat tuberous head with many fibres.—*Culm* erect, from one to six feet high, mostly naked, smooth, three-sided.—*Leaves* sheathing as in the other species, their margins slightly armed with minute prickles.—*Umbel* compound, or decompound, from two to four inches high.—*Umbellets* sessile, and peduncled, composed of erect, compound spikes, of small diverging, linear, acute, from four to six-flowered spikelets.—*Involucre* many-leaved, unequal, most of them longer by far than the umbel, margins armed like the leaves.—*Involucel* many-leaved, length of the umbellets.—*Seed* three-sided, oblong, with grooved sides.

Obs. In Bengal I have met with this species six feet high, with a decompound umbel, a foot or more every way, and with from twelve to twenty-four flowers to the spikelets, the culm between round and three-sided, and in the largest almost entirely round.

38. C. *verticillatus. R.*

Culms four feet high, three-cornered, leafy at the base. *Umbel* decompound; *involucre* many-leaved, *involucels* filiform, spikes of the umbellets verticelled; *spikelets* lanceolate. *Scales* with a blunt point. *Seed* linear, oblong, three-sided.

Beng. Burethi.

Found during the rains, in wet low places, in the vicinity of Calcutta.

Root perennial, somewhat tuberous, with numerous dark-coloured, thick, spongy fibres.—*Culms* naked, except just at the base, from two to six feet high, and about as thick as a stout quill, absolutely three-sided, smooth, shining, deep green.—*Leaves*, one, two, or three at the base of each culm, and about the same length, keeled, somewhat spongy, very deep green, whitish underneath, margins armed with minute, short, sharp, aciculæ, pointing forwards. —*Umbel*, compound and decompound. *Umbellets* many on smooth somewhat angular peduncles of various lengths, composed of from five to eight cylindric, spreading spikes in a verticel, with an erect, sessile, simple or pedicelled compound one in the centre.—*Spikelets* most numerous, imbricated, lanceolate, from eight to twelve-flowered.—*Involucre* many-leaved; some of them nearly as long as the culm, and like the radical leaves keeled and armed on the margins. —*Involucels* scarcely so long as their respective umbellets.—*Scales* of the spikelets boat-shaped, with membranaceous margins, and sharp thickened points.—*Stamens* three.—*Seed* oblong, obsoletely three-sided.

Obs. This species comes near Rottboell's *Cyperus alopecuroides.*

39. C. *distans. Linn. Sp. Pl. ed. Willd.* 1. 288. *Vahl. Enum. Pl.* 2. 362.

Culms about three feet high, three-cornered. *Umbel* decompound, with the *involucre* and *involucels* many-leaved; *spikelets* linear, alternate; *flowers* distant, with obtuse scales. *Seeds* oblong, three-sided.

C. *elatus. Rottb. gram.* 37. *t.* 10.

Beng. Panee-mulunga.

Is a native of wet places around Samulcota.

Root fibrous.—*Culms* erect, two feet high, four-fifths or more naked, three-sided, smooth; angles blunt.—*Leaves* from two to four, embracing the base of the culm, and nearly as long, sheathing, deeply grooved above, and keeled below, smooth.—*Umbel* decompound, from four to six inches high. *Umbellets* some sessile, and from four to eight, with erect peduncles of unequal lengths. *Partial* umbellets oblong, raceme-like, being composed of alternate, needle-like, diverging spikes.—*Involucre* from three to six-leaved, unequal; the largest about a foot, or fifteen inches long, all are smooth.—*Involucels* three-leaved, length of the umbellets.—*Flowers* remote, scales obtuse.—*Seed* three-sided.

40. C. *tegetum.* *R.*

Leafless, culms from three to six feet high, angles rounded, umbel decompound. *Involucre* as long, and longer than the umbel; *involucels* chaffy. *Spikelets* linear-lanced, alternate. *Seed* clavate, three-sided.

Hind. and *Beng.* Madoor-kati.

Is common in ditches, borders of lakes, &c. in the vicinity of Calcutta during the rains.

Root creeping under ground, perennial.—*Culms* naked, generally from three to six feet high, obsoletely three-sided, smooth.—*Leaves* no other than two or three sheaths embracing the base of the culms.—*Umbel* decompound; umbellets sub-sessile, and on peduncles of various lengths, sub-erect.—*Involucre* about four-leaved, one or two of them longer than the umbel.—*Involucels* minute.—*Spikelets* alternate, linear-lanceolate, many-flowered.—*Seeds* elliptically triangular.

Obs. Those elegant, useful, durable, large mats so common on the floors of rooms in and about Calcutta, are made of the culm of this plant. When green they are split into three or four pieces, which in drying contract so much as to bring the margins in contact, or to overlap each other; in this state they are wove.

41. C. *alopecuroidus.* *Rottb. gram.* 38. *t.* 8. *f.* 2. *Vahl. Enum. Pl.* 2. 368.

Culms from four to five feet high. *Umbel* decompound; *involucre* three-leaved, *spikes* nodding, *spikelets* alternate, many-flowered.

Wara-pulla. *Rheed. mal.* 12. *p.* 77. *t.* 42.

Cyperus *glomeratus.* *Linn. Sp. Pl. ed. Willd.* 12. 277.

This is also a large species, and a native of the same places.

Root fibrous.—*Culms* erect, from four to five feet high, except at the bases three-sided, smooth.—*Leaves* as in the other species.—*Umbel* decompound, drooping, from six to twelve inches high. *Umbellets* one or two sub-sessile, and from four to eight with peduncles of various lengths. *Partial umbellets* peduncled and composed of beautiful, linear, bowing, compound racemes of lanceolate, many-flowered, chesnut-coloured spikelets.—*Involucre* three-leaved, unequal. *Involucels*, small, frequently only chaffy scales.—*Seed* three-sided, smooth.

Obs. I have seen small plants with the umbellets small, dense and globular.

42. C. *nudus*. *R.*

Leafless. *Culms* from four to six feet high, three-cornered. *Umbels* super-decompound, naked. *Involucres*, and *involucels* chaffy; *spikelets* subulate, alternate. *Seed* oblong, compressed without angles.

Beng. Goola-Methhi.

A native of low wet places over Bengal. Flowers during the rains.

Root perennial, somewhat tuberous.—*Leaves* no other than a short sheath or two, which envelope the base of the culm, the uppermost one sometimes ends in a short ensiform leaflet.—*Culm* about five feet high, naked, smooth, bright-green, three-cornered, with the angles rounded. *Involucres* three or four, sub lanceolate, erect, about one-fourth the length of the umbel.—*Umbels* decompound and super-decompound, with the sub-divisions, and slender, smooth, compressed peduncles of various length.—*Spikelets* alternate, subulate, of a bright brown colour, each bearing from twenty to thirty flowers.—*Stamens* three.—*Style* three-cleft.—*Seed* oblong, compressed, without angles.

Obs. This species is very like C. *tegetum*, and about the same size, though I am informed, never used for mats, as that is. To know it from C. *tegetum* attend to the involucre, which in this is only about one-fourth the length of the umbel, but in that as long or longer.

43. C. *Haspan. Linn. Sp. Pl. ed. Willd.* 1. 287.

Culms from four to sixteen inches high. *Umbel* super-decompound, involucre two-leaved. *Seeds* globular, white.

Gramen cyperoides madasarpatanum, &c. *Pluck. alm.* 192. *f.* 2.

Teling. Poo-cushalic.

An elegant small species, a native of cultivated moist places.

Root fibrous.—*Culms* erect, from four to twelve inches high, three-fourths naked, obsoletely three-sided, smooth.—*Leaves* sheathing, often as long as the culm, very slender.—*Umbel* super-decompound, in small plants as long as the culm. *Umbellets* one or two sessile, and from four to six peduncled. *Partial* umbellets of the first and second degrees pedicelled.—*Involucre* generally two-leaved, unequal, the longest rather longer than the culm; *involucel* a few chaffy scales only.—*Stamens* two or three.—*Seed* sub-globular, pure white, smooth.

SCIRPUS.

Glumes chaffy, imbricated on all sides. *Corol* none. *Seed* one.

Sect. 1*st.* *Without leaves.*

1. S. *tuberosus. R.*

Root tuberous. *Culms* culumnar, jointed, leafless. *Spikes* terminal, cylindric, naked. *Scales* oblong. *Base* of the style cordate. *Seed* obcordate, beset with glochidate bristles.

Pi-tsi, Maa-tai, Pu-tsai, or Pe-tsi, or water chesnut of the Chinese.

It is mentioned by Abbe Grosier under the first name, and under the other appellations was transmitted from Canton, by Mr. Duncan, at the desire of the Governor General, for the Company's Botanic Garden, where it blossoms about the close of the rains, in September.

Root fibrous, with stoloniferous shoots and round turnip-shaped tubers.—*Culms* erect, naked, slightly twisted to the right, inwardly intercepted by slender membranes, distant a line, or a line and a half from each other, every eight or tenth of these partitions, is of a firmer texture than the others and visible on the outside, even when the plant is growing.—*Leaves* none except a slender short sheath or two at the base of each culm.—*Spike* terminal, solitary, cylindric, naked. —*Scales* oblong, membranaceous margined.—*Stamens* three.—*Style* two or three-cleft, at the base swelled out into an ovate-cordate, compressed shape.—*Seed* obcordate, surrounded with glochidate bristles.

Obs. For the œconomical uses of the tuberous roots of this plant, I beg leave to refer to Abbe Grosier, and other historians who have had opportunities of observing in what manner they are employed by the Chinese, and to the following extract from Mr. Duncan's letter which accompanied the plants.

"The *Maa-tai, Pee-tai* or *Pi-tse* of Abbe Grosier, the *water-ches-* "*nut*, grows in ponds, which are manured for its reception about "the end of March. A pond being drained of its water, small pits "are dug in its bottom, these are filled with human manure, and "exposed to the sun for a fortnight; their contents are then inti- "mately blended with the slimy bottom of the pond, and the slips "or roots of the plant deposited therein; the water is now return- "ed to the pond, and the new crop of tubers come to perfection "by the beginning of September.

"This nut is in high estimation among all ranks of Chinese, "not only as a pot root, but also a medicine. It is eat either boiled "or raw. I will not vouch for the truth of the singular virtues as- "cribed to the Maa-tai, but just relate one of the most plausible. "The children here often play with cash in their mouths, which "sometimes slip down into the stomach, and bring on alarming "symptoms. The nut is immediately given in quantities, either raw "or boiled, and they say it never fails to give relief being consider- "ed as specific which has the effect of decomposing the metal."

2. S. *plantaginus. Linn. Sp. Pl. ed. Willd.* 1. 291. *Rottb. gram.* 45. *t.* 15. *f.* 2. *Vahl. Enum. Pl.* 2. 251. *Retz. Obs.* 5. 14.

Culms, columnar, from two to four feet high, intercepted. *Spike* terminal, naked, sub-cylindric ; *scales* oblong, obtuse. *Seeds* beset with bristles.

A leafless, rush-like species ; it grows in shallow standing sweet water.

Root stoloniferous, with thick, spongy, and capillary fibres.—*Culm* erect, round, smooth, naked, except the lower part, which is covered with a membranaceous sheath or two; inwardly intercepted with transverse, rigid septums, at every inch and quarter, besides numberless, irregular, smaller ones between.—*Spike* terminal, cylindric, two inches long, naked.—*Scales* obovate, smooth, a little keeled, margins membranaceous.—*Stamens* three, with bristles intermixed round the base of the germ.

3. S. *tumidus. R.*

Culms from two to three feet high, round, regularly intercepted. *Spike* terminal, cylindric. *Scales* broad-lanceolate. *Seeds* beset with simple bristles.

This species is very much like the last described, and a native of similar places, but much rarer.

Root, the same.—*Culm* erect, three feet high, round, smooth, naked, seemingly swelled, intercepted with a conspicuous transverse partition every sixth of an inch.—*Leaves* none, but sheathes as in the former.—*Spike* single, terminal, erect, cylindric, an inch long. —*Scales* broad-lanceolate, acute, a little keeled, otherwise smooth. —*Stamens, bristles,* &c. as in the other species.

4. S. *spiralis*. *Linn. Sp. Pl. ed. Willd.* 1. 290. *Rottb. gram.* 45. *t.* 15. *f.* 1. *Vahl.* 2. 252.

Culms from two to three feet high, three-sided. *Spike* terminal, naked, cylindric, obtuse. *Scales* lapped. *Seeds* beset with bristles.

This is a native of the same places with the last described.

Root stoloniferous, &c. as in S. *plantaginus.*—*Culm* erect, from two to three feet high, naked, three-sided, angles sharp, smooth, not articulated, nor interrupted inwardly.—*Leaves* none, but a sheath or two embracing the base of the culm.—*Spike* as in S. *plantagi-*

nus, but the flowers are more numerous.—*Scales* imbricated seemingly without order, smooth, lapped, margins broad, and membranaceous.—*Stamens, bristles,* &c. as in the two preceding species.

5. S. *acutangulus. R.*

Culms from two to three feet high, triangular. *Spike* terminal, tapering, acute; *scales* somewhat acute; *seeds* bristly.

Grows in similar places to those which produce the two last species.

Root as in the last species.—*Culm* three-sided, from two to three feet high, naked, smooth; angles very sharp, even almost winged.—*Leaves* no other than a sheath or two involving the base of each culm. —*Spike* terminal, &c. as in the two former.—*Scales* disposed spirally, oval, somewhat acute, striated, margins membranaceous.—*Filaments* membranaceous. *Bristles,* &c. as in the former two.

Obs. It differs from S. *spiralis* in having the angles of the culm sharp. In having the flowers few and disposed spirally. In having the scales striated, oblong and somewhat pointed.

6. S. *medius. R.*

Culms from one to two feet high, three-cornered. *Spike* terminal, sub-cylindric; *scales* obtuse. *Seeds* bristly.

A species resembling the last, and growing in similar places.

Root as in the species already mentioned.—*Culm* erect, one or two feet high, smooth, three-sided, angles rounded.—*Spike* as in the former.—*Scales* smooth, obovate.—*Stamens, bristles* and *pistil* as in the last four species.

7. S. *incurvatus. R.*

Culms in tufts, culumnar, from six to nine inches high, incurved, leafless, intercepted. *Spikes* in a globular, naked *verticil* near the base of the culm. *Stigma* trifid. *Seed* three-sided.

A native of moist places near Calcutta. Flowers during the cool season.

Root of rust-coloured, capillary fibres.—*Culms* numerous, from their incurved shape forming nearly a globe, leafless, round, smooth, intercepted at every half inch or less; general length from six to eight inches, and thicker than a crow quill.—*Sheathes* a few between the spikes and root.—*Spikes* numerous, sessile, ovate-oblong, forming a round *verticil* about an inch and a half above the root.—*Scales* round, ovate-obtuse.—*Seed* ovate, triangular.

8. S. *articulatus. Linn. Sp. Pl. ed. Willd.* i. 290. *Vahl. Enum.* ii. 258.

Culms from two to four feet high, culumnar, intercepted with conspicuous partitions. *Spikes* pointed, crowded into a head near the base. *Seed* naked, three-sided.

Tsjelli. *Rheed. mal.* 12. *p.* 135. *t.* 71.

Hind. and *Beng.* Putputi-chechka.

Teling. Lutti-pitta-allie.

This is also a native of standing sweet water.

Root stoloniferous, and fibrous as in the former.—*Culm* erect, from two to four feet high, round, smooth, naked, intercepted inwardly by transverse septums at every inch, or half inch; they are conspicuous on the outside, even while the plant is growing.—*Leaves* no other than a short sheath or two, involving the base of the culm.—*Head* lateral, near the base of the culm sessile, sub-globular, composed of many, (fifty to sixty) oblong-pointed, sessile, many-flowered spikes.—*Scales* broad-cordate, a little pointed.—*Stamens* three; the bristles of the foregoing five species are here wanting.—*Style* slightly two-cleft.—*Seed* three-sided, smooth, white.

9. S. *subarticulatus. R.*

Culms from two to three feet high; culumnar, intercepted with inconspicuous partitions. *Spikes* obtuse, crowded into a head near the base. *Seed* three-sided, pointed.

A native of the same places with the former.

Root resembling that of the preceding species.—*Culm* erect, from two to three feet high, round, smooth, naked, intercepted every quarter of an inch by a slight membrane; their places do not appear externally even when the plant is dry.—*Leaves* no other than a sheath or two.—*Head* lateral, near the base of the culm, sessile, globular, composed of many, (about one hundred) small, oval, obtuse, sessile, many-flowered spikes.—*Scales* oval, membranaceous.—*Seed* three-sided, white, smooth.

10. S. *dubius*. *R.*

Root tuberous. *Culms* naked, culumnar; intercepted with numerous partitions.

Teling. Allike.

Roots tuberous, with stolones and numerous fibres issuing from them, outwardly of a dark dusky colour, inwardly white.—*Leaves* erect, cylindric, smooth, about a foot long.

Obs. I have never met with flowers of this plant which is a native of wet sandy pasture ground, and was formerly considered as a species of *Isoetes*. The natives boil and eat the roots which they say they are as good as yams.

11. S. *capitatus*. *Linn. Sp. Pl. ed. Willd.* i. 294. *Vahl.* ii. 250.

Culms naked, from four to eight inches high, culumnar; *spikes*, terminal, globular. *Seed* black, roundish, bristle-beset.

S. *caribæus*. *Rottb. gram.* 46. *t.* 15. *f.* 3.

A native of barren, sandy, moist places.

Root fibrous.—*Culms* erect, from four to eight inches high, round, smooth, naked.—*Leaves* no other than a sheath or two.—*Spike* terminal, naked, sub globular; permanent bristles surround the germ in this species, as in those with single, terminal, cylindric spikes.—*Style* two-cleft.—*Seed* obcordate, compressed, smooth, shining black.

12. S. *juncoides*. *R.*

Culms from one to three feet high, culumnar. *Spikes* oval, from three to five in a sessile head, far below the apex. *Seed* roundish, beset with bristles.

A native of wet and marshy places.

Root fibrous. *Culms* erect, from one to three feet high, naked, round, and slender in proportion to the height, striated, otherwise smooth. *Leaves* none, except two sheaths embracing the base of each culm. *Spikes* from three to five, lateral, sessile, oval, naked, from four to six inches, below the apex of the culm, which is erect, and like that below the spikes. *Involucre* none. *Stamens* three, intermixed with bristles. *Stigma* two-cleft. *Seed* obcordate, compressed, smooth, white.

13. S. *mucronatus. Linn. Sp. Pl. ed. Willd.* i. 303. *Vahl. Enum.* ii. 256.

Culms from six to twelve inches high, three-cornered. *Spikes* peduncled and sessile below the apex. *Seed* black, scabrous, subtriangular.

A native of sandy ground.

Root fibrous.—*Culms* erect, from six to twelve inches high, firm, naked, three-sided, striated, with rounded angles, above the spikes half round, with a groove on the flat side. *Spikes* lateral, sessile and peduncled, naked, from six to twelve, oblong, many flowered. *Nectarial bristles* of the germ wanting.—*Seed* somewhat three-sided, black, with scabrous spots.

14. S. *triangulatus. R.*

Culms in tufts, erect, two feet high, naked, acutely three-angled. *Spikes* from eight to twelve, sessile just under the straight three-winged apex. *Germ* embraced by six long, slender bristles, besides the three-broad filaments.

Native of Chittagong, where like most of the order, it delights in low wet ground.

15. S. *supinus. Vahl. Enum. Pl.* ii. 257.

Culms in tufts, curved, three-cornered, generally naked. *Spikes* several, under the grooved apex, sessile and peduncled. *Involucre* one-leaved, besides the apex of the culm. *Seed* three-sided, transversely rugose on the back, black.

S. *lateralis. Linn. Sp. Pl. ed. Willd.* i. 302. *Retz. Obs.* iv. 12.

Found growing in tufts, in wet places, in the vicinity of Calcutta, during the cool season.

Root of brown capillary fibres.—*Culms* numerous, about a foot high, curved, three-cornered, angles rounded, above the spikes incurved, and deeply channelled.—*Leaves* rarely any other than a few short sheaths, investing the base of the culms, and they generally end in an ensiform point, when present solitary and like the culm above the spikes.—*Spikes* many, peduncled and sessile, considerably under the apex, oblong.—*Scales* ovate, mucronate, with brown chaffy margins.—*Seed* three-sided, black, transversely rugose.

16. S. *globulosus. Linn. Sp. Pl. ed. Willd.* i. 301.

Culms from one to two feet high, nearly culumnar. *Umbel* compound; *involucre* shorter than the umbel. *Seed* long, obovate, smooth.

Teling. Billapa.

A native of wet sandy places.

Root fibrous.—*Culms* erect, from one to two feet high, as thick as a crow quill, a little compressed, smooth, except near the base which is involved in, from three to five, large, pointed sheaths.—*Umbel* terminal, small, compound. *Spikes* oval.—*Involucre* two-leaved, unequal, erect, shorter than the umbel.—*Filaments* three, broad, and membranaceous.—*Style* broad, margins fringed. *Stigma* two-cleft. —*Seed* obcordate, compressed, smooth, light brown.

17. S. *pectinatus. R.*

Culms from three to four feet high, culumnar. *Panicle* lateral, just below the straight subulate apex of the culm. *Scales* daggered. *Stigma* two-cleft. *Seed* roundish, beset with four pinnate bristles.

Grows chiefly in marshy places, where the soil is sandy.

Root fibrous.—*Culms* erect, from three to four feet high, as thick as a quill, naked, round, smooth.—*Leaves* no other than a sheath or two round the base of each culm.—*Umbel* lateral, about its own length below the apex of the culm, compound, rarely decompound, bowing.—*Spikes* oblong.—*Involucre* no other than the termination of the culm, which is erect and about as long as the umbel.--*Scales* chaffy, daggered.—*Stigma* two-cleft.—*Nectarial bristles* two on each side of the germ they are beautifully pinnate and permanent.--*Seed* obovate, compressed, smooth, brown, embraced by the nectarial bristles.

18. S. *pentagonus*. *R.*

Culms, from six to nine inches high, somewhat five-sided. *Umbel* terminal, compound; *involucre* chaffy; *flowers* monandrous. *Seed* obovate, triangular.

A native of pasture ground.

Root fibrous.—*Culms* erect, from six to nine inches high, naked, somewhat five-sided, slightly striated, otherwise smooth.—*Leaves* no other than a sheath or two.—*Umbel* terminal, compound, small, erect.—*Involucre*, a few small chaffy scales, with long tapering points. —*Stamens* single.—*Stigma* three-cleft.—*Seed* somewhat three-sided, white, scabrous.

Sect. 2d. With leaves.

19. S. *atropurpureus*. *Linn. Sp. Pl. ed. Willd.* i. 294.

Grows in small tufts. *Culms* from three to four inches high, setaceous, erect, culumnar, longer than the leaves. *Spikes* terminal, naked, solitary, oval.

A native of poor, wet sandy, pasture land.

20. S. *minimus*. *R.*

Grows in tufts. *Culms* from one to two inches long, setaceous, curved, shorter than the leaves. *Involucre* three-leaved. *Spikes* terminal, globular, squarrose, about as long as the involucre.

Found with S. *atropurpureus* on poor, wet sandy lands.

21. S. *monostachyos*. *Kön.*

Culms straight, from six to twelve inches long, culumnar. *Spike* terminal, solitary, naked, scales roundish. *Seed* obcordate, smooth.

Delights in barren, sandy pasture lands.

Root fibrous, purple-coloured.—*Culms* erect, from six to twelve inches high, round, smooth, three-fourths or more naked.—*Leaves* sheathing, nearly as long as the culm, filiform.—*Spikes* terminal, single, naked, oval; flowers few.—*Scales* sub-orbicular.—*Style* broad, with downy margins. *Stigma* two-cleft.—*Seed* obcordate, compressed, smooth, light brown.

22. S. *strobilinus. R.*

Culms two feet high, three-sided, base leafy. *Leaves* triangular. *Spike* lateral, solitary. *Scales* cordate, cuspidate. *Stigmas* two-cleft. *Seed* roundish, beset with bristles.

A native of the soft, wet banks of the Ganges. Flowering ime the cool season.

Root stoloniferous, with pale-coloured fibres.—*Culms* simple, straight, about two feet high, three-sided.—*Leaves* two or three, of different sizes, embracing the base of each culm, the longest nearly the length of the culm, keeled and deeply channelled, smooth.—*Spike* solitary, lateral, one-third part of the culm below its apex, ovate-oblong, about one inch in length. *Involucre* one-leaved, besides the continuation of the culm, length of the spikes.—*Scales* cordate, cuspidate, smooth.—*Stamina* three, with about double the number of bristles between them.—*Anthers* linear, when old spiral.—*Germ* obovate. *Stigma* two-cleft.—*Seed* flat, particularly on the inside, roundish, obcordate, crowned with a minute portion of the remaining style, and embraced by about six slender, soft, permanent bristles which are about the same length.

23. S. *scaber. R.*

Culms erect, from four to eight inches high, angular. *Spikes* terminal. *Scales* oblong, diandrous. *Stigma* bifid. *Seed* obcordate, scabrous.

Compare with S. *polytrichoides. Retz. Obs.* iv. *p.* 11.

A native of dry, sandy pasture ground on the coast of Coromandel.

Root composed of dark purple capillary fibres. *Culms* erect, from four to eight inches high, four-fifths naked, angular, smooth.—*Leaves* sheathing, mostly radical, filiform, half the length of the culm.—*Spike* single, erect, oblong, many-flowered.—*Involucre* two or three caducous scales, shorter than the spike.—*Scales* oblong, apex rounded, falling as the seeds ripen.—*Stamens* generally two.—*Stigma* two-cleft.—*Seed* obcordate, compressed, lead-coloured, rugose.

24. S. *bispicatus. Kön.*

Culms erect, from six to twenty-four inches high, filiform. *Spikes* from two to three, terminal, sessile, and peduncled, involucred; *scales* roundish. *Seed* obcordate, with a callous margin.

Teling. Khooshi-*tunga.*

Beng. Kesoori-*mulunga.*

Delights in a moist rich soil, grows in large distinct tufts.

Root fibrous.—*Culms* sub-erect, four-fifths naked, very slender, filiform, somewhat five-sided, smooth.—*Leaves* sheathing, in large plants long, slender, and often twisted a little, in small plants more rigid.—*Spikes* two or three terminal, ovate; one is always sessile, the other one or two peduncled.—*Involucre* one or two-leaved, about as long as the peduncled spikes.—*Scales* nearly orbicular.—*Stamens* and *style* as in S. *monostachyos*.—*Seed* obcordate, compressed, white, with a thick callous margin.

25. S. *schœnoides. Kön. Retz. Obs.* v. 14.

Culms erect, from six to twelve inches long. *Spikes* one or two, terminal, naked, acute. *Scales* pointed. *Seed* obcordate, angular, dotted.

Beng. Moormoori.

A native of dry pasture ground, generally amongst bushes.

Root fibrous.—*Culms* sub-erect, from six to twelve inches high, three-fourths naked, filiform, a little compressed, particularly towards the tops, and striated.—*Leaves* as in the former.—*Spike* generally single, though there are sometimes two, in which case one is peduncled, few-flowered.—*Involucre* entirely wanting.—*Scales* large, with a large keel and sharp point.—*Style* enlarged at the base. *Stigma* three-cleft.—*Seed* three-sided, very scabrous, pearl-coloured.

26. S. *tristachyos. Vahl. Enum.* ii. 254.

Culms from six to eight inches long, slender, curved, striated. *Leaves* shorter than the culms. *Spikes* from two to four, oblong. *Involucre* two-leaved, one longer than the spikes, apparently a continuation of the culms. *Style* two-cleft, villous. *Seed* obcordate.

A native of Bengal, appears during the cool season in low moist places.

Root fibrous, brown.—*Culms* numerous, in tufts. slender, variously bent, though nearly erect, slender, grooved, particularly deep on the opposite sides, from six to eight inches high.—*Leaves* few, subulate, hard, half the length of the culms.—*Spikes* generally three, varying however from one to five, sometimes short-peduncled, oblong.—*Involucre*, two-leaved; one erect, longer than the spikes, seems a continuation of the culm.—*Scales* oval, somewhat mucronate.—*Germ* cordate. *Style* villous. *Stigma* two-cleft.

Obs. It differs from Rottboell's plant in having the *scales* mucronate, and the *stigma* two-cleft; yet I think they must be the same.

27. S. *squarrosus. Linn. Sp. Pl. ed. Willd.* i. 308. *Rottb. gram.* 49. *t.* 17. *f.* 5.

Culms from three to six inches long. *Spikes* three; *involucre* two-leaved. *Scales* ending in a long recurved hairy horn. *Seeds* angular.

Motta-pullu. *Rheed. mal.* 12. *p.* 72. *t.* 38.

Beng. Goori.

A small delicate species, growing in distinct tufts on dry sandy pasture ground.

Root fibrous.—*Culms* numerous, bent in various directions, from three to six inches long, four-fifths naked, round, smooth, filiform. —*Leaves* a few only, sheathing, shorter considerably than the culm. —*Spikes* terminal, generally three, sessile, oval.—*Involucre* two-leaved, the largest about an inch long; being nearly erect, it appears a continuation of the culm.—*Scales* numerous, terminating in a long, hairy, recurved point.—*Seed* three-sided, oblong, smooth.

28. S. *monander. Linn. Sp. Pl. ed. Willd.* i. 311. *Rottb. gram.* 50. *t.* 14. *f.* 3.

Culms erect, from three to six inches high, angular. *Head* terminal, of three to five spikelets ; *involucre* from two to three-leaved; *scales* acute, monandrous. *Seed* roundish, angular.

A native of sandy pasture ground.

Root consisting of capillary fibres.—*Culms* naked, straight, from three to six inches high, angular, smooth.—*Leaves* filiform, as long as the culm, woolly near the mouth of the sheaths.—*Head* terminal, sessile, composed of from three to five small, sessile, few-flowered spikes.—*Involucre* from two to three-leaved. *Scales* pointed.—*Stamen* single.—*Stigma* three-cleft.—*Seed* three-sided, short, thick, smooth, white.

29. S. *antarcticus. Linn. Sp. Pl. ed. Willd.* i. 310. *Vahl. Enum.* ii. 261.

Culms erect, from three to six inches high, round. *Head* terminal, of many spikelets. *Involucres* from two to three-pointed. *Seed* three-sided.

Grows on sandy pasture ground.

Root fibrous.—*Culms* erect, from three to six inches high, round, smooth.—*Leaves* radical, embracing the base of the culms, sheathing, half the length of the culm.—*Head* or *Umbellet* terminal, globular, composed of numerous small, sessile, few-flowered, conical, pointed spikes.—*Involucre* from two to three-leaved, scarcely the length of the head. *Scales* pointed.—*Seed* three-sided, smooth.*

* This is *Isolepis barbata*, R. Brown, prodr. nov. holl. I. 222.—I have had specimens from Nepala, where it is common during the rainy season, forming extensive tufts. The mouth of the sheaths, as well as their ligula, are bearded with long hairs. —N. W.

30. S. *argenteus. Kön. Rottb. gram.* 51. *t.* 17. *f.* 6. *Linn. Sp. Pl. ed. Willd.* i. 311.

Culms sub-erect, from four to eight inches high, semi-culumnar. *Head* terminal, of many sub sessile spikelets; *involucres* from two to three, *scales* obtuse, monandrous. *Seed* roundish, compressed.

Fimbristylis argentea. *Vahl. Enum.* ii. 294.

Found on sandy pasture ground, growing in large tufts, with the exterior culms of the tuft curved.

Root fibrous.—*Culms* erect, or ascending, mostly naked, from four to eight inches high, round on one side, and flattened on the other, sometimes with a groove on the flat side.—*Leaves* as in the former species.—*Head* or *Umbellet* terminal, composed of from fifteen to twenty cylindrical, obtuse, sessile, silver-coloured, many-flowered spikes.—*Involucre* two, three, or four-leaved, unequal.—*Scales* oblong, obtuse; *keel* green, the rest silver-coloured.—*Stamen* single.—*Stigma* two-cleft.—*Seed* obcordate, compressed, smooth, pearl-coloured.

31. S. *glomeratus. Linn. Sp. Pl. ed. Willd.* i. 296.

Culms six inches high, semi-culumnar. *Leaves* numerous, from one to two inches long. *Umbel* composed of from six to twelve ovate spikelets. *Involucres* and *involucels* the length of the spikelets.

Hind. Bura Nirbishee.

Native of Coromandel and Bengal.

32. S. *arvensis. Retz. Obs.* iv. *p.* 11.

Culms from six to twenty-four inches high, compressed, furrowed. *Leaves* tortuose. *Umbel* simple. *Involucre* one or two-leaved, shorter than the umbel. *Scales* obtuse, diandrous. *Seed* obcordate, striated and dotted.

A native of various parts of India, appearing in low moist places during the rainy season.

Culms about six inches high, compressed, and furrowed.—*Leaves* radical, or surrounding the lower portion of the culms, narrow, a little twisted, and nearly as long as the culms.—*Umbel* generally simple, of from four to six sessile, or pedicelled, ovate-oblong brown spikes; sometimes one of the pedicels supports two spikes.—*Involucre* one or two-leaved, small and shorter than the umbel.—*Scales* of the spikes ovate, chaffy.—*Stamens* two.—*Style* thick at the base; *apex* two-cleft, and villous.—*Seed* abcordate, furrowed longitudinally, and dotted with white on a brownish ground.

33. S. *tenellus. R.*

Culms twelve inches high, leafy, sub-cylindric. *Umbels* thin, scarcely compound. *Involucres* minute. *Style* with enlarged base. *Stigma* three-cleft. *Seed* obovate, dotted, pearl-coloured.

A native of Coromandel.

Culms many, uncommonly slender, erect, about a foot high, toward the base invested with the sheaths of a few slender leaves.—*Umbel* very thin, composed of one sessile, oblong spike, and of from six to twelve or about half the number of pedicels of various lengths.—*Involucres* minute, the longest scarcely exceeding the sessile spike.—*Scales* boat-shaped, scariose, with a short point.—*Style* swelled at the base to the full size of the germ; this swelled part, however, is not more permanent than the rest of the style. *Stigma* trifid.—*Seed* obovate, obscurely three-sided, much dotted, pearl-coloured.

34. S. (*Rhynchospora?*) *mollis. Wall.*

Culms roundish, erect, longer than the narrow, setaceous, villous leaves; *umbel* composed of about seven unequal, one to three-spiked rays; *involucre* about five-leaved, shorter than the umbel, villous; *spikes* oblong *scales;* ovate, sub-aristate, smooth.

A native of Nepala, where it is found during the rainy season.

Root creeping, brown.—*Culms* one or two feet high, jointless, slightly furrowed and hairy, somewhat compressed at the top.—*Leaves* a few towards the base of the culm, five or six inches long, channelled above, with long membrane-margined, very hairy sheaths.—

Rays of the umbel from half an inch, to one and a half inches long, with one or two sessile spikes among them.—*Leaves* of the *involucre* setaceous, hairy, one or two, as long as the rays; the rest as well as the *involucels* shorter than the brown, acute, from three to six lines-long spikes.—*Scales* smooth, broad-ovate, with a greenish rib ending in a short round point; the lowest smaller, two or three of them abortive.—*Stamens* three, short.—*Ovarium* minute, surrounded by three long, hairy bristles.—*Style* compressed, hairy and ciliated, with a broad cordate articulate base.

Obs. The habit of this grass is that of a *Fimbristylis,* from which however it differs in having several long, curved bristles attached to the base of the ovarium; it seems therefore to be a legitimate species of *Rhynchospora.*—N. W.

35. S. *tridentatus. R.*

Culms straight, from two to five feet high, angular. *Umbel* compound; *involucres* from two to three; *spikelets* pointed; *scales* lapped, jagged and daggered.. *Seed* roundish, three-angled, and bristly.

Compare with S. *maritimus.*

Teling Goorapa-shaka-tunga.

Grows in water courses, ditches, &c. where the water is fresh; never in salt, or brackish water.

Root stoloniferous with some small tubers, and many fibres. —*Culms* erect, rigid, leafy, jointed, from two to five feet high, three-sided, smooth, angles sharp.—*Leaves* numerous, sheathing, sub-erect, rigid, keeled, smooth.—*Umbel* compound, small for the size of the plant.—*Involucre* from two to three-leaved, erect, unequal, the longest being three times the length of the umbel, and the shortest equalling the latter.—*Spikes* lanceolate or cylindric, large; some sessile, some peduncled.—*Scales* three-toothed, the lateral ones torn, the middle one long and subulate, keel green, the rest rust-coloured. —*Seed* three-sided, short, thick, smooth, with bristles round the base.

Obs. I believe König called it S. *maritimus indicus,* but the form of the seeds and jointed culm, will, I think, render it a very well marked and distinct species.

36. S. *strictus. R.*

Culms straight, from twelve to eighteen inches high, culumnar. *Umbel* compound. *Involucres* chaffy; *scales* obtuse, diandrous and roundish, compressed.

Grows on a moderately dry, hard, barren, sandy soil.

Root fibrous.--*Culms* straight, naked, rigid, from twelve to eighteen inches high, round, very smooth.—*Leaves* radical, short, rigid, but smooth.—*Umbel* terminal, sometimes decompound, contracted, seldom more than two inches long, and one broad.—*Involucre* from two to four, small, chaffy, tapering sheaths.—*Spikes* ovate, small, many-flowered.—*Scales* with a broad, white, membranaceous border.—*Stamens* two.—*Stigma* two-cleft.—*Seed* much compressed, obcordate, dotted with brown.

37. S. *brevifolius. R.*

Culms erect, from six to eight inches high, semi-culumnar. *Umbel* decompound; *involucels* chaffy; *scales* cuspidate, monandrous. *Stigma* bifid. *Seeds* obcordate, striated.

Is found generally on low sandy pasture ground.

Root fibrous.—*Culms,* erect, mostly naked, from six to eight inches high, flat on one side and rounded on the other.—*Leaves* sheathing, short.—*Umbel* terminal, decompound,—*Involucre* from two to three-leaved, shorter than the umbel.—*Spikes* small, oblong; while young sessile, but when old appearing peduncled, the scales and seeds gradually dropping off and leaving the rachis naked.—*Scales* keeled and pointed.—*S amen* single.—*Style* broad, enlarged at the base; above, the margins are hairy. *Stigma* two-cleft.—*Seeds* obcordate, convex on the outside, flat on the inner, white, striated.

Obs. The single stamen and striated seed are the best specific marks. It much resembles Rottboel's figure of S. *dichotomus,* but that species is described by Vahl as having three stamina: (see *Fimbristylis dichotoma, Enum.* ii. *p.* 287,) consequently they must be different.*

38. S. *astivalis. Retz. Obs.* iv. 12.

Spikes rounded, subsquarrose; *involucre* six-leaved, longer than the compound umbel; *scales* monandrous. *Vahl. Enum.* ii. 288. (*under Fimbristylis.*)

I received this elegant little grass from Nepal*a* where it grows in low fields during the rainy season.

Obs. My specimens agree perfectly with the descriptions quoted above. They are three inches high; spikes oblong, obtuse, measuring about three lines in length, with villous scales which end in a long, recurved, scabrous awn as in S. *squarrosus. Ovarium* lenticular, obovate, smooth, shining, pearl-coloured, covered almost to its base with numerous loosely adhering, white, flat fascicles of villi, which proceed like rays from the enlarged base of the smooth, bifid style, and serve to keep the latter attached to the seed. This is a second instance of a singular structure, observed by the immortal author of the Prodromus floræ novæ hollandiæ, in another species of Fimbristylis, (vol. i. p. 225).—N. W.

39. S. *diphyllus. Linn. Sp. Pl. ed. Willd.* i. 301.

Culms erect, from one to two feet high, semi-culumnar. *Umbel* decompound; *scales* obtuse, diandrous. *Seeds* roundish, obcordate, smooth.

Fimbristylis diphylla. *Vahl. Enum.* ii. 249.

Delights in a rich wet soil.

Root fibrous.—*Culms* erect, seven-eighths naked, from one to two feet high, sometimes a little compressed.—*Leaves* chiefly radical, half the length of the culm, rigid.—*Umbel* terminal, decompound. —*Involucre* two or four-leaved, the larger about the length of the um-

* I have received specimens of this grass, which undoubtedly is a *Fimbristylis*, from Nepal*a*; it is intermediate between F. *dichotoma* and *glauca*, Vahl, differing from both in being monandrous.—N. W.

bel, *partial involucres* small.—*Spikes* numerous, oblong, some sessile, some peduncled, exclusive of the naked rachis of those that are old, dark brown.—*Scales* oblong, obtuse, dark brown.—*Stamens* two.—*Stigmas* two-cleft.—*Seed* obcordate, much compressed, smooth, brown.

40. S. *(Isolepis) densus. Wall.*

Culms tufted, setaceous as well as the short leaves; sheaths bearded; *umbel* decompound with unequal capillary rays; *involucres* squarrose, awned; *spikelets* oval, long-peduncled, except the central ones.

Native of Nepala; in vigour during the rains.

A most slender, glaucous, capillary, spreading, smooth grass, which forms extensive and dense tufts. *Root* grayish, consisting of innumerable long fibres. *Culms* from six to twelve inches high, three-cornered.—*Leaves* three or four near the base of the grass, and a great many radical ones, three or four times shorter than the culm; sheaths short, flaccid, brown, with a number of long curved hairs at their back and mouth. Rays about five, unequal, mostly an inch long; some of them bearing a few-spiked umbellet, others supporting only a single spikelet. *Scales* of the involucre chaffy, brown, imbricating, one of them often ending in a capillary leaflet.—*Spikelets* very small, brown, somewhat angular, consisting of oval, keeled, pointed scales.—*Seed* obovate, sharply three-cornered.—*Bristles* none.

Obs. The root, culm, leaves and sheaths of this elegant little grass are exactly like those of Isolepis barbata, (see above;) the inflorescence, however, is totally different.—N. W.

41. S. *miliaceus. Linn. Sp. Pl. ed. Willd.* i. 305. *Rottb. gram.* 57. *t.* 5. *f.* 2. *Retz. Obs.* 5. 16.

Culms erect, from one to two feet high, semi-culumnar. *Umbel* decompound; *involucres* and *involucels* shorter than the umbels and umbellets. *Seed* obcordate, striated.

Fimbristylis miliacea. *Vahl. Enum.* ii. 287.

This species delights in wet pasture ground.

Root fibrous.—*Culms* erect, from one to two feet high, seven-eighths naked, generally round, though sometimes a little compressed, and chiefly near the apex striated, otherwise smooth.—*Umbel* decompound, very thin.—*Involucre* three or five-leaved, unequal, but the largest is seldom so long as the umbel. *Involucels* like the involucre, but much smaller.—*Spikes* sessile, or peduncled, oval, rust-coloured, many flowered.—*Scales* broad, a little pointed, very concave.—*Stamens* two or three.—*Style* enlarged at the base, broad, with hairy margins. *Stigma* two-cleft.—*Seed* obcordate, compressed, pearl-coloured, striated, without bristles.

42. S. *complanatus. Vahl. Enum.* ii. 279. *Retz. Obs.* 5. 14.

Culms sub-erect, from one to two feet high, edged. *Umbel* from compound to super-decompound *Involucre* two-leaved, erect; *scales* lanceolate. *Seed* obcordate, dotted.

Hind. Kuratiya-juvani.

A native of moist or wet places.

Root fibrous.—*Culms* nearly erect, but waving a little, from one to two feet high, naked, except at the base, compressed, two-edged; edges sharp, and turned different ways, striated, smooth, often a little twisted.—*Leaves* radical, or embracing the base of the culm, nearly as long as the culm, linear, waved, often a little twisted, apex abrupt.—*Umbel* super-decompound, dense.—*Involucre* two-leaved, opposite, erect, unequal, the longest about the length of the umbel. *Involucels* chaffy, ending like bristles.—*Spikes* sessile, or short-pedicelled, oblong, few-flowered.—*Scales* rust-coloured, acute.—*Seeds* obcordate, rugose, white.

43. S. *tetragonus. R.*

Culms from one to two feet high, four-angled. *Umbel* decompound and super-decompound. *Involucre* and *involucels* bristly, *scales* roundish; diandrous. *Seeds* round.

Hind. Bura-juvani.

Teling. Murcushalee.

Grows with the other species on moist pasture ground.

Root fibrous.—*Culms* nearly erect, from one to two feet high, naked, except at the base, four-sided, smooth.—*Leaves* radical, and about the base of the culm, bifarious, tapering, nearly as long as the culm.—*Umbel* super-decompound, rather thin, and of an irregular form.—*Involucre* and *involucels* small, bristle-like.—*Spikes* minute, globular, many-flowered.—*Stamens*, from one to two, rarely three.—*Stigma* two or three-cleft.—*Seed* nearly globular, compressed on the inside, scabrous, white.

44. Sc. *quinquangularis. Vahl. Enum.* ii. 279.

Culms from one to two feet high, five-angled. *Umbel* super-decompound; *involucre* much shorter than the umbel; *scales* oval, monandrous. *Seed* three-sided, rugose.

A native of similar places with the last species.

Root fibrous.—*Culms* as in *complanatus*, and *tetragonus*, and of the same height, but in this species always five-sided. The angles sharp and the sides concave.—*Leaves* as in Sc. *complanatus.*—*Umbel* super-decompound, rather thin, and unequal as in Sc. *tetragonus.*—*Involucre* three-leaved, the largest about the length of the umbel.—*Spikes* mostly pedicelled, oval, very small, few, from six to eight-flowered.—*Scales* oval.—*Stamen* one.—*Stigma* three-cleft.—*Seed* oblong, three-sided, rugose.

45. Sc. *pallescens. R.*

Culms from six to ten inches high, culumnar. *Umbel* super-decompound, rather shorter than the involucre, monandrous *Scales* obtuse. *Seed* obcordate, without angles, streaked, and wrinkled.

A native of low pasture ground.

Root fibrous.—*Culms* nearly erect, from six to ten inches high, mostly naked, nearly round and striated.—*Leaves* sheathing, as long as the culm, concave; edges slightly serrulate, somewhat woolly.—*Umbel* terminal, super-decompound, large in proportion to the rest of the plant.—*Involucre* two-leaved, nearly erect, opposite margins slightly serulate, longer than the umbel, a few small scales intermixed.—*Spikes* very minute.—*Scales* oval, concave.—*Stamens* solitary. —*Style* two-cleft.—*Seed* obcordate, a little compressed, striated longitudinally, and wrinkled transversely.

46. Sc. *anceps. R.*

Culm erect, twelve inches high, two-edged. *Involucre* two-leaved, shorter than the super-decompound, four or five times-divided umbel.

Beng. Joopi.

A native of Bengal.

47. Sc. *comosus. Wall.*

Glaucous. *Leaves* radical, together with those of the involucrum very long, channelled, ending in a triangular capillary acumen. *Corymb* super-decompound, nodding. *Spikes* geminate or ternate, sessile, oblong, crowned with the long straight silvery hairs of the seeds.

A native of Nepala, where it is found among rocks during the rains.

Root fibrous —*Culms* many, tufted, erect, from six to twelve, or more inches long, very slender, scarcely thicker than a crow-quill, obscurely three-cornered, smooth, naked.—*Leaves* radical, numerous, embracing the base of the culm with their short brownish, membranaceous sheaths, which burst on one side ; generally twice as long as the culm, very narrow, slightly keeled, channelled above, the margins and keel scabrous, ending in a long three-cornered point.—*Corymb* terminal unilateral and slightly nodding, composed of about twelve unequal, slender, somewhat flattened, smooth rays from three to six inches long, surrounded at the base with a short, brown, narrow, two-keeled *ochrea ;* they end in partial, more or less compound umbellets, consisting in general of four or five simple or divided short capillary peduncles.—*Involucrum* consisting of five very long leaves, resembling the radical ones, but somewhat broader at their base, embracing the apex of the culm with their brown, membrane-winged insertions ; the largest measuring from twelve to eighteen inches.

—*Involucels* consisting of a few chaffy, brown, subulate scales.—*Spikes* very numerous, about three lines long, terminal, or lateral, sessile, geminate, sometimes ternate; some single pedicelled ones are crowded in the centre of the general and partial umbels.—*Scales* lanceolate, keeled, brown, with short subulate points; the lowermost broader and sub-aristate, barren.—*Ovarium* linear, oblong, surrounded at the base by about twelve silvery, shining straight hairs, which are twice as long as the scales. *Style* trifid, scabrous, shorter than the hairs.

Obs. This beautiful grass, which I have received both from Katmandu and *Gossein-Than*, an uninhabited place at the foot of the Himal*uya* mountains, from whence Mr. Gardner, the Honourable Company's Resident at Nepal*a*, has sent me the most rare plants, comes near Sc. *eriophorus* and like that connects the genus *Scirpus* with *Eriophorum*. It differs in its shorter and smaller stature, and in having oblong sub-cylindric spikes. Perhaps it is only a variety of that species.—N. W.

48. Sc. *Kysoor. R.*

Culms from five to six feet high, triangular, and hispid. *Umbel* super-decompound. *Spikes* ovate. *Seed* oblong, three-sided, crowned, and beset with five villous bristles.

Beng. Kesoor or Kesooree.

Common in Bengal, growing on the borders of lakes, ponds, &c. of fresh water, in flower during the latter part of the rains.

Root tuberous, with numerous ferruginous fibres.—*Culms* straight, from four to six feet high, three-angled, the sides deeply grooved and the angles very sharp, and backwardly hispid.—*Leaves* several to each culm, and about their length, sheathing at the base, above that part deeply channelled, with the margins and keel while young, somewhat hispid, and the points long and subulate.—*Umbel*, terminal, super-decompound.—*Involucres* about three, very unequal, the largest from one to two feet long, and the shortest as many inches; resembling the leaves in form.—*Spikes* ovate, brown.—*Scales* roundish oval, concave, chaffy, brown, smooth; the rib or nerve ending in a subulate point beyond the rounded apex. —*Stamens* three on the exterior side of the germ, one-bearded bristle between each filament, and three on the back, making in all five. —*Germ* ovate!—*Style* with swelled, permanent, cordate base.—*Stigma* three-cleft.—*Seed* oblong, three-sided, surrounded by the five-bearded bristles.

Obs. It approaches Sc. *grossus. Vahl. Enum.* ii. 270; the best marks to distinguish them by, are the bristles which embrace the germ, and continue with the seed; here they are five in number and villous, there six, and barbed, or armed on the margin, with sharp bristles pointing backward. My Sc. *maximus* is a much larger species, and the germ destitute of bristles.

49. Sc. *grossus. Vahl. Enum.* ii. 270.

Culms erect, from six to ten feet high, triangular. *Umbel* super-decompound. *Seed* three-sided, beset with six, backwardly hispid bristles.

Exclude Sc. *grossus. Retz. Obs.* v. 15. which I presume is my *Sc. maximus;* the germ and seed being there destitute of the bristly involucre. To which of these *Sc. grossus,* Linn. *Suppl.* ought to be referred, I cannot say.

This, next to Sc. *maximus,* is by far the largest species I have ever met with. It grows in standing sweet water.

Culm erect, mostly naked, from six to ten feet high, three-sided; *angles* sharp, *sides* concave.—*Leaves* radical, and round the base of the culms sheathing, much keeled, almost as long as the culms; edges slightly scabrous.--*Umbel* super-decompound, five or eight inches each way.—*Involucre* three-leaved, the largest about two feet, the shortest six inches, they taper to a long, acute point; margins pretty smooth.—*Involucels* chaffy.—*Spikes* oblong, dark-brown, many-flowered.—*Stamens* three, with two bristles between each pair. —*Style* three-cleft.—*Seed* three-sided, embraced by the six enlarged bristles, which are armed on the edges with stiff, bent back, short hairs.

50. Sc. *maximus. R.*

Culms straight, from six to fifteen feet high, triangular. *Umbel* decompound; *spikelets* roundish. *Seed* obcordate, three-sided, without bristles.

Sc. *grossus. Retz. Obs.* v. 15. and probably Linn. *Suppl. p.* 104.

Teling. Booda-*tunga.*

Is found only in pretty deep, standing, sweet water.

Root fibrous, and stoloniferous.—*Culms* erect, most rigid, naked, except at the base, from six to fifteen feet high, and from one to two inches in diameter, three-sided; *angles* very sharp; *sides* somewhat concave.—*Leaves* mostly radical, almost as long as the culm, much keeled, (a transverse section appears like the letter V,) smooth, inside only a little striated.—*Umbel* super-decompound, generally about a foot long.—*Involucre,* from three to four-leaved, the largest from three to four feet long, the smallest from three to four inches, smooth.—*Involucel* chaffy.—*Spikes* minute, ovate, few-flowered.—*Scales* broad, and short.—*Stamens* three.—*Stigma* three-cleft.—*Seed* three-sided, without bristles, which easily distinguishes it from the last Sc. *grossus.*

Obs. I have not in any of the foregoing genera of *Cyperoideæ* taken notice of the sheaths which embrace the base of the peduncles and pedicels of the umbels, and their sub-divisions; because they are common to all, and so much alike, that I do not think they can convey any information. Nor have I attended to the form of those peduncles and pedicels; because I have not found it uniform. The culm, inflorescence, involucre, and seed, are I think, the best marks to discriminate the species by, particularly the culm and seed. The number of stamens and divisions of stigma, sometimes vary even in the same plant, much more in different plants of the same species.

Nearly the whole of the plants belonging to the six foregoing *Genera* (called *Calamariæ* by Linneus, and *Cyperoideæ* by Jussieu) are natives of low, barren, moist places, and borders of rice fields on the coast of Coromandel. They appear, blossom, and ripen their seed during the rains, and the beginning of the cool season. They are in general rejected by cattle.

DIGYNIA.

PEROTIS. Schreb. Gen. N. 1714.

Calyx none. *Corol* two-valved, *valves* equal, and awned.

1. P. *latifolia. Linn. Sp. Pl. ed. Willd.* i. 324.

Culms piped, from nine to eighteen inches high. *Leaves* short, waved, and ciliate. *Raceme* cylindrical.

Tsjeria-kuren-pullu. *Rheed. Mal.* 12. 117. *t.* 62.

Anthoxanthum indicum. Linn. Sp. Pl. ed. Willd. i. 157.

Saccharum spicatum. Linn. Sp. Pl. 79.

Is a native of hign, dry sandy barren ground, over the coast of Coromandel.

Root fibrous.—*Culms* at the base resting on the ground, and there branchy, above erect, leafy, jointed, from nine to eighteen inches high, smooth.—*Leaves* sheathing, short, broad at the base, margins waved and hispid; sheaths about half the length of the joints. —*Racemes* (not spikes) terminal, erect, cylindrical, lead-coloured.—*Flowers* numerous, diverging.—*Calyx*, two-valved, both awned, and are totally destitute of hairs round their base, or any where else:—*Corol* two-valved, very minute. *Seed*......

Obs. Cattle are not fond of this grass.

CENCHRUS. Schreb. Gen. N. 1574.

Involucre echinate, and four-flowered. *Calyx* from two to three-valved, one or two-flowered.

1. C. *biflorus. R.*

Spikes simple, cylindric. *Involucres* two-flowered; *segments* subulate, hooked, and hairy within.

A native of the interior, elevated, dry parts of the Coromandel coast.

Culms straight, from six to twenty-four inches high, smooth.—*Leaves* long, slender, somewhat scabrous; mouths of their sheaths a little hairy.—*Spikes* terminal, solitary, simple, cylindric, from three to four inches long. *Rachis* flexuose, three-sided.—*Flowers* numerous, solitary, alternate.—*Involucre* two, rarely one-flowered, multifid, divisions longer than the flowers, subulate, grooved on the back, inside hairy, apex slightly hooked. Both flowers hermaphrodite.—*Calyx* three-valved, one-flowered, the exterior one smaller, all are thin, smooth, nerved.—*Corol* two-valved, valves hard, acute, but not awned, in length nearly equal, and scarcely longer than the calyx.—*Stamina* three.—*Styles* two, plumose.

SACCHARUM. *Schreb. Gen. N.* 104.

Calyx one-flowered, two-valved, girt at the base with wool; *corol* uncertain.

1. S. *cylindricum. Linn. Sp. Pl. ed. Willd.* i. 323.

Culms from one to two feet high, shorter than the leaves. *Panicle* linear, white with the wool of the calyx. *Corol* and *calyx* two-valved and diandrous.

Lagurus *cylindrica. Sp. Pl.* 120.

S. *Thunbergi. Linn. Sp. Pl. ed. Willd.* i. 323. *Retz. Obs.* v. *p.* 17.

Gramen caricosum. *Rumph. amb.* 6. 17. *t.* 7. *f.* 2.

Beng. Ooloo.

Teling. Baroomhiss.

Is a native of moist, stiff, pasture ground, and particularly common over Bengal, where the fields are white with it when in flower, after the first rains in April and May.

Root creeping at some depth under the surface.—*Culms* erect, simple, from one to two feet high, leafy, jointed, joints bearded.—*Leaves*, those of the root or lower part of the culm are much longer than the culm, tapering equally towards each end, smooth, those of the culm are little else than sheaths.—*Panicles* terminal, cylindric, from three to four inches long, composed of small, erect branchlets, of a bright cotton-like whiteness, from the quantity of wool that surrounds the insertion of the flowers.—*Flowers* pedicelled, generally simple.—*Calyx* two-valved, with here and there a long white hair.—*Corol,* two-valved, minute.—*Stamens* two.

Obs. Cattle are not fond of it, particularly when old. It is used in the marriage ceremonies of the Telingas. In Bengal it is much used for thatch.

2. S. *spontaneum. Linn. Sp Pl. ed. Willd.* i. 321.

Culms from five to fifteen feet high. *Leaves* involute when dry. *Panicle* linear, spreading, verticilled; *ramification* simple, *calyx* ciliate. *Corol* one-valved.

Sans. Kasha.

Hind. Kagara.

Teling. Relloo-gaddy.

Grows on the banks of rivers, in hedges, and on moist, uncultivated land; in a good soil, it is frequently from ten to fifteen feet high, in a poorer soil, from five to ten.

Root perennial, creeping very deep in the ground.—*Culms* annual, erect, leafy, round, lower parts perennial, and as thick as the finger, annual shoots about as thick as a common quill, solid, height various as above observed.—*Leaves* sheathing, remarkably long and narrow, but firm; margins hispid; a transverse section appears crescent; sheaths with their mouths woolly.—*Panicles* terminal, spreading, erect, oblong, from one to two feet long, composed of verticilled, filiform, simple, (except the lower verticil or two,) spiked racemes; *spiked,* because half the flowers are sessile, and half pedicelled. The immense quantity of long, bright, silver-coloured wool which surrounds the base of the flowers gives this species a most conspicuous, gaudy appearance.—*Flowers* paired, one-pedicelled, the other sessile.—*Calyx* two-leaved; margins ciliate, acute, surrounded with many soft, very long, silver-coloured hairs.—*Corol* one-valved, ciliate, acute, membranaceous.—*Stigma* feathery, purple.

Obs. I have not seen the ripe seed. It is so very coarse that cattle do not eat it, except while very young. It makes good mats for various purposes, and is used to thatch houses.

3. S. *fuscum. R.*

Erect, from five to eight feet high. *Leaves* sub-lanceolate, smooth. *Panicle* spreading, verticilled; *ramifications* compound. *Flowers* paired, both pedicelled, wool of the calyx short and scanty. *Corols* three-valved.

Beng. Khuree or Pata-Khuree.

A native of damp places over Bengal. Flowering-time the rainy season.

Culms erect, from five to eight feet high, as thick as the little finger, hairy near the top.—*Leaves* sheathing, linear-lanceolate, smooth in every part, except the margins of the sheaths, which are fringed with much soft hair; length about three or four feet, breadth about, or under two inches.—*Panicles* from one to two feet long, erect, linear, verticilled, except toward the apex; *ramifications* compound; *ramule* nodding.—*Flowers* paired, one short-pedicelled, the other longer, both hermaphrodite; wool of the pedicels, &c. short and in small quantity.—*Calyx* with ciliate margins, otherwise smooth and shining.—*Corol* three-valved, of which two are equal and nearly as long as the calyx, the third minute, all much ciliated.—*Seed* long, obovate, brown, smooth.

Obs. The natives make their pens of the culms of this and the following species, and use it for screens and light fences.

4. S. *semidecumbens.*

Lower portion of the culms procumbent, and perennial. *Leaves* narrow, channelled. *Panicle* oblong, lax, with compound verticilled ramifications, *corol* one-valved.

Beng. Khuree.

A native of Bengal, where it delights in low wet places, blossoms about the close of the rains, and the beginning of the cold season.

Root perennial.—*Culms* of great length, i. e, from eight to sixteen or more feet long, one half or more, generally resting on the ground, and striking roots at the joints; this portion is often as thick as the little finger, and perennial; the extreme portion erect, straight, and taper to the thickness of a crow-quill at the panicle; filled with pith in every part.—*Leaves* from one to six feet long, very slender, pointed, channelled, bottom of the channel white, margins most acutely hispid, mouths of the sheaths and for some way above them bearded with much soft white hair.—*Panicles* ovate, or oval, erect, large. *Ramifications* coloured, verticilled, compound.—*Flowers* approximate, paired; one sessile, the other pedicelled, both girt with much, pure white, soft hair.—*Calyx* two-valved; inner valve ciliate.—*Corol* one-valved, ciliate.

5. S. *Officinarum. Linn. Sp. Pl. ed. Willd.* i. 321.

Culms from six to twelve feet high. *Leaves* flat. *Panicle* ovate, spreading; *branches* alternate, decompound. *Corol* one-valved.

Sans. इक्षुः, Ikshoo, रसालः, R*u*sal*a*; the light-coloured varieties; पुंड्रः, Poondr*a*, and कांगरकः, Kang*uruka*, the red.

Beng Ik, Ook, Ak. Kooshiar, also P*oori* and K*u*ll*oo*a the two pale varieties, and Kaj*ooli* the red.

Teling Cherukoo-bodi, or Cherukoo-duboo.

Where wild, I do not know.—*Panicles* terminal, spreading, erect, oblong, from one to three feet long, of a grey-colour, from the large quantity of long soft hair that surrounds the flower. *Ramifications* alternate, very ramous, expanding.—*Rachis* striated.—*Flowers* hermaphrodite, in pairs; one sessile the other pedicelled.—*Calyx* two-leaved, smooth.—*Corol* one-valved, smooth, membranaceous, rose-coloured.—*Stamens* and *Pistil* as in the genus. I have not seen the seed.

Obs. It is much cultivated in the Rajamundr*i* Circar, where they only make a coarse sort of brown raw sugar, which is sold on the spot for about three half-pence per Pound. The cane is cultivated on their best lands, which can be easily watered; but, unless the farmer pleases, not easily flooded during violent rains.

The land is first well ploughed during the month of April and beginning of May. The field is then flooded from the river, if there is not sufficient rain; the upper part of the cane is then cut into two lengths of one or two joints each, (the lower part of the same canes are employed to make sugar from) these are placed over the wet field, at about fifteen or eighteen inches asunder, in rows, the rows about four feet from one another, and trode under the soft wet surface with the foot; in six days after the planting, the field is again flooded, if there has not been rain. In about eight days more the shoots appear; the land is soon after slightly hoed, and weeded. A month after the planting some rotten chaff, or other such manure, is scattered about the young plants. Every ten or fifteen days, if there be not sufficient rain, the field is watered; two months from the planting, some stronger manure is strewed about the plants; and every fifteen or twenty days the field is slightly hoed, and the weeds rooted out.

During the wet season, drains must be made, to carry off the superabundant water. By August or September, the cane will be from three to five feet high; in each shoot, the produce of every cutting, which may contain from three to six canes, a straight bamboo is stuck into the earth, in the centre; to this the canes are tied by their leaves. In this country the leaves are never stript from the cane, but as they wither are tied round them. This must impede the free circulation of air, which may be conceived hurtful. In January, viz. between nine and ten months from the time they were planted, the cane, when stript of its leaves, and the useless top cut off, will be about as thick as a good stout walking cane, and from four to six feet long; they then begin to cut the cane, express the juice, and boil the sugar, which is with the natives here, a very simple process; a small mill turned by cattle squeezes the cane, and one boiler boils it.

In Bengal there are three varieties cultivated, viz. The above described or common yellow cane, called by the Bengalees *Poori.* Secondly the purple cane called by them *Kajooli,* which is said to yield juice one-eighth part richer than the yellow cane; but the sugar thereof is always of a dark-colour. The third and last is a very large, light-coloured cane, called *Kullooa,* which grows in a low swampy soil, where neither of the other two will succeed; its juice is still weaker than that of the yellow cane; but it has these advantages, that it grows to a much larger size, and where neither of the other two will thrive. It is therefore much cultivated, because the other sorts planted on higher situations are apt to suffer from drought.

6. S. *sinensis. R.*

Culms from six to ten feet high. *Leaves* flat, with margins hispid, panicles ovate, with simple and compound verticilled branches. *Corol* of two valves on the same side.

Stem erect, jointed, whole height from ten to fifteen feet; the greatest part covered with the sheaths of the leaves; *joints* from four to eight inches long, and from two to three inches in circumference, colour pale brownish yellow.—*Leaves* sub-bifarious, tapering from the base, to a long, fine, point; plain, smooth on both sides; margins armed with numerous, small, very acute spines pointing forward, length from two to three feet, and about an inch and a half broad at the base.—*Sheaths* smooth, with a small ligula, or stipulary ring round the inside of the mouth.—*Panicle* ovate, erect; branches, simple and compound, sub-verticilled, reclinate, long and slender. In S. *officinarum* the branches of the panicle are scattered over the common rachis and are decompound, and super-decompound; on this I rest the most obvious specific deference, independent of the additional small inner scale, or valve of the corol.—*Calyx,* and its surrounding wool, as in the genus.—*Corol* of the two valves on the same (anterior) side, the inner one very small.—*Nectary* of two, large, broad, cuneate, crenulate, fleshy scales, which occupy the two posterior sides of the germ, opposite to the two valves of the corol.

Obs. This new species, for such it appears to me, was introduced from China into the Botanic Garden, at the close of the year 1796, in hope of finding it in some respects better than the common cane cultivated over India. It promises considerable advantage; particularly from its being so solid, and hard, as to resist the forceps of the white ants, and the teeth of the jackall, two great enemies to our East Indian sugar plantations. At the same time it bears drought much better than the sorts in general cultivation. It produces a profitable crop even to the third year; while the common cane of India must be renewed every year. It is also said to yield juice of a richer quality.

This seems to be the sort employed over China for making their sugars, as appears from the following extract of a letter from Mr. A. Duncan, surgeon to the factory at Canton, to Richard Hall, Esq. President, &c. of the Select Committee, dated Cantón 26th October, 1796, which accompanied plants of this cane from China; explaining the mode of cultivation and manufacturing the sugar in the province of Canton.

" *Sugar canes.* A first and second sort are cultivated in the pro-" vince of Canton; the former being small, compact and highly Sac-" charine, is manufactured into sugar, whereas the latter, being of a " larger size, and looser texture, is generally cut into joints, and eat " by the natives, either raw, or gently heated.

" The culture of sugar canes commences about the middle of " March or beginning of April, in proportion to the temperature of " the weather, when a good, light, sandy soil is thrown up into beds " of about four feet diameter leaving intermediate spaces of one " and a half, which serve as drains and passes for the husbandmen. " A composition of human manure and water is thrown over the beds, " after which the tops of sugar canes, fifteen inches long, (the Chi-" nese are here economical) are planted in a slanting position, and " occasionally watered at sun set, till the young shoots rise a cubit " high, when the husk and refuse of the ground nut, or of some oleagi-" nus seeds after the oil is extracted, are well beaten into a kind of

" powder and put round the root of the plants, in small quantities and " afterwards covered over with earth. The canes are now left to the " fertility of the soil till they arrive at maturity, except when it proves " such a dry season, that watering becomes requisite, and are cut " down about the 1st of November. The first sort planted in the " foregoing manner, generally yields three crops, while the second " requires fresh planting.

" The sugar canes are now removed from the ground to the mill " for expression. The juice, being extracted, is carefully strained, " mixed with a small quantity of lime water, then poured into large " warm iron pans, placed on a plat-form of brick work, with fire " places below; it is now boiled for a length of time, and being " removed from the fire, and allowed to cool, it is a second time " strained, and being again returned to the pans, the boiling is re- " peated for once, or twice, till it acquires a saccharine consistence. " The sugar, still very moist, is now put into cylindrical jars, which " have a small hole in each of their bottoms, for the purpose of " draining; and their tops being covered over with from half an " inch to an inch of clayey moist earth, these are placed on stands, " and remain for three or four days; when the old earthen covering " is removed, and a fresh one put on. They continue in that situ- " ation for three or four days more, the covering is then taken off, " and the sugar being sufficiently clarified is removed from the jars, " and spread out on flat baskets, exposed to the sun till dry, when " it is fit for use.

" The mill for expressing the juice; of which I have sent a very " exact miniature model, appears so far as I recollect, preferable to " those used in India. The machine is wrought by a buffalo " yoked to the small end of the crooked beam, while its thick end is " fastened to the long axle of one of the cylinders. The canes hav- " ing undergone one expression, and being consequently in a split state, " are not thrust a second time through till all the irons are greased. " The cylinders are not always of the same dimensions; those I ex- " amined were two cubits high, and one and two-thirds in diameter, "these are the common size; I am however told they are sometimes "much larger."

To point out the benefit that may accrue to the cultivators of this sugar cane, I need only add the following copy of a letter from Mr. Richard Carden, superintendant of the Honourable Company's Rum and Sugar works at Mirzapore Culna in Bengal, to me, dated 13th August, 1801, soon after the introduction of the cane.

"With respect to the produce of the common Bengal sugar canes, "I have never been able to collect an account that can be depended "upon; the natives generally manufacture the juice into Jaggery in "my neighbourhood, which yields them nearly 14 cutcha maunds, "of 60 sicca weight, per *bigha* on an average; and a profit of about "11 or 12 rupees per *bigha*. Neither the white ants nor jackalls, "have committed any depredations on the China canes that I have "planted, although the latter have often been seen among them, "which certainly gives these canes a decided preference to the Ben-"gal sugar canes. I do not think the China ones degenerate in the "least, nor do they improve; they appear to me to remain nearly "in the same state. If planted at the same time the natives put "their canes into the ground, they will not make such good returns "as the Bengal sort, but planted in the West India mode, in the "month of September or October, and suffered to remain on the "ground till the December or January twelve months, they will then "yield double the returns of what the Bengal canes do, which is "owing in part to the length of time they are in the ground, and "principally to the ants and jackalls not destroying them, whereas "if the Bengal cane was to remain so long on the ground, the na-"tives would have great difficulty to prevent the greatest part of "them from being destroyed, and the young shoots would suffer very "much from the hot winds, and ants, which I witnessed the second "year I came to Mirzapore; but the shoots from the China canes "I cut last January, stood the last hot season uncommonly well, "and will next January, I have reason to believe, from the present "appearance, make half, or nearly three-fourths the quantity of "sugar they did last January; and that with the trifling expence "of clearing the ground twice; cutting, and manufacturing the juice. "I should be extremely happy to have sent you a more particular "account, but that is not in my power, as the native statement is "not to be depended upon. I did purchase twelve cottas of the "best Bengal canes last January, merely as an experiment, which "yielded just half the quantity of sugar my China canes did."

7. S. *procerum*. *R.*

Perennial, from ten to twenty feet high, erect. *Leaves* ensiform, with a white rib, and hispid margins. *Panicle* diffuse, with verticilled, compound and decompound branches. *Corol* of the pedicelled flower two, and of the sessile three-valved, with the inner one retuse.

Beng. Teng.

A native of Bengal, and by far the most beautiful of the genus I have yet met with. It comes nearest in appearance to S. *Officinarum*, but is a taller and much more elegant plant.

Culms perennial, straight, simple till the second or third year, then branchy, about as thick as a slender walking cane; joints from six to twelve inches long, and filled with insipid pith; height of the whole plant, when in flower from ten to twenty feet.—*Leaves* from three to five feet long, tapering to a long and very fine point, the greatest breadth is at one or two feet above the sheath, and is there from one to two inches; nerve white, margins hispid.—*Sheaths* bearded round the mouth, and at their insertion on the outside.—*Panicles* large, from one to two feet long, ovate, erect, composed of numerous expanding, subverticilled, compound, woolly branches, when in blossom much expanded, afterwards they become erect, and pressed in on the common rachis, forming a dense cylindrical panicle, and I have repeatedly observed this variation in the form of the panicle at different periods, to run through the whole genus.—*Flowers* in pairs, one sessile, the other pedicelled.—*Calyx* purple, woolly.—*Corol* of the sessile flower three, of the pedicelled two-valved, all are exceedingly delicate, and the third valve of the sessile flower is retuse.

Obs. The seeds or culms are long, strong and straight, and employed by the natives for skreens, and various other economical purposes.

8. S. *sara. R.*

Erect, from eight to twelve feet high. *Leaves* flat with prickly margins. *Panicle* dense, sub-verticilled; *ramifications* decompound. *Flowers* paired, one of the pair sessile. *Corol* three-valved.

Sans. गुंद्रः, Goon*dra*, तेजनकः, Te*junuka*, शरः, Sh*ura.*

Beng. Sh*ur*.

Sara. *Asiat. Res.* iv. 247.

Found in the vicinity of Calcutta but rather rare, whereas S. *spontaneum* (which Sir William Jones mistook for Sara) is very common every where. It is readily distinguished by being a stronger reed, the leaves much larger, with very hispid margins, the ramifications of the panicle decompound, and a corol of three-valves. *Culms* perennial, erect, from six to sixteen feet high, often near the base as thick as the little finger, smooth, remarkably strong.—*Leaves*, the lower ones from four to eight feet long, and narrow; the superior ones shorter, broader, tapering from the base to a most fine point, strong, and rigid; concave above, margins hispid.—*Sheaths* from twelve to eighteen inches long, with a tuft of hair above their mouths on the inside.—*Panicle* dense, when in flower open, when in seed condensed and of a lanceolate shape, from one to three feet long; *branches* decompound, or more; the inferior alternate, the superior sub-verticilled with generally three sharp angles armed with small stiff bristles besides long white silky hairs.—*Flowers* paired, one sessile, the other pedicelled.—*Calyx* two-valved, clothed with long, soft, white, silky hairs.—*Corol* three-valved, fringed.

9. S. *exaltatum. R.*

Culms from ten to sixteen feet high. *Leaves* flat with prickly margins. *Panicles* linear, crowded. *Calyces* woolly. *Corol,* two and three-valved.

Kerpa. *Hort. mal.* 12. 85. *t.* 46. seems to be this, rather than S. *spontaneum*.

Teling. Konda-Kanamoo.

A very high beautiful species, a native of vallies; compare with S. *Arundinaceum*.

Root creeping.—*Culms* many, erect, from ten to sixteen feet high, towards the base perennial, ligneous, inwardly spongy, (not piped,) smooth, round, without any groove, or hollow any where, about an inch in circumference at the thickest part which is near the middle. —*Leaves* alternate, sheathing, remote, remarkably long and narrow, but firm, tapering from the middle towards each end; margins armed with very sharp spines, mouth of the sheaths woolly.—*Panicles* terminal, crowded, when the flowers are on the decay linear, oblong, erect, eighteen inches long, three in diameter, and composed of beautiful cream-coloured, downy, short, sub-erect, ramous, sub-verticelled spiked racemes.—*Rachis* round, striated.—*Flowers* in pairs, one sessile, the other pedicelled, both hermaphrodite.—*Calyx* one-flowered, two-valved, both valves pointed, and covered with much long, cream-coloured soft hair as well as their pedicels, and particularly round their insertion.—*Corol* of the sessile flower three-valved, of the pedicelled flower two-valved, membranaceous, white. I could not observe any *nectary*.—*Stamens*, &c. as in the genus.—*Seed* clavate, smooth, clear whitish yellow.

Obs. Cattle do not eat it.

10. S. *Munja. R.*

Straight, from eight to ten feet high, every part smooth, except the inside of the base of the hispid, margined, long, linear, white-nerved, channelled leaves. *Panicles* large, oblong, spreading; *branches* thereof verticelled, and super-decompound. *Flowers* all hermaphrodite. *Corols* two-valved.

Hind. Munja. *See Asiat. Res.* iv. 248.

A native of the countries about Benares where the natives make ropes of it. In the Botanic Garden it blossoms in November. I never saw it, nor, so far as I can find, is it known so low down as Bengal, otherwise I might have thought it König's S. *bengalense*.

11. S. *canaliculatum*. *R.*

Perennial, erect, (from eight to twelve feet high.) *Leaves* subulate, deeply channelled, very long. *Panicles* linear-oblong, *ramifications* simple, and verticelled; calycine valves with thick callous base and fringed margins, *corol* one-valved.

A most beautiful stately species, a native of Bengal in most thickest, &c. where the soil is rich, flowering in August and September.

Culms about as thick as a common ratan, incrusted with a white substance under the sheaths; filled with pith.—*Leaves* from five to seven feet long, semi-cylindric, no where thicker than a pack thread, the whole being only a perfect channel; their sheaths simply bearded at the mouth.—*Panicle* sub-cylindric, about two feet long, waving with its immense quantity of silver-coloured long, soft hair, to every breeze.—*Flowers* paired, one sessile, the other pedicelled. *Calycine valves* are remarkable on account of their large thick smooth callous base.

ANTHISTIRIA. *Schreb. Gen. N.* 1567.

Polygamous. *Involucre* one-valved, seven-flowered, hermaphrodite, solitary, with a two-valved, one-flowered calyx, and corol of one valve and an awn. *Accessary florets* male or neuter, six; four sessile, resembling a calyx, and two pedicelled, their structure various.

1. A. *ciliata*. *Linn. Sp. Pl. ed. Willd.* iv. 899.

Erect. *Leaves* taper from the ciliate base. *Accessary florets* two or three-valved, neuter, exterior valve of the sessile flower papillose, ciliate near the apex.

A. ciliata. *Gaert. Carp.* ii. *p.* 465. *t.* 75.

An erect annual species, a native of various parts of India, appears about the close of the rains, in October and November, and continues during the cool season.

Culms erect, about two feet high, in our rich soil of Bengal, more slender than a crow quill, smooth, and often coloured.—*Leaves* few, ensiform; broadest at the base and there more ciliate, particularly the small floral leaves.—*Sheaths* much shorter than the joints, smooth, and carinated on the back.—*Ligula* short and chaffy.—*Panicles* sometimes drooping, though in general erect and composed of a few, rather remote, axillary branches.—*Involucres* longer than the flowers, smooth, cuspidate.

HERMAPHRODITE FLOWERS bearded at the base.—*Calyx* two valved; valves hard, obtuse, and somewhat hairy, by age changing from straw colour to dark brown.—*Corol* one small, pellucid valve, and a very strong, long, thick, brown, hairy awn on the opposite side.—*Stigma* three.—*Styles* two, with ochraceous plumose stigmas, which become brown by age.—*Nectarial*, scales large, with crenate, truncate apices.

ACCESSARY FLORETS six, all neuter.—*Calyces* two-valved, those of the sessile four, papillose, ciliate near the apex; those of the upper pedicelled two, smaller and smooth.—*Corols* none, or of one, small, hyaline valve.—*Stamina* none.

2. A. *polystachya. R.*

Culms numerous, sub-erect *Leaves* taper from their sub-ciliate base. *Accessary florets* three-valved, male, exterior valves of the sessile four, papillose, ciliate near the apex.

A native of Bengal, and with the other species which are natives of the same country, appears at the close of the rains, in October and November.

Culms many, annual, erect, or ascending, smooth, from one to two feet high.—*Leaves* about the length of the joints of the culm, taper from the base, and there sometimes sub-ciliate.—*Panicles* slender, erect, or nearly so, composed of a few alternate, axillary branches.—*Involucres* longer than the flowers, smooth, sub-aristate.

HERMAPHRODITE FLOWERS bearded at the base.—*Calyx* two-valved, both hard, a little hairy awn on the opposite side.—*Corol* one-valved with a long, strong, hairy awn on the opposite side.

ACCESSARY FLORETS six, all male.—*Calyces* two-valved, the exterior one of the four sessile florets papillose, ciliate near the apex. —*Corol* of one transparent valve.—*Stamens* three.—*Pistil* none.

3. A. *scandens. R.*

Perennial, ramous, scandent, geniculate, emitting roots from the joints. *Floral leaves* with the base ciliate. *Panicles* copious, shaggy. *Involucres* smooth. *Accessary florets* two-valved, neuter; exterior valve of the sessile four, papillose, ciliate near the apex.

A native of Bengal; flowers during the cool season.

Root perennial.—*Culms* scandent, running to an extent of many feet, smooth, ramous, jointed, and coloured, from the joints numerous roots issue, and enter the ground if not too far removed from it.—*Leaves* ensiform, the floral ones much shorter, and ciliate at the base.—*Panicle* terminal, composed of erect ramifications of numerous, rather remote fascicles of axillary, short pedicelled umbellets, each of which is furnished with a boat shaped spathe or bracte, and consists of four, sessile, neuter, two-valved florets in a verticil; above these are two more, but much smaller, on short pedicels, with one sessile, awned, hermaphrodite one in the centre, on a short scaly receptacle.—*Calyx* two-valved, the exterior valves of the lower four neuter florets bearded, of the hermaphrodite ones boat-shaped, and very obtuse.—*Corol* of the hermaphrodite flower two-valved, with a long, ferruginous, straight arista.

Obs. It differs from *A. ciliata* in being an extensively scandent, perennial plant.

4. A. *heteroclita. R.*

Culms leaning, geniculate. *Leaves* ciliate. *Involucres* five-flowered, two awned hermaphrodite ones, and three ciliate male ones.

This is one of the smallest of the genus, a native of newly made pasture land in the vicinity of Calcutta, where it appears about the close of the rains, and during the cool season.

Culms annual, erect, or leaning, geniculate, smooth, and coloured; from twelve to twenty-four inches high.—*Leaves* ensiform, soft and hairy. *Sheaths* much shorter than the joints.—*Panicles* copious, hirsute, composed of numerous small umbells and these in general of five or more umbellets, the former with a common cymbiform, bristly involucre, and the latter with a small, proper characteristic involucre, equally bristly; all the bristles have a round bulbous base.—*Umbellets* five-flowered; the two hermaphrodite ones awned, and the three male ones pedicelled, and awnless; one of the former is sessile below, with a pedicelled male one by it; the other elevated on a common peduncle, embraced by the other two male ones on their proper pedicels, in all of them the calyx is two-valved, no corol, the arista in the hermaphrodite flowers excepted.

Obs. This species may be immediately distinguished by its umbellets being composed of two hermaphrodite awned florets, and three male ones. In all the rest, which I have met with there is one awned hermaphrodite, and six male, or neuter florets in the umbellet.

5. A. *prostrata. Linn. Sp. Pl. ed. Willd.* iv. 901.

Creeping. *Leaves* scarcely longer than their sheaths. *Involucres* smooth. *Accessary florets* male, with a two-valved hairy calyx and one-valved awnless corol.

A common grass, delighting in rich pasture ground.

Culms creeping, or leaning on other plants, filiform, ramous, winding from joint to joint, round, smooth, from one to three feet long.—*Leaves* below, and on leafy shoots bifarious, small, particularly the floral ones, smooth.—*Sheaths* short, compressed.—*Panicles* thin, leafy.—*Flowers* numerous, collected in little pedicelled, involucred fascicles from the axils of the exterior leaves. *Fascicles*, or *umbellets*, their peduncles are jointed near the apex, and from thence upwards embraced by a long, boat-shaped sheath or involucre. Flowers, of the umbellet seven; one hermaphrodite or female, and six male; four of the male florets on their proper short pedicels surround a common pedicel which elevates the hermaphrodite or female one, and the other two male florets; these last two have also their proper pedicels: all their insertions are surrounded with hair.

HERMAPHRODITE TWO, OR FEMALE THREE FLOWERS. *Calyx* two-valved, pointed.—*Corol* one-valved, a pretty long twisted awn occupying the place of the other.—*Stamens* three, or none.—*Male flowers*, four at the base, and two embracing laterally the hermaphrodite, or female flowers, each of these two has its proper pedicel.—*Calyx* two-valved, lanceolate, hairy.—*Corol* one-valved.—*Stamens* three.

Obs. It often happens that there are two hermaphrodite, and seven male flowers in the fascicle; in that case one of the usual lateral pedicels supports the additional two flowers.

6. A. *cymbaria. R.*

Erect, ramous. *Leaves* many times longer than their sheaths.—*Involucres* smooth.—*Accessary florets* male, with a two-valved, ciliate calyx, and a one-valved, awnless corol.

Andropogon cymbarius. *Linn. mant.* 303. *Sp. Pl. ed. Willd.* iv. 909.

Teling. Pedda Ghella.

This is a large species. I have generally found it amongst bushes on mountains.

Culms many, erect, from three to six feet high, very ramous, smooth, solid, ligneous, joints smooth.—*Leaves* large in proportion to the size of the plant.—*Panicles* linear, leafy, erect, composed of innumerable bracted, fascicles like those of the last described.—*Bractes* boat-shaped, ending in a very long, subulate point, sometimes coloured, and in luxuriant plants fringed with many long rigid hairs.—*Flowers* seven in each fascicle or umbellet as in the last, and disposed exactly in the same manner, only here the four male flowers that surround the base, of the common pedicel are sessile.

HERMAPHRODITE FLOWERS. *Calyx* two-valved.—*Valvelets* nearly equal, obtuse, smooth, of a firm texture.—*Corol* one-valved, the place of the second is occupied by a long twisted awn.—*Stamens* three, &c. as in the genus.

MALE FLOWERS. *Calyx* two-valved, the exterior valve fringed with many strong hairs about the apex.—*Corol* one-valved.—*Stamens* three.

7. A. *arundinacea* R.

Perennial, straight, from six to twelve feet high; *leaves* from four to eight feet long, *accessary florets* male, with both *calyces* and *corols* two-valved; the exterior valvelet of the former shaggy.

A native of Bengal. Flowering time the beginning of the cool season.

Root perennial.—*Culms* straight; jointed, glaucous, between the joints much contracted, round and very smooth, the lower joints of the large plant as thick as the fore finger; every part filled with white spongy pith; height of the whole plant when in flower, from six to sixteen feet. In the Botanic Garden where the soil is rich some plants are full sixteen feet high.—*Leaves* from six to eight feet long, slender, deeply channelled toward the base, recurvate, margins hispid. The radical ones are bifarious, the floral ones alternate, sheaths smooth, compressed, much shorter than the joints of the culm.—*Panicle*, the whole upper half of each culm may be so called, and is composed of numerous, axillary, very ramous branches, bending elegantly to one side, particularly with a breeze of wind.—*Bractes* one to each umbellet of flowers, boat-shaped.—*Umbellets* numerous, long, slender, pedicelled, drooping, each one is composed of four, sessile, awnless accessary male flowers surrounding, and inserted on a short clavate pedicel, which supports an hermaphrodite awned flower, and two pedicelled awnless ones.

MALE FLOWERS. *Calyx, glume*, two-valved, the exterior valve clothed with much diverging fulvous hair, making it quite shaggy.—*Corol, glume*, two-valved, linear, membranaceous.—*Nectary* of two cuneate, truncated scales.—*Stamens* three —*Pistil* none.

HERMAPHRODITE FLOWERS. *Calyx* two-valved, villous.—*Corol* one-valved, a long arista occupies the place of the usual second valve.—*Nectary* and *stamen* as in the male.—*Germ* oblong; *styles* two, villous.

ANDROPOGON. *Schreb. Gen. N.* 1566.

Polygamous. *Flowers* in pairs, hermaphrodite, sessile.—*Calyx* two-valved, one-flowered.—*Corol* for the most part two-valved, and generally awned. *Male*, or *neuter* pedicelled.—*Calyx* as in the hermaphrodite.—*Corol* for the most part two valved, generally awnless.

Sect. 1*st.* *With terminal spikes.*

1. A. *serratus.* *Retz. Obs.* v. 21. *Linn. Sp. Pl. ed. Willd.* iv. 903. *Spikes* solitary, or paired, imbricated with sessile, hermaphrodite, awned flowers on the outside, and two rows of awnless pedicelled male ones on the inside.

Sans. वेणी, Venee, खरा, Kh*u*ra, गरी, G*u*ree, अगरी, *U*g*u*ree, खरागरी, Kh*u*rag*u*ree, गरागरी, G*u*rag*u*ree, देवताडः, Dev*u*tar*a*, जीमूतः, Jeemoot*a*.

Beng. Detar*a*, or Dett*a*.

A native of pasture ground. Flowers during the cold season.

Culms at and near the base creeping, branchy, with only about a foot of their flower-bearing extremities erect, and as thick as a sparrow's quill, round, smooth, joints slightly villous.—*Leaves* small, inside a little hairy, mouth of the sheaths hairy.—*Spike* terminal, generally solitary, though sometimes there are two or even three; when single, long-peduncled, when two, each has a very short, erect pedicel; the spike or spikes are about an inch or an inch and a half long.—*Flowers* in pairs, numerous, imbricated, on one side are two rows of sessile, hermaphrodite ones, and in the other two rows of pedicelled ones; *male rachis*, waved, jointed and hairy. Hermaphrodite flowers sessile.—*Calyces* two-valved, the exterior one broad, compressed, striated, a little hairy, rounded at the apex; *margins* fringed with hairs; *involucre* smooth, glossy, keeled, pointed.—*Corol* of the hermaphrodite flower one-valved; the arista which is pretty long and twisted, occupies the place of a second; that of the male two-valved, the inner one slender and bifid.

Obs. Cattle will eat it, but not greedily.

2. A. *contortus. Linn. Sp. Pl. ed. Willd.* iv. 904.

Monoicous. *Spikes* solitary: *inferior flowers* male or neuter and awnless; *superior flowers* one female and awned, the other male and awnless.

Teling. Yeddi.

Grows in tufts on rich pasture ground.

Culms near the base spreading a little, there compressed and ramous, the rest from one to two feet high, erect, round, and smooth; joints smooth.—*Leaves* below bifarious, middle-sized, upper side scattered with a few hairs, margins and mouths of the sheaths hairy.—*Spikes* always single, terminal, bowing, cylindric; *rachis* jointed, waved, round, below where it is occupied by the male and neuter flowers smooth; above where the female flowers are, it is covered with much, short, brown hair.—*Flowers* numerous, imbricated in pairs; on the lower half of the spike they are all neuter, or male, and awnless, one sessile, one pedicelled; on the upper half the sessile flower is female, and the pedicelled one male.

MALE or neuter FLOWERS. *Calyx* two-valved, one-flowered, exterior valvelet green, and covered with stiff white hair, inner valve membranaceous, and smooth, except the margins, which are slightly ciliate. *Corol* two-valved, membranaceous, margins slightly ciliate. *Stamens* three or none.

FEMALE FLOWERS. *Calyx* one-valved, one-flowered, rigid, embracing the corol on all sides, brown, a little hairy.—*Corol* one-valved, small, the place of the other is supplied by a very long, twisted arista.

3. A. *tenellus. R.*

Filiform, *spikes* paired, *flowers* paired, the sessile ones herma phrodite, and awned; the pedicelled ones neuter, and awnless; exterior valves of both calyces oblong, striate, hairy, and three-toothed, neuter corol of one minute valve.

A most delicate tall grass, a native of Bengal, where it appears in hedges, &c. during the cool season.

Culms from two to three feet long; at the base resting on the ground, ramous, scarcely so thick as a pack thread; *joints* villous.—*Leaves* slender, a few hairs toward the base.—*Spikes* paired, erect, short-pedicelled; *outside* imbricated with two rows of sessile, awned, hermaphrodite flowers; *inside* with rows of pedicelled, awnless, neuter ones.—*Rachis* jointed and hairy.—*Calyces* two-valved; *exterior* oblong, three-toothed, striated, hairy; *interior* acute, three-nerved, that of the neuter flower shorter.—*Corol* of the hermaphrodite flower one-valved, with a long brown awn on the opposite, or inner side, that of the neuter flower single, and very minute.

4. A. *binatus. Linn. Sp. Pl. ed. Willd.* iv. 918. *Retz. Obs.* v. 24.

Slender, towards the base creeping; *spikes* paired, incurved, *flowers* paired, the sessile ones hermaphrodite, and awned, the pedicelled ones male, with an awnless, two-valved corol.

A native of Bengal, appears during the cool season.

Culms very slender, toward the base ramous, and creeping; *joints* with a scanty short beard.—*Leaves* slender, somewhat hairy toward the base.—*Spikes* paired, on a long filiform peduncle, short pedicelled, somewhat incurvate; *outside* imbricated with two rows of sessile, awned, hermaphrodite flowers, *inside* with two very distinct (for they are separated by a deep groove) rows of pedicelled, awnless, male ones. *Rachis* jointed, and hairy.—*Calyces* of both two-valved; *exterior* oblong, hairy, ciliate, five to seven-nerved, edges thin, apex of the male tridentate, of the hermaphrodite obtuse. *Inner valves* acute, three-nerved.—*Corol* of the hermaphrodite flower of one hyaline valve on the outside and an arista for the other; of the male one of two very unequal hyaline valves.—*Styles* two ; *stigmas* purple, villous.—*Seed* oblong, smooth.

5. A. *conjugatus. R.*

Straight, very ramous, filiform. *Spikes* conjugate, both flowers hermaphrodite, and both awned ; *exterior* valves of the calyces lanceolar, and copiously ciliate.

A native of Bengal, and the smallest of the genus I have yet met with. Flowers during the cool season.

Root a few very slender, brown fibres.—*Culms* in tufts, and very ramous, filiform, erect, smooth, even the joints ; from six to twelve inches high.—*Leaves* minute, a few hairs near the base.—*Spikes* always exactly paired as in most species of *Ischæmum*, erect, flat on the inside, where they touch each other. *Rachis* winding, and clothed with long, silver-coloured hairs, length about an inch.—*Flowers* in pairs, one sessile, one pedicelled, both hermaphrodite, and both awned.—*Calyces* two-valved, one flowered ; *exterior* valve in both lanceolate, with the margins copiously ciliate.—*Corol* two-valved ; *valves* transparent, and brown, that (the inner) which gives support to the arista, bifid.—*Stamens* three. *Styles* two. *Stigmas* long and plumose.

6. A. *filiformis. R.*

Filiform, annual, erect. *Spikes* paired. *Flowers* all sessile, solitary ; hermaphrodite, diandrous, with a corol of one valve, which ends in a long awn.

This is the most delicate species I have yet met with ; appears in small tufts, on pasture land near Calcutta, in the months of October, November, and December.

Culms numerous, filiform, smooth, erect, from nine to twelve inches high.—*Leaves* small, slender, and mooth.—*Spikes* two, terminal, filiform, scarcely two inches long.—*Flowers* all hermaphrodite, solitary, sessile, no pedicelled flower being found, no pubescence about the calyx, no rachis.—*Calyx* two-valved, one-flowered.—*Corol* of one large, awned valve.—*Stamina* two, one of the anthers larger.—*Germ* ovate.—*Styles* two.

7. A. *tristachyos. R.*

Spikes three, both flowers hermaphrodite, and both awned. *Calyces* lanceolate, and ciliate.

A delicate, beautiful species, found on newly laid down pasture ground in the vicinity of Calcutta, and in flower in October, November, and December.

Annual, with numerous, nearly straight, erect, smooth, slender, pithy culms, of from one to two feet in height.—*Leaves* taper from the base, most acute, about as long as the joints of the culms; a few long soft hairs on the inside near the base.—*Sheaths* smooth, shorter than the joints.—*Spikes* two, rarely three, and I have not seen four, one sessile the other two a little more elevated on a common pedicel, expanding; from two to three inches long. *Rachis* jointed, winding, and hairy.—*Flowers* in pairs, one sessile, the other pedicelled, both hermaphrodite, and both awned.—*Calyces* two-valved; *valvelets* lanceolate, ciliate; backs smooth, and brown.—*Corol* of two, small, pellucid, brownish valves, the inner of each, ends in a long, brown arista, proceeding from the bifid apex of the valvelet.

8. A. *tridentatus. R.*

Spikes paired, hirsute, both flowers hermaphrodite, and awned. *Calyces* shaggy, exterior valves cuneate, truncate, tridentate. *Corols* four-valved.

A delicate, rather tall species: a native of Coromandel.

Culms straight.—*Leaves* small.—*Spikes* terminal, paired, sessile, lanceolate, shaggy.—*Flowers* in pairs; one sessile, the other short-pedicelled, both hermaphrodite, and both awned.—*Calyx*, exterior valves in both flowers cuneiform, with apices truncate, tridentate, back striated; middle part, as well as the margins from thence down, and the base bearded with fulvous hairs; inner valve longer than the exterior, ending in a pretty long arista, margins amply fringed with long fulvous hairs.—*Corol* in both flowers four-valved; the exterior one on the inside bifid, and awned.

9. A. *lanceolatus. R.*

Leaves lanceolate, with cordate stem-clasping base, bristle-ciliate. *Spikes* tern, peduncled. *Sessile* or hermaphrodite flower awned, with the margin of the exterior valve of the calyx acutely pectinate, serrate.

A native of Coromandel, and very much like *Panicum setigerum.*

Culms scandent, or creeping, emitting roots from the joints, seems perennial.—*Leaves* lanceolate, base cordate, and stem-clasping, apex acute, smooth on both sides, but the edges from the middle down, bristle-ciliate.—*Spikes* terminal, two, three, or four, peduncled. *Rachis* and *pedicel* of the male flower somewhat woolly.—*Calyces* lanceolate; *exterior valve,* and chiefly that of the hermaphrodite sessile flower with the margins minutely and acutely pectinate-serrate.—*Corol* in both two-valved, with an arista in the hermaphrodite one.

10. A. *scandens. R.*

Climbing, *spikes* from three to six, digitate; *calyx* lanceolate, male corols one-valved, and awnless, hermaphrodite ones with an one-valved corol and an awn.

Grows commonly in hedges, and on that account scandent; appears and flowers during the rains.

Culms long, ramous, creeping, or climbing over bushes, &c. emitting long roots from the joints; smooth and deeply grooved on one side; flower-bearing extremities erect, and about one or two feet long; *joints* woolly.—*Leaves* sometimes a little hairy on the upper side, mouths of the sheaths membrane-stipuled, and hairy.—*Spikes* generally from three to six, terminal, short pedicelled, sub-panicled flowers in approximate pairs; one hermaphrodite, and sessile, the other male, and pedicelled. In both the calyx is of two lanceolate, somewhat hairy valves. In the hermaphrodite one the corol is of one valve, and the place of the second valve is occupied by a long twisted arista. In the male one it is two-valved.

Obs. It is a coarse grass. Cattle are not fond of it.

11. A. *pertusus. Linn. Sp. Pl. ed. Willd.* iv. 922.

Near the root creeping. *Spikes* from four to eight, digitate, pedicelled, exterior valve of the sessile, awned, hermaphrodite calyx with a pit on the back; male or neuter corol one-valved

Holcus pertusus. *Kön. Linn. Mant.* 301.

Is found on old pasture ground, which is generally shaded by trees.

Culms creeping near the base, &c. as in the last two species; joints bearded.—*Leaves* small, ciliate; *mouths* of the sheaths bearded.—*Spikes* from four to eight, in other respects exactly as in the last species.—*Flowers,* the specific mark is to be found in the pit of the exterior valves of the calyx of the hermaphrodite flowers, as in A. *strictus,* and A. *punctatus,* to be described; an arista occupies the place of the second valve of the corol; the valvelets of the calyx are both awnless.

Obs. I should doubt this being König's *Holcus pertusus* if he had not pointed it out to me himself.

12. A. *Bladhii. Linn. Sp. Pl. ed. Willd.* iv. 920.

Sub-scandent; joints pubescent. *Spikes* from three to six, sub-panicled. *Calyces* lanceolate; male corols two-valved; hermaphrodite ones one-valved with an arista; *flowers* paired.

Beng. Loari.

A native of hedges, and road sides, but chiefly of old pasture ground.

Culms as in the last described species; joints much bearded.—*Leaves* the same.—*Spikes* from four to eight, short-pedicelled, inserted round the last half in each of the culms; in other respects they are as in the last described species.—The flowers resemble those of that species, but are a little more remote.—*Calyx* and *Corol* the same as in A. *scandens.*

13. A. *Ischæmum. Linn. Sp. Pl. ed. Willd.* iv. 921.

Spikes from six to eight, fascicled, peduncled ; *rachis* hairy. *Flowers* lanceolate ; hermaphrodite, sessile, and awned ; male pedicelled and awnless ; *calyces* acute ; exterior valve hairy.

A native of Coromandel, a stout, erect species, with a branch from the axil of each of the superior smooth short leaves, having much the appearance and habit of *Bladhii* and *pertusus* ; from the former it differs most conspicuously in the valves of the calyces being acute, whereas in *Bladhii* they are broad and rounded ; and from *pertusus* in the want of the pit on the back of the exterior valve of the hermaphrodite flower.

Sect. 2d. With long peduncled, axillary spikes.

14. A. *monandrus. R.*

Erect, filiform, with erect, filiform, branches at the top. *Spikes* solitary, flowers hermaphrodite, and female ; *exterior* valvelet of the calyces and corols of both awned, hermaphrodite, monandrous.

A native of mountainous countries.

Culms many, ramous, filiform, straight ; about a foot high, round and smooth.—*Leaves* small, base and mouths of the sheaths hairy.—*Spikes* generally solitary, terminal, and from the exterior axils, long-peduncled, straight, about an inch long.—*Peduncles,* filiform, jointed, and spathed at the middle. *Rachis* jointed, winding, hairy —*Flowers* numerous, imbricated in pairs ; one, the hermaphrodite, sessile and the other, female, pedicelled.

HERMAPHRODITE FLOWERS sessile. *Calyx* two-valved, nearly equal, exterior valvelet ending in a long slender arista, the inner one with its apex obtuse, and bearded.—*Corol,* exterior valve ending in a long slender awn.—*Stamen* one.—*Styles* two, &c. as in the genus.

FEMALE FLOWERS pedicelled. *Calyx* and *corol* as in the hermaphrodite ones ; only the inner valve of the calyx is pointed, and bearded.

15. A. *strictus. R.*

Straight, ramous at the top; base of the leaves ciliate. *Spikes* fascicled on their own proper pedicels; exterior valvelets of both calyces pitted; inner glume of the hermaphrodite corol awned.

A native of mountains, and their vicinity.

Culms straight, ramous, about two feet high, and as thick as a sparrow's quill, round and smooth; joints bearded.—*Leaves* slender, a little hairy, and ciliate near the base; mouths of the sheaths stipuled.—*Spikes* from three to six, long-pedicelled, and collected on a common peduncle, one of which is terminal, and one or two more from the next one or two exterior axils; very slender, perfectly erect, and about an inch long. *Rachis* as in the former species.—*Pedicels* most fine, straight, two inches long, jointed near the base, where a fine slender spathe begins.—*Flowers* in pairs; one hermaphrodite and sessile, the other small and pedicelled.

HERMAPHRODITE FLOWERS sessile. *Calyx* as in the genus, except that the exterior large valvelet has a large pit in it, as in *Andropogon pertusus.*—*Corol,* inner valve slender, ending in a long twisted awn.

MALE FLOWER pedicelled. *Calyx* one-valved, and pitted as in the Hermaphrodite flower.—*Corol* two-valved, awnless.

16. A. *polystachyos. R.*

Straight, *spikes* numerous, axillary, simple, each elevated on a long jointed, sheathed peduncle. *Flowers,* inferior pairs male and awnless; one of the superior ones is female and amply awned, the other male and awnless.

A native of the Peninsula of India. The flowers greatly resemble those of A. *contortus,* but the habit of the grasses is very different.

Culms straight, ramous.—*Leaves* ensiform, straight, acute, ciliate at the base.—*Spikes* simple, many together on the axil of the leaves, each elevated high on its long, filiform, jointed, sheathed peduncle, length from one to two inches.—*Flowers* paired, those on the lower two-thirds of the spike are all male, and awnless, and, as usual, one of them sessile, the other short-pedicelled; *superior pairs*, the sessile one of each of these is a slender female, amply bearded at the base with much dark brown hair, which is the more conspicuous as every other part of the spike is smooth; the other short-pedicelled, male, and awnless, like the inferior pairs.

Male Flowers. *Calyx* two-valved.—*Corol* two-valved.—*Stamina* three.

Female Flowers. *Calyx* of one linear, striated, hard, dark-brown valve.—*Corol* one-valved with a very long, twisted awn on the other side.—*Stamina* none. *Styles* two.

Sect. 3d. Panicles naked, and terminal.

17. A. *aciculatus*. *Linn. Sp. Pl. ed. Willd.* iv. 906.

Creeping. *Leaves* shorter than their sheaths. *Panicle* linear-oblong, erect, *ramifications* three-flowered, one hermaphrodite, sessile, and two male, peduncled.

Gramen aciculatum. *Rumph. Amb.* 6. *t.* 5. *f.* 1. good.

Raphis trivalvis. Lour. *Cochinch.* 676.

Kudira-pullu. *Rheed. Mal.* 12. *p.* 97. *t.* 43.

Sans. संखिनी, Sh*u*nkh*i*nee, चोरपुस्पी, Chor*a*-pooshpee, केषिनी, Kesh*i*nee.

Beng. Chor*a*-kanta.

Grows on barren, moist, pasture ground; its seeds are exceedingly troublesome to those who walk where it grows, as they stick in the stockings and produce a disagreeable itching.

Culms straight, simple, from one to two feet high, round and smooth.—*Leaves* small, waved, with the margins prickle-ciliate.—*Panicle* terminal, erect, linear-oblong, from two to three inches long. *Ramifications* (or rather peduncles) verticelled, simple, three-flowered, nearly erect; apex armed with short stiff hairs.—*Flowers*, one sessile, hermaphrodite, and two pedicelled male ones on each peduncle.

Hermaphrodite Flower. *Calyx* two-valved, nearly equal; the inner one ends in a short arista, and has the keel scabrous; base involved in brown short hair.—*Corol* two-valved, the inner one ending in a short rigid arista.—*Stamens* three, &c. as in the genus.

MALE FLOWERS. *Calyx* and *Corol* as in the male, except that there is no *arista* to the corol.

Obs. It is of a very coarse nature. I never found it touched by cattle.

18. A. *verticillatus.* *R.*

Erect. *Leaves* long, and very hairy; *Panicle* conical, *ramifications* verticelled, three-flowered; one hermaphrodite, sessile, with two awns; and two-pedicelled, male, with one awn each.

A native of mountains, flowering about the end of the wet season.

Root woody, perennial.—*Culms* erect, simple, very leafy, smooth, round, towards the base solid, as thick as a small goose quill; from four to six feet high.—*Leaves* approximated, eighteen inches long, and three quarters of an inch broad; margins set with minute spines, and particularly hairy on the inside near the base.—*Panicles* erect, conical, lax, from eight to ten inches long, composed of many verticelled, most simple, filiform, waved, bowing, three-flowered branches, as in A. *aciculatus.*

HERMAPHRODITE FLOWERS sessile. *Calyx*, glume one-flowered, two-valved, inner valve awned, both hairy, having the base surrounded with much brown hair.—*Corol* one-valved, a long twisted awn occupying the place of the other.—*Stamens* three.—*Style* yellow.

MALE FLOWERS pedicelled, one on each side the hermaphrodite one; pedicels covered with brown hairs.—*Calyx*, exterior valve a little hairy, and awned, base involved in short brown hair.—*Corol* two-valved.—*Stamens* three.

19. A. *saccharoides.* *Linn. Sp. Pl. ed. Willd.* iv. 912.

Sub-erect. *Leaves* remarkably long. *Panicle* oblong, composed of simple, alternate, filiform, short-pedicelled, erect, patent branches; neuter florets two-valved and pedicelled.

A large, leafy, sub-erect species; a native of the mountainous parts of the northern Circars.

I am not possessed of any other description of Swartz's A. *saccharoides* than the specific character, as it stands in Willdenow's edition of the species, which agrees with my plant; I therefore conclude they are the same.

20. A. *Ischæmum. Linn. Sp. Pl. ed. Willd.* iv. 921.

Spikes from six to eight, fascicled, peduncled; *rachis* hairy. *Flowers* lanceolate, hermaphrodite, sessile and awned; *male* pedicelled and awnless; *calyces* acute, exterior valve hairy.

A native of Coromandel. It is a stout erect species, with a branch from the axil of each of the superior smooth short leaves; it has much the appearance and habit of *Bladhii*, and *pertusus*; from the former it differs most conspicuously in the valves of the calyces being acute, whereas in *Bladhii* they are broad and rounded; and from *pertusus* in the want of the pit on the back of the exterior valve of the hermaphrodite flowers.

21. A. *punctatus. R.*

Erect, simple. *Leaves* large, long, and numerous. *Panicle* of numerous, simple, secund ramifications. *Exterior* valves of the calyces of both hermaphrodite and neuter flowers pitted.

This is a mountain grass.

Culms from two to four feet high, without branches, below bent towards the earth for half a foot, often erect, entirely surrounded with the sheaths of the leaves; not piped, but replete with spongy pith.—*Leaves* numerous, large, with some long white hairs scattered over them near the base; mouths of the sheaths stipuled, bearded, and woolly.—*Panicle* erect, oblong, from four to five inches high, composed of many, erect, short-peduncled, filiform, secund, spiked racemes; *rachis* and flowers exactly as in A. *fascicularis*; except that here the outer valve of the calyx of both flowers has a pit in the middle.—The *corol* of the hermaphrodite flower wants the inner valve, in its place is a pretty long twisted awn.

22. A. *fascicularis. R.*

Erect, ramous. *Panicle* composed of fascicled, simple, or divided branches. *Flowers* paired; hermaphrodite ones awned; neuter ones awnless and without a corol.

Teling. Kunda-jeri.

A native of mountains.

Culms erect, jointed, smooth, rather thicker than a crow's quill, from three to five feet high, sulcated, mostly naked, rush-like, not piped.—*Leaves* from three to four, small, remote, retrorso-scabrous; mouths of the sheath hairy; from each sheath springs a short panicle-bearing branch.—*Panicle* terminal, six inches long, erect, fascicled; *fascicles* remote, alternate, composed of several erect, filiform, often ramous, secund spikes, or spiked-racemes.—*Flowers* in pairs; the hermaphrodite ones sessile; the neuter peduncled. *Common rachis* round, smooth; *partial one* undulated, jointed, compressed, woolly.

HERMAPHRODITE FLOWER. *Calyx,* glume one-flowered, two-valved, smooth, awnless, base surrounded with wool.—*Corol, exterior valve* the length of the calyx; *inner* short, bifid, with a long twisted awn from the fissure. *Neuter floret* small, its pedicel woolly, length of the hermaphrodite flower.—*Calyx* two-valved.—*Corol* none.

23. A. *muricatus. Retz. Obs.* iii. *N.* 95. and v. 20.

Erect in all situations. *Leaves* sub-bifarious, long, narrow, and rigid. *Panicle* verticelled, *ramifications* simple, expanding, hermaphrodite and male flowers awnless, with muricate calyces, and corols of three valves.

Sans. वीरणं, Veer*u*n*u*ng, वीरतरं, Veer*u*t*u*r*u*ng.

Beng. Bena, the name of the plant, and *Khuskhus* the root.

Tam. Woetiwear.

Virana. *Asiat Res. vol.* iv. *p.* 306.

It is very common in every part of the Coast and Bengal where it meets with a low, moist, rich soil, especially on the banks of water courses, &c.

Root of many perennial, long, spongy, brown fibres.—*Culms* many, smooth, simple, rigidly erect, a little compressed at the base, jointed, smooth, inwardly spongy; from four to six feet high and as thick as a goose quill.—*Leaves* near the base bifarious, narrow, erect, from two to three feet long, pretty smooth, rigid.—*Panicle* from six to twelve inches long, conical, composed of numerous spreading, short-peduncled, simple, linear, verticelled *spiked-racemes*. *Ruchis* of the *spiked-racemes* jointed, winding, naked.—*Flowers* paired, awnless, one hermaphrodite, sessile; the other male, and pedicelled; insertion nearly naked.

HERMAPHRODITE FLOWER. *Calyx* two-valved, nearly equal, muricated.—*Corol* three-valved, membranaceous.—*Nectary*, two obcordate scales embracing the germ and insertions of the filaments.—*Stamens* three.—*Stigmas* feathery.

MALE FLOWER. *Calyx* as in the hermaphrodite.—*Corol* two-valved. *Nectary* as above.—*Stamens* three.

Obs. The roots, when dry, and then gently moistened, emit a pleasant kind of fragrance; they are employed to make large fans, commonly called *Vissarees;* and also to make screens which are placed before windows and doors, which being kept moist during the hot winds render the air that passes through them, both cool and fragrant. See Observation on *Aristida setacea*. The grass is often employed for thatch.

Dr. König was too accurate a Botanist to describe this very conspicuous plant under two names, viz. A. *squarrosus*. Linn. Sp. Pl. Willd. iv. 908, and A. *muricatus;* the former is evidently a very different species, and found by him in Ceylon, floating on pools of water; whereas A. *muricatus* (which I formerly called *aromaticus)* is always rigidly erect, whether growing in water, or on dry land. Its root is delightfully aromatic, as mentioned by König himself, and by Sir William Jones, particularly when moistened with water.

I am inclined to think König's *Andropogon squarrosus* is my *Panicum paludosum;* a species with thick spongy culms which is generally found swimming on pools of sweet water.

24. A. *glaber*. R.

Erect, ramous, smooth. *Panicle* ovate, verticelled; *ramification* simple, two or three-cleft. *Flowers* paired, the hermaphrodite ones awned, with the calyx pitted; the male ones neither pitted nor awned, with a two-valved corol.

Beng. Gundha goorana.

Found thinly scattered on rather elevated spots over Bengal.

Root perennial.—*Culms* sub-erect, very ramous, smooth, from three to four feet high.—*Leaves* glossy, smooth in every part.—*Panicles* ovate, verticelled; *ramifications* simple, and two-cleft, or even three-cleft sometimes.—*Flowers* paired; the *hermaphrodite* ones sessile and awned; the *male* ones pedicelled, and awnless.

HERMAPHRODITE FLOWERS. *Calyx* two-valved, smooth, purple-coloured; exterior valve pitted on the back.—*Corol* one-valved, a long awn occupying the place of the second.

MALE FLOWERS nearly as large as the female.—*Calyx* not pitted. —*Corol* two-valved.

25. A. *montanus*. R.

Shrubby near the base, creeping. *Leaves* long and slender. *Panicle* sub-verticelled; *ramifications* ramous; *flowers* paired, the hermaphrodite ones awned, the neuter awnless, with a corol of one valve.

Teling. Auroots hangali.

A native of mountainous countries.

Culms many, below resting on the ground and rooting, round, smooth, from three to four feet high, and as thick as a large crow-quill.—*Leaves* long and narrow, a little hairy on the upper side near the base only; mouths of the sheaths bearded.—*Panicle* terminal, conical, from four to eight inches long, composed of many, ramous, sub-verticilled spikes, or rather spiked-racemes; *rachis* winding, jointed, a little hairy.—*Flowers* as usual, in pairs, an hermaphrodite one sessile; and a neuter one pedicelled.

HERMAPHRODITE FLOWERS. *Calyx* two-valved, the exterior

one a little hairy.—*Corol* one-valved, a twisted awn two or three times the length of the corol supplies the place of the other.—*Neuter* flowers like the hermaphrodite ones, but without arista.

26. A. *bicolor. R.*

Erect. *Panicle* diffuse, with expanding, decompound and superdecompound branches; *hermaphrodite* calyces with smooth backs, and fringed margins; *corol* three-valved, and awned, calyx of the *neuter flower* hairy with a two-valved corol.

Holcus bicolor. *Linn. Sp. Pl. ed. Willd.* iv. 929.

Beng. Kal*a*-deodhan.

Hind. Kal*a*-jooar.

Teling. M*u*ka-janoo, and M*u*ka janoolar the grain.

I have only found it in a cultivated state.

Culms generally single, straight, round, smooth, from four to ten feet high, and as thick as the finger or thumb, mostly covered by the sheath of the leaves.—*Leaves* from one to three feet long, and from one to three inches broad, smooth; *mouths* of the sheaths bearded.—*Panicle* erect, oval, dense, composed of many ramous, expanding, sub-verticelled branches, the various ramifications angular and very hispid when the finger is drawn backwards over them, their flower-bearing extremities hairy.—*Flowers* paired, and, as in the genus, the sessile one is hermaphrodite and the pedicelled one neuter.

Hermaphrodite Flowers sessile. *Calyx, valvelets* nearly equal, with the back smooth, and as they ripen becoming of a shining black, with many white hairs round the margins (ciliated.)—*Corol* three-valved, fringed, the second or solitary valve ending in a short twisted arista.—*Seeds* oval, brown, smooth, free, with an impression on the back near the base.

Neuter Flowers pedicelled. *Calyx* striated, hairy.—*Corol* two-valved, awnless.—*Stamens* none.

Obs. This and the following species A. *Sorghum* seem to me to possess more of the character of *Andropogon* than of *Holcus.* I have therefore changed their places from the latter to the former.

27. A. *Sorghum. R.*

Erect. *Panicle* contracted, dense; hermaphrodite calyces hairy. *Corols* three-valved, awned.

Holcus Sorghum. *Linn. Sp. Pl. ed. Willd.* iv. 929.

Beng. and *Hind.* Jooar.

Teling. Tella, or Konda (white) jan*oo* the name of the plant, and Tella-jan*oo*lar the grain.

Obs. This also I have only found in a cultivated state. It differs from the former.

1*st.* In having the panicle contracted, very dense, and less regularly verticelled.

2*d.* In having the glumes of the calyx of a whitish green colour and very woolly.

3*d.* In having three-valves to the neuter flowers.

There are two other sorts, which I take to be only varieties of the last; their names in Telinga are *Pacha* (green) *Janoo* and *Yerra* (red) *Janoo.*

These last two with their varieties are much cultivated on a rather elevated, good soil. Seed time October; and harvest time January. The produce in good soil is often upwards of an hundred-fold. The grain is much used for food. It is probable that through the whole of Southern Asia as many of the inhabitants live on the various kinds of dry or small grain, as upon rice, and they are reckoned fully as wholesome as that is.

All kinds of cattle are remarkably fond of the straw notwithstanding its thickness, and solidity.

28. A. *cernuum. R.*

Erect, from five to fifteen feet high; lower half with verticels of roots from the joints. *Panicles* oval, with numerous, long, compound, cernuous branches; *glumes* villous and fringed; *corol* three-valved and fringed, the inner scarce awned.

Holcus cernuus. *Linn. Sp. Pl. ed. Willd.* iv. 930.

This species, (or variety of *Sorghum;*) is the kind cultivated by the inhabitants of the *Munipoora,* Koonkee, and other mountainous districts immediately east of Bengal. The grain is milk white; some of it was sown in the Botanic Garden, during the early part of the rainy season of 1812, and in November the plants were from ten to fifteen feet high, several ramous stems arising from the same root, or grain of seed; all the joints of the lower half of the original stems which are as thick as a slender walking cane, throw out copiously verticels of roots from all the joint, the lower ones near the soil enter it, and give additional support and nourishment to the plants, which are of two, or more years duration, if suffered to remain; the leaves are from twenty to forty inches long, by two or three broad, soft and smooth, the rib white on both sides. Panicles large, oval on the more slender branches, and in those, while in flower erect; but in such as terminate the primary stems the form is obscure, from the drooping habit of their branches. The flowers agree with those of *Sorghum*, except that the awn is so small and short as to be hid within the glume of the calyx; the neuter flowers are very minute, and consist of only one or two slender, villous, calycine glumes. The grain of this plant is the staff of life of those wild savage mountaineers, who inhabit the above mentioned countries, where it is one of the few articles cultivated by them. Cattle are fond of the straw, or rather canes.

29. A. *sacharatus*. *R.*

Erect. *Panicles* verticelled, with ramifications drooping. *Calyx* hairy, corols awnless; hermaphrodite three-valved, neuter one-valved.

Holcus saccharatus. *Linn. Sp. Pl. ed. Willd.* iv. 930.

Hind. and *Beng.* Deo-dhan.

Is much cultivated over various parts of India during the rainy and cold seasons upon land which is too high for rice.

Root ramous.—*Culms* erect, round, jointed, sometimes ramous, lower joints radicating, from eight to twelve feet high, three or four inches in circumference near the base.—*Leaves* bifarious, sheathing sword-shaped, margins hispid, the rest smooth, with a white nerve on the upper side, three feet long, and three or four inches broad. *Sheath* as long as the panicle is when young, afterwards the whole is completely bent down from the weight of the seed.—*Flowers* paired, one hermaphrodite, sessile; the other neuter, and pedicelled.

HERMAPHRODITE FLOWERS sessile. *Calyx* one-flowered, two-valved, both valves hairy.—*Corol* three-valved, the solitary valve daggered; dagger, shorter than the calyx.—*Seed* oval, white.

NEUTER FLOWERS pedicelled. *Calyx* two-valved, hairy.—*Corol* one-valved, awnless.—*Stamens* none.

Obs. The only circumstance that renders me uncertain whether this is *Holcus saccharatus* of Linneus is the total want of the arista in the hermaphrodite flowers; in other respects they agree.

30. A. *laxus. Linn. Sp. Pl. ed. Willd.* iv. 907.

Erect, from one to fifteen feet high. *Panicle* ovate, lax; *ramifications* sub-verticelled, compound and decompound. *Calyx* hairy; *corols* fringed, those of the hermaphrodite three-valved, and awned.

Beng. Kula-moocha.

Teling. Gaddi-janoo.

Grows in hedges, on banks of water courses, and on land that has lately been in cultivation.

Root somewhat woody, perennial.—*Culm* near the base bending to the earth, and there rooting at the joints, the rest erect; the whole is of a firm ligneous texture, inwardly spongy, jointed, round and smooth; below naked, above covered by the sheaths of the leaves. In hedges on the banks of the rivers, &c. where the soil is good, it is often from ten to fifteen feet high; while on a barren dry soil, not more than one or two.—*Leaves* long, smooth, except the margins, which are scabrous.—*Panicle* very large, oval, often bowing, rather thin, composed of sub-verticelled, ramous, spreading branches, with woolly insertions; the branchlets of the branches are composed of four or six pedicelled, spiked racemes. *Rachis* of the spikes jointed, &c.—*Flowers* paired, &c. as in the other species.

HERMAPHRODITE FLOWERS sessile. *Calyx* two-valved; *valve-*

lets equal, hairy.—*Corol* three-valved, membranaceous, margins fringed with hairs; the second one retuse, with a pretty long twisted arista; the inner one very small.—*Nectary* as in the last species.

MALE FLOWERS pedicelled. *Calyx* as in the hermaphrodite.—*Corol* two-valved.

31. A. *miliaceus. R.*

Erect, from six to ten feet high. *Panicle* ample, lax; *ramifications* verticelled, expanding, or drooping. *Corol* three-valved, in both hermaphrodite and male flowers.

The seeds of this most beautiful stately grass were sent me from Lucknow, by the late Gen. Claude Martin, under the name of hill grass. It being a native of the mountains north of Oude. It blossoms in this garden during the latter part of the rains.

Culms erect, ramous, smooth, the largest of them nearly as thick as the little finger, general height (in the rich soil of the Botanic Garden) about ten feet.--*Leaves* ensiform-lanceolate, soft, and smooth, except the margins, which are armed with minute sharp bristles, pointing forward, general length from two to three feet, and from one to two inches broad near the base, where broadest.—*Sheaths* smooth, except at the mouths, and there bearded on both sides.—*Panicle* very large, erect, with long, slender, smooth, elegantly drooping, verticelled, compound, and decompound ramification.—*Flowers* in pairs, as in the genus; one hermaphrodite, and sessile; the other male, and pedicelled.—*Calyx* two-valved, one-flowered, oval, rather obtuse, and hairy, but without the characteristic woolly, or hairy involucre round the base of the flowers.—*Corol* in both sorts three-valved, membranaceous, and fringed, sometimes the hermaphrodite flower is awned.—*Nectary* two, fleshy, broad, truncated, fringed scales.

Obs. In habit this beautiful species comes near my *Andropogon tumbackianus,* from the Cape of Good Hope, but is more elegant in the panicle.

Sect. 4th. Panicles foliaceous.

32. A. *pumilus.* R.

Erect; twelve inches high; *panicles* composed of numerous axillary, and terminal, conjugate spikes, on long, jointed, sheathed peduncles. *Calyx* of the sessile hermaphrodite, flower awned, cuspidate.

A native of Coromandel, and one of the smallest of the genus.

Culms ramous, erect, smooth.—*Leaves* rather small, particularly the floral ones, which are little more than large sheaths.—*Panicle* composed of numerous, axillary, and terminal, conjugate, hirsute, secund spikes, elevated on slender, jointed peduncles, embraced by many delicate, chaffy bractes at the base, and by a sheath from the joint upwards.—*Flowers* in pairs on the joints of the hairy rachis, one sessile and hermaphrodite, the other peduncled and male.—*Calyx* two-valved, that of the hermaphrodite flower cuspidate.—*Corol* one-valved, an arista occupies the place of a second in the hermaphrodite flower.

33. A. *parviflorus.* R.

Ascending. *Leaves* linear. *Panicle* leafy, thin; *spikelets* solitary, with proper, jointed, leafy spathes; both flowers awned, the pedicelled one a rudiment only.

A very beautiful most delicate species, a native of pasture land up amongst the Circar mountains and also of the mountains themselves.

Culms very branchy, ascending, filiform, about a foot, or a foot and a half high.—*Leaves* numerous, small, linear, rather obtuse, mouths of the sheaths stipuled.—*Spikes* axillary, minute, peduncled, generally one or two, rarely three. *Rachis* jointed and waved as in the other species, but here it is smooth, except at the insertions of the flowers.—*Peduncles* most slender, jointed at the middle, from whence a small spathe ascends, which generally hides a part of its spike.—*Flowers* in pairs; one hermaphrodite and sessile, the other merely the rudiment of a floret on a long clubbed pedicel.

34. A. *Schœnanthus. Linn. Sp. Pl. ed. Willd.* iv. 915.

Perennial, erect. *Panicle* sub-secund, linear, leafy; *spikelets* thereof paired, on a common, spathed pedicel, with proper pedicels and spathes; *flowers* of both sorts awnless, the male corol one-valved.

Sans. मालातृणकं, Malatr*i*n*u*k*u*ng, भूस्तृणं, Bhoostr*i*n*u*ng.

Beng. G*u*ndh*a*-ben*a*.

Ramacciam. *Rheed. Hort. mal.* 12. *p.* 137. *t.* 72.

Schœnanthum amboinicum. *Rumph. amb.* 5. *p.* 181. *t.* 72. *f.* 2.

Schœnanthus, or Juncus odoratus of the *Materia Medica.*

On the Coast I have only found this elegant valuable species in a state of cultivation, few gardens being without it; I have but twice found it in flower and both times during the rainy season.*

Root perennial, young propagating-shoots issue from the axils of the leaves that surround a short, sub-ligneous leaf-bearing culm.—*Culms* from five to seven feet high, erect, simple, smooth, about as thick as a goose quill.—*Leaves* many, near the root bifarious, few on the upper part of the culm, of a soft texture, pale green colour, slightly scabrous on the margins when drawn backward against the fingers, otherwise smooth; from three to four feet long including their sheaths, and about three-quarters of an inch broad. *Floral leaves* small.—*Panicle* linear, a little bent to one side, composed of many fascicles of spikes that are both terminal, and from the exterior axils.—*Spikes* generally paired on a common peduncle, with a common boat-like spathe, or involucre at the division; each has also its proper pedicel, and both spathe-shaped. *Rachis* articulated, much waved, hairy.—*Flowers* in pairs, one hermaphrodite and sessile, the other male and pedicelled; the last hermaphrodite flower of each spike has two males; below, only one male, as the *rachis* occupies the place of the other; this is the case with most of the species of the genus.

Hermaphrodite Flowers sessile. *Glume* girt at the base with wool, &c. as in the family.—*Corol* two-valved, awnless. *Nec-*

* In the North of Bengal large tracts of waste land are covered with it. *Ed.*

tary, two, broad, short, wedge-formed, obliquely lobed, crenulated bodies embrace the insertions of the filaments and the forepart of the germ.

MALE FLOWERS pedicelled. *Calyx, glumes* as in the hermaphrodite ones.—*Corol* one-valved, awnless. *Nectary* as in the hermaphrodite.—*Stamens* three.

Obs. The fresh leaves are much used as a substitute for tea. The white succulent centre of the leaf-bearing culms is often put into curries, to give them an agreeable flavour.*

35. A. *Iwarancusa. Philosoph. Transact. Vol.* 80. 284. *t.* 16.

Perennial, erect. *Leaves* long, and very slender. *Panicles* linear; *spikelets* paired, five-jointed, with common spathes; hermaphrodite flowers awned; male one-valved and awnless.

Beng. and *Hind.* Ibharank*u*sh*a*, Iwarank*u*sh*a*, K*u*rank*u*sh*a*.

Iwarancussa. *Asiat. Res.* iv. 109.

A native of the skirts of the northern mountains of India. Dr. Blane found it between the river Raptee and the mountains, and Dr. Boyd about Hurdwar; it is therefore probably a native of a very extensive tract along the skirts of that range of mountains which bounds the plains of Bengal, Bahar, Oude, &c. Flowering time the rainy season.

It grows in large tufts, each tuft composed of a number of plants adhering together by their roots.

Root perennial, the principal parts thereof are thicker than a crow's quill, and marked with circular cicatrices, from these issue many small, firm, aromatic fibres.—*Culms* erect, generally simple; from three to six feet high, smooth in every part, not hollow, but filled with a light spongy substance.—*Leaves*, those near the root longer than the joints of the culm, margins hispid when the finger is drawn backwards over them, otherwise smooth in every part. —*Panicle* axillary, and terminal, or the whole may be reckoned a

* The natives of the Molucca Islands extract a pleasant-tasted essential oil from the leaves. *Ed.*

long, linear, erect, or drooping, intercepted panicle, composed of numerous, (many of them proliferous,) fascicles of slender, pedicelled, thin spikes of five joints; each fascicle is furnished with its own proper, boat-shaped spathe, besides many chaffy bractes within it among the insertions of the pedicels; and a proper spathe to each pair of spikes.—*Flowers* on the body of the rachis paired; one awned, hermaphrodite, and sessile; the other awnless, male, and pedicelled, the terminal florets are three, one hermaphrodite, sessile, and awned; the other two male, pedicelled, and awnless.

HERMAPHRODITE. *Calyx* one-flowered, two-valved, base girt with wool, as is also the rachis, and proper pedicels.—*Corol* two-valved, and fringed; inner valvelets two-parted, with a short arista in the fissure. *Nectary* two, minute, leaflets embracing the germ laterally.—*Stamens,* pistil, and seed, as in the genus.

MALE. *Calyx* as in the hermaphrodite. *Corol* one-valved, *nectary* and *stamens* as in the hermaphrodite, no pistil.

Obs. This species comes remarkably near to A. *Schœnanthus* both in habit and taste.

For the virtues of the roots of this plant, see Dr. Blane's account thereof in the second part of the 80 volume of the Philosophical Transactions of the Royal Society of London; also Asiatic Researches iv. p. 109. The virtues seem to reside almost entirely in the larger parts of the root, marked with annular cicatrices, the taste is agreeably aromatic, with some degree of bitterness.

36. A. *Martini. R.*

Perennial, erect. *Panicles* linear, sub-secund; *spikelets* paired, three-jointed, with common spathes only; hermaphrodite corols of one valve, and an awn; male awnless.

A native of the high lands of Balla-ghat, General Martin collected the seeds while there with the army, during the last war with Tippoo Sultan, and has reared abundance of it at Lucknow.

It is also now plentiful in the Company's Botanic Garden, raised from seed sent from thence by that gentleman, whose name I have applied as a specific one for this elegant plant.

Root perennial, with long wiry fibres.—*Culms* erect, from three to six feet high, often ramous, smooth, filled with a spongy pith.—*Leaves* very long, tapering to a very fine point, smooth in every part and of a soft delicate texture.—*Sheaths* shorter than the joints on full grown plants, with a membranaceous stipulary process at the mouth.—*Panicles* as in A. *Iwarancusa; spikelets* paired, but with only three joints.—*Flowers* also paired, &c. as in the former species, only there the lowermost pair on the most sessile of the two spikelets are both male; and one of them rests upon a smooth, convex, callous receptacle instead of a pedicel.—*Rachis* jointed, and woolly.—*Calyx* as in A. *Iwarancusa.*—*Corol* one-valved, a long black awn occupies the place of the other, which has two small filaments near its base. *Nectary,* &c. as in the foregoing species.

Obs. General Martin, who sent me this grass from Lucknow, writes to the following effect:

"I took particular notice of a sort of long grass which the cattle were voraciously fond of, which is of so strong an aromatic and pungent taste, that the flesh of the animals, as also the milk and butter, have a very strong scent of it. Of this grass I send you a small stalk, some roots, and seed ; if you taste the latter, though old, you will find it of a very pungent aromatic taste."

PASPALUM. Schreb. Gen. N. 81.

Calyx one-flowered, two-valved, valvelets rounded, equal. *Corol* of two-valves, and like the calyx. *Seed* adhering to the corol.

1. P. *scrobiculatum. Linn. Sp. Pl. ed. Willd.* i 330.

Spikes, axillary and terminal, sessile, alternate, erect. *Flowers* alternate, in two rows, or crowded, smooth. *Calyces* from three to five-nerved. *Culms* erect, two feet high, smooth. *Seed* round, smooth.

Sans. कोरदूष, Korudoosh*a,* कोद्रव, Kodr*uva.*
Beng. Kod*a.*
Hind. Koda-ka-choul.
Teling. Aruga.

This is cultivated by the natives over many parts of India. It delights in a light, dry, loose soil, but will grow in a very barren one. Time of cultivation the rainy season.

Root fibrous.—*Culms* erect, ramous, jointed, smooth ; about two feet high, involved in the sheaths of the leaves.—*Leaves* sheathing, bifarious, longer than the culm, every part smooth. *Sheaths* longer than the joints, often embracing the spikes like a spathe.—*Spikes* axillary and terminal, from two to four, alternate, sessile, erect, secund.—*Rachis* broad, membranaceous, with a waved keel on the inside.—*Flowers* oval, alternate, in two rows ; in luxuriant plants the flowers are crowded without order.—*Calyx, exterior* valve three-nerved, *interior* valve five-nerved.—*Seeds* brown, smooth, size of that of hemp.

Obs. The seed is an article of diet with the Hindoos, particularly with those who inhabit the mountains and most barren parts of the country, for it is in such countries only where it is cultivated, it being an unprofitable crop, and not sown where others more beneficial will thrive. I have eaten of the boiled grain, and think it as palatable as rice.

2. P. *Kora. Linn. Sp. Pl. ed. Willd.* i. 332.

Spikes terminal, alternate ; *flowers* alternate, in two-rows, smooth. *Calyces* from three to five-nerved. *Culms* below procumbent, from one to eight feet high.

P. *orbiculare. Forst. prodr. N.* 35.

Hind. Kodu.

Teling. Neer (i. e. water) Aruga.

A native of the borders of water courses.

Root fibrous, annual.—*Culms* resting on the ground near the base, above ascending, branchy, jointed, smooth ; from one to three feet long.—*Leaves* as in P. *scrobiculatum,* but shorter. *Spikes* generally two, or three, though in very luxuriant plants I have seen seven or eight, terminal, sessile, horizontal, and erect, the rest as in the last described species.

Obs. Cattle are very fond of these two species, whether green or dry.

3. P. *longiflorum. Linn. Sp. Pl. ed. Willd.* i. 332. *Retz. Obs.* iv. 15.

Spikes paired, terminal, horizontal. *Calyces* oblong, obtuse, smooth. *Culms* creeping.

Teretu-pullu. *Rheed. mal.* 12. *p.* 81. *t.* 44.

Paspalum *longiflorum. Retz. Obs.* iv. *p.* 15.

A native of low pasture ground. Compare it with P. *distichum. Linn. Sp. Pl. ed. Willd.* i. 322.

Root creeping.—*Culms* creeping, with their extremities ascending six or eight inches high, jointed, smooth.—*Leaves* sheathing, alternate, bifarious, short; mouths of the sheaths bearded.—*Spikes* paired, terminal, sessile, spreading, secund.—*Flowers* in two rows, alternate, oblong, somewhat paired, smooth.—*Calyx* and *corol* without nerves.

4. P. *longifolium. R.*

Erect, simple. *Leaves* as long as the culms. *Panicle* of many simple, alternate, diverging spikes scattered round a three-sided rachis. *Valvelets* of the calyx three-nerved. *Seed* oval, lucid, and marked with minute pits.

Of what country this is a native is uncertain. It appeared in the Botanic Garden in 1807, in a place where plants from Sumatra had been planted, it is therefore more than probable that the seeds were in the earth.

Culms several from one root, simple, straight, jointed, nearly erect, round and smooth, height of the whole plant when in flower above three feet.—*Leaves* long, viz. from one to three feet, sheaths included, slender, acute, smooth, except the edges when rubbed backward. *Sheaths* longer than the joints of the culm, and smooth, except at the top, where there are a few, long, soft hairs near the short scariose ligula.—*Panicle* composed of from twelve to twenty-four, simple, diverging spikes, scattered alternately round a three-sided rachis, nearly a foot in length.—*Spikes* sessile, with a few, long, straight, white hairs round their insertions, about three inches long. *Rachis* flat, with a waved keel on the underside, and coloured, waved margins.—*Flowers* in numerous, alternate, imbricated pairs on the underside of the flat rachis, on unequal, short-pedicels.—*Calyx* of two, very equal, oval leaflets with a nerve or rib round the margins, and one down the middle.—*Seeds* oval, dotted with innumerable, small pits, shining.

LAPPAGO. Schreb. Gen. N. 131.

Calyx from two to three-valved.—*Corol* two-valved, resupine.

1. L. *biflora. R.*

Culms creeping. *Pedicels* two-flowered; inner valvelets of the calyx armed with hooked bristles.

Phalaris muricata. Forsk. descript. 202.

It is a small, prostrate grass, a native of the dryest pasture ground on the Coromandel Coast.

Culms pressing on the earth, and striking root from the joints, from six to twelve inches long, ramous.—*Leaves* short, margins ciliate, and waved; mouths of the sheaths bearded.—*Racemes* terminal, simple, from one to two inches long.—*Pedicels* short, diverging, two-flowered.—*Flowers* lanceolate, echinated on the back, or outside, diverging on all sides round the rachis, which is waved, angular, striated, and downy.—*Calyx* one-flowered, two-valved; the exterior valve very minute, second or inner valve, (for there is no other,) lanceolate and completely embracing the corol on all sides, on the back and sides covered with many, strong, hooked bristles.—*Corol* two-valved, membranaceous. *Nectarial* scales large.—*Stamens, Pistil,* &c. as in the genus.

PANICUM. Schreb. Gen. N. 107.

Calyx, Glume one or two-flowered, three-valved : exterior valvelets minute; *seed* adhering to the corol.

Sect. 1*st. Spikes simple.*

1. P. *indicum. Mant.* 184. *Retz. Obs.* iii. *p.* 9.

Culms from four to eight inches, ascending from a repent base. *Spikes* sub-cylindric, naked. *Calyx* smooth, no involucres.

P. *conglomeratum. Linn. Sp. Pl. ed. Willd.* i. 341. does not agree with our plant in having cylindrical spikes with flowers equally disposed on all sides. This is a very small species, a native of Coromandel.

2. P. *barbatum. R.*

Culms from three to five feet high ; very ramous at top. *Leaves* lanceolate. *Spikes* simple, cylindric. *Involucrets* simple, and bearded, surrounding on every side a single flower ; *seed* smooth.

This species sprung up in beds, in the Botanic Garden where earth had been thrown that came from Sumatra, or Amboyna. Compare with P. *polystachyon. Linn. Sp. Pl. ed. Willd.* i. 335.

Root annual in Bengal.—*Culms* erect, about five feet high, simple, till near the top, there very ramous.—*Leaves* lanceolate, often a foot and a half long, hispid on the margins, otherwise smooth ; the sheaths about half the length of the joints, with a few single hairs at the mouth.—*Spikes* terminal, and solitary (though from the numerous *ramification* of the top of the culms the whole may be called a large leafy panicle) sub-cylindric, about three or four inches long.—*Involucrets* of two sorts, the longest about twice the length of the flowers with their lower half bearded, (hence the specific name, which for the same reason is also applicable to my *Panicum holcoides,*) the innermost one is in this species about double the length of the other five, six, or seven ; all are of a reddish brown colour from the bearded part to the point. The other sort of involucrets are simple and about the length of the flowers. The two sorts surround on all sides a single flower.—*Calyx* one-flowered, exterior valve very minute.—*Seeds* smooth, slender and pointed.

Obs. It differs from my P. *holcoides* in the culms being very ramous at the top only, and in the leaves being lanceolate; from *polystachyon* it differs, I think, in having two sorts of involucrets equally disposed round the flower, and in the leaves being lanceolate and smooth.

3. P. *spicatum. R.*

Erect. *Spikes* cylindric. *Involucres* hairy, surrounding from one to three awnless polygamous flowers. *Calyces* two-valved, both shorter than the corol, the inner longer and retuse.

Holcus spicatus. *Linn. Sp. Pl. ed. Willd.* iv. 928.

Gramen paniceum, *Pluck.* &c. *alm. t.* 32. *f.* 4. good.

Hind. Bujera, or Bujra.

Pedda-Gantee is the Telinga name of the plant, and Gantiloo, the grain.

I have only found this in a cultivated state. It is sown about the beginning of the rains, viz. the end of June, and the beginning of July, and ripe in September.

Culms several, if the soil is good, from the same grain of seed, erect, with roots from the lowermost joint or two, round, smooth, from three to six feet high, and nearly as thick as the little finger. —*Leaves* alternate, sheathing, broad and long; mouths of the sheaths bearded.—*Spikes* or rather *racemes*, terminal, cylindric, erect, as thick as a man's thumb, or more, and from six to nine inches long. *Pedicels* generally two-flowered, though sometimes only one and sometimes as many as four.—*Flowers* surrounded with many, woolly, hispid purple bristles or *involucres*, about the length of the flower.—*Calyx* two-flowered, one hermaphrodite the other male, two-valved; exterior *valvelet* minute, *interior* nearly as long as the corol, retuse, both awnless.—*Corol* of the hermaphrodite flower two-valved, of the male one-valved.—*Style* single. *Stigma* two-cleft, feathery.—*Seed* obovate, pearl-coloured, smooth.

Obs. It agrees perfectly well with our Indian species of *Panicum,* on that account I have transferred it to that genus.

This species is much cultivated over the higher lands on the Coast of Coromandel. The soil it likes is one that is loose and rich; in such it yields upwards of an hundred fold, the same ground will yield a second crop of this or some other sort of dry grain during October, November, December, and January.

The Hindoo farmer knows four other varieties of this species, all of which he cultivates. Their Telinga names are 1*st*. *Pitta* (birds) *Gantee; 2d. Munda-boda-Gantee; 3d. Palla-boda-Gantee;* and 4*th*. *Yerra-Gantee.*

Cattle are fond of the straw, and the grain is a very essential article of diet amongst the natives of these parts.

4. P. *involucratum. R.*

Erect. *Spikes* cylindric, numerous, scattered, two-flowered, alternately longer and ciliate; shorter and smooth. *Calyces* two-valved, the exterior one minute, the inner one shorter than the corol, and emarginate.

A native of mountains chiefly, where it grows wild.

Culms as in the last species, from two to four feet high; joints woolly.—*Leaves,* mouths of their sheaths bearded.—*Spikes* as in P. *spicatum,* but the pedicels smaller, two-flowered.—*Involucre,* many bristles surrounding on all sides the flowers, they are of two sorts, simple ones awled, and longer ones with fringed margins.—*Calyx* one or two-flowered; valvelets as in the last species.—*Corol,* when there is only one to the calyx it is hermaphrodite, when two, one hermaphrodite the other male as in the last species.—*Styles* two.

Obs. I know of no use this is put to at present; it may be *Panicum spicatum* in its wild state, though it is more like my *Panicum holcoides* in its present wild state.

5. P. *glaucum. Linn. Sp. Pl. ed. Willd.* i. 335.

Culms erect, from one to three feet high; *involucel* one bundle of hairy bristles, one-flowered. *Flowers* generally polygamous. *Seed* transversely rugose.

Beng. Pingi-natchi.

Teling. Naka-kora.

This species is generally found on cultivated lands, growing amongst the different sorts of small or dry grain.

Root fibrous.—*Culms* sometimes near the base resting on the ground, and striking root; the rest erect, ramous, smooth, a little compressed; from one to three feet high.—*Leaves* sheathing, nearly bifarious, smooth; mouths of the sheaths hairy.—*Spikes* terminal, columnar, from two to six inches long.—*Flowers* solitary, or joined to the rudiments of one or two more, short-pedicelled, numerous.—*Involucels* longer than the flowers, consisting of one bundle of hairy bristles which issue from the middle of the underside of the pedicel.—*Calyx* generally two-flowered, in which case one is *male*, the other *hermaphrodite*, with its corol, as in the family. The male corol consists of one membranaceous glume between the inner valve of the *corol* and inner glume of the *calyx*, it has three stamens only.

Obs. It may probably be P. *helvolum* of the *Supplementum Plantarum.* P. 107.

In India we have two or three varieties of this coarse grass, can any of them be *Alopecurus indicus?* the small reddish one comes very near it.

6. P. *holcoides. R.*

Erect. *Involucels* numerous, scattered, one-flowered, alternately one longer and woolly, and another shorter and smooth; *flowers* sometimes polygamous. *Seed* oblong, smooth, resembling a small grain of common oats.

Beng. Swati.

Grows, with the former, on cultivated lands, but chiefly up amongst the mountains.

Root fibrous.—*Culms* erect, ramous, from two to four feet high, round, smooth.—*Leaves* sheathing; *sheaths* half the length of the joints; mouths bearded.—*Spikes* as in P. *glaucum*, but larger.—*Flowers* solitary, short-pedicelled, without order.—*Involucels*, numerous, bristles entirely surrounding the flower; of two sorts, the largest twice the length of the flower, and woolly from the middle down, the other sort are smaller, and without wool or hairy.—*Calyx* one or two-flowered, as in the last; exterior valve minute, the other two as long as the corol, of which the apex of the exterior one is somewhat three-toothed.—*Seed* smooth, very like common oats, but smaller.

Obs. It differs from P. *polystachyon* in having the involucels scattered round the pedicels and flower. It may therefore be *Rumph's gramen caricosum. vol.* vi. *t.* 7. *f.* 2. A.

7. P. *interruptum. Linn. Sp. Pl. ed. Willd.* i. 341.

Culms above water erect. *Spikes* simple, cylindric; *flowers* crowded, smooth, polygamous; *seed* smooth.

Beng. Nardula.

Teling. Wolam.

A very large species, grows in deep standing water.

Culms ramous, those parts under the water swelled, as thick as the little finger, with many fibres from the joints; the parts above the water erect, about as thick as a common quill, from two to four feet high, smooth, involved in the sheaths of the leaves.—*Leaves*, sheaths longer than the joints, smooth.—*Racemes* or *spikes* simple, erect, columnar, from four to six inches high.—*Flowers* small, oval, pedicelled, numerous, generally many, from nearly the same place.—*Involucre* entirely wanting.—*Calyx* two, inner glumes striated, from six to seven-nerved.—*Corol*, the small flower is present here.—*Seed* oblong, smooth, shining white.

8. P. *curvatum. Linn. Sp. Pl. ed. Willd.* i. 352.

Culms erect. *Spikes* curved; *flowers* crowded, smooth, hermaphrodite; *seed* smooth.

Grows about the borders of lakes.

Culms numerous, ramous, erect, as thick as a crow quill, from two to three feet high, smooth.—*Leaves* numerous, very long, slender, and smooth.—*Racemes,* or *spikes* columnar; lower part somewhat ramous; branchlets adpressed.—*Flowers* numerous, covering most completely every part of the racemes, pedicelled, oval.—*Calyx,* two inner glumes striated.—*Corol* no neuter, only one hermaphrodite.—*Seed* smooth, shining, white.

Obs. This may be only a variety of P. *interruptum.*

9. P. *dimidiatum. Linn. Sp. Pl. ed. Willd.* i. 339.

Spikes secund, jointed; joints excavated, daggered on alternate sides of the apex, from one to five-flowered. *Calyx* two-flowered, one hermaphrodite, the other male.

A native of the Peninsula of India, scarce.

Culms short, oblique, smooth; whole height from six to twelve inches.—*Leaves* slender, smooth, obtuse, rather broad toward the apex.—*Spikes* solitary, secund, jointed, smooth, from three to five inches long. *Joints* of the *rachis* from eight to twelve rigid, excavated for the reception of the flowers, except the lower one or two, which are longer; in all except these a long sharp point or dagger proceeds from the apex, on the alternate sides of the articulation.—*Flowers* lodged in the concave joints of the rachis, on very unequal pedicels, the lowermost having each three or more, while those about the middle of the spikes have uniformly two, and the last two or three only one.—*Calyx* two-flowered, two-valved; flowers both sessile, one hermaphrodite, the other male; valves unequal, the exterior one smaller; the inner one about as long as the valve of the corol, or it may be described with three valves as in the genus; but then the male flower will have but one glume.—*Corols* two-valved; valves nearly equal, those of the male flower much firmer, the sides of the exterior one (which may be called the inner valve of the calyx, if it be allowed, to have three), with the sides bent in at right angles.—*Stamina* three.—*Germ* in the hermaphrodite flower oblong. *Style* undivided for a short distance. *Stigma* plumose.—*Seed* oblong, dark brown, dotted; on the anterior side near the base is an oblong elevated mark.

Obs. Dr. Rottler, of Madras, an excellent Botanist, obliged me with the only specimens of this rare grass which I have yet met with, and had he not informed me that this was considered to be P. *dimidiatum* I should have been inclined to have taken it for an *Ischaemum.*

Sect. 2d. Spikes paired.

10. P. *conjugatum.* *R.*

Spikes conjugate, secund. *Flowers* solitary, sessile, awnless. *Corol,* with an accessary neuter valve.

A slender, soft, villous, half creeping species; a native of Coromandel. It differs from P. *distachyon* in the number of the spikes being constantly two, and the flowers always solitary and sessile. To these marks of distinction may be added that the valves of the calyx are three-nerved, and the accessary one particularly large.

11. P. *squarrosum.* *Linn. Sp. Pl. ed. Willd.* i. 345. *Retz. Obs.* iv. 15. and v. *t.* 1.

Culms creeping, from ten to twenty inches long. *Spikes* paired, horizontal, *rachis* articulate; *flowers* in a fascicle on the upper end of each joint of the rachis.

Cenchrus muricatus. *Mant.* 302.

P. *dimidiatum.* *Burm. Ind.* 25. *t.* 8. *f.* 2.

A native of dry sandy ground near the sea.

Culms branchy, creeping, with their flower-bearing extremities sub-erect; from ten to twenty inches long.—*Leaves* short, but rather broad, and covered with soft hair; *sheath* large, downy, involving most of the culms.—*Spikes* two, terminal, spreading, horizontal, or ascending like a pair of horns, secund. *Rachis* composed of from four to eight, oblong joints, divided by a waved ridge; on each side of the ridge membranaceous.—*Flowers* collected in sessile bundles of from four to eight, alternately disposed on the upper end of each joint.—*Calyx*, the *exterior* one minute, and lanceolate; the *second* large, embracing loosely the corol, pointed, and striated; the *inner* one nearly as small as the *exterior*, tapering from the base to a fine point, three-nerved; margins fringed, of a chaffy texture.—*Corol* as in the genus, and rather longer than the inner glume of the calyx, here is a third neutral membranaceous valve, as in many species of this genus.

Sect. 3d. Spikes fascicled.

12. P. *dactylon. Linn. Sp. Pl. ed. Willd.* i. 342.

Smooth, creeping. *Spikes* digitate, secund, corol gibbous on one side, and twice as long as the calyces.

Agrostis linearis. *Linn. Sp. Pl. ed. Willd.* i. 375. *Retz. Obs.* iv. *N.* 51. Sir W. Jones. *Asiat. Res.* iv. *p.* 248.

Sans. दुर्व्वा, Doorva, शतपर्व्विका, Shutupurvika, सहस्रवीर्य्या, Suhusruveerya, भार्गवी, Bharguvee, उरुहा, Ourooha, अनन्ता, *Ununta.*

Beng. Doorba,

Teling. Ghericha.

Tam. Arugam-pilla.

This is by far the most common and useful grass in India. It grows every where abundantly, and flowers all the year.

Root creeping.—*Culms* creeping, with their flower-bearing branchlets erect, from six to twelve inches high, smooth.—*Leaves* small, and smooth.—*Spikes* from three to five, terminal, sessile, filiform, expanding, secund, from one to two inches long. *Rachis* waved.—*Flowers*, alternate, single, disposed in two rows on the underside.—*Calyx* much smaller than the corol.—*Corol*, the large or exterior valve boat-shaped, keel slightly ciliate.—*Stigmas* villous, purple.

Obs. This most valuable grass forms three-fourths of the food of our horses and cows in India. It is by the brahmuns of the Coasts held sacred to *Gunesha,* (the *Janus* of the ancients,) under the name of *Doorwall.*

13. P. *aegyptiacum. Linn. Sp. Pl. ed. Willd.* i. 343.

Creeping at the base. *Spikes* from four to eight; corymbed, smooth. *Flowers* paired on unequal pedicels; accessary valve of the calyx minute or wanting, the other two very unequal, nerved, and ciliate on the margins only.

Obs. Is rather a scarce grass in India, and grows in tufts. Cattle do not seem fond of it, whereas all are fond of P. *dactylon.*

14. P. *ciliare. Linn. Sp. Pl. ed. Willd.* i. 344.

Culms creeping at the base. *Spikes* sub-digitate. *Flowers* paired, one sessile the other pedicelled; inner two valves of the calyces equally long, and bearded with four woolly ridges; third minute.

Hind. Mak*ur*-jalee.

Teling. Shangal*i*-gadd*i*.

It delights most in newly laid down pasture ground.

Culms creeping, with one or two feet of the extremities erect, these ramous, round and smooth.—*Leaves* sheathing, short for the size of the grass; margins ciliate near the base; *sheaths* sometimes a little hairy, shorter than the joints of the culm, their mouths rise above the insertion of the leaf, stipule-like, as in Dr. Smith's *Erharta calycina,* but here it is entire.—*Spikes* or rather spiked-racemes, from four to ten, digitate, expanding, secund.—*Flowers* paired, one sessile, one peduncled.—*Rachis* three-sided, waved.—*Calyx, exterior* valve most minute. *Interior* two many-nerved, four of the nerves are clothed with very long, white, soft hairs.

Obs. Small plants on a poor soil, have much the appearance of *Agrostis radiata.* Cattle are very fond of this grass.

15. P. *filiforme. Linn. Sp. Pl. ed. Willd.* i. 343.

Creeping, filiform, smooth. *Spikes,* from two to four, sub-digitate, filiform, secund; *flowers* paired, one sessile the other pedicelled. *Calyx* with the accessary valve, minute; inner one half the length of the corol or middle one; these last two are three-nerved, and villous on the margin.

From China it was brought to the Botanic Garden, where it thrives well, and blossoms the whole year, but chiefly during the rains.

Culms ramous, creeping, very delicate, and smooth, a large portion next the spikes erect, naked, coloured and filiform.—*Leaves* soft and small, ciliate, with a few long hairs at the base.—*Spikes* generally from two to four issuing from the same point, very slender, secund. *Rachis* flexuose.—*Flowers,* in a pair, one sessile.—*Calyx* one-flowered, flower hermaphrodite; *accessary* valve exceedingly minute, but by careful observations may always be found; second valve as long as the co ol, the inner one only half their length, which is one of the best specific marks; these two ha e three smooth nerves on the back, and their margins are slightly bearded.—*Corol* two-valved.

16. P. *lineare. Linn. Sp. Pl. ed. Willd.* i. 344.

Sub-erect, smooth. *Spikes* from two to six, digitate, filiform. *Flowers* all hermaphrodite, awnless, and pedicelled. *Calycine* valves, the accessary one minute; the inner two equalling the corol, and three-nerved.

P. *lineare. Burm. Ind. p.* 25. *t.* 10. *f.* 3, is a tolerable representation of this pretty delicate grass.

A native of China, accidently introduced from thence into the Botanic Garden, where it is in flower the whole year; but like most other grasses most luxuriant during the rains. It has a great resemblance to P. *dactylon, ciliare, filiforme,* and *aegypticum,* but does not creep like them. The flowers are less regularly paired, more minute, smooth, and obtuse, the two inner valves of the calyx equal, and as long as the corol, and the accessary one very minute, or not to be found.

17. P. *cimicinum. Linn. Sp. Pl. ed. Willd.* i. 344.

Culms erect, hairy, from one to two feet high. *Racemes* peduncled, corymbed; *flowers* polygamous, in pairs; exterior valvelets of the calyces fringed; *corol* awned.

Milium cimicinum. *Mant.* 184.

A rare plant, native of ground that has been lately in cultivation.

Culms from one to two feet high, sub-erect, hairy.—*Leaves* short and broad, hairy, and fringed round the margin; *sheaths* also hairy.—*Racemes* from four to ten; when more than four or five, the rest are elevated on a common peduncle above the point from whence the first four or five issue; all are pedicelled, bowing, secund.—*Flowers* in two rows and in pairs, scarcely ever single.—*Calyx*, border of the inner valve fringed, with red hairs.—*Corol*, the arista is sometimes wanting, and its valves three-nerved; in mine there is a second, or male floret as in many species of *Panicum.*

18. P. *corymbosum. R.*

Sheaths of the leaves bristly. *Spikes* numerous, corymbed. *Calyces* two-valved. *Seeds* oblong, smooth.

Teling. Cheepa-bora-gaddi.

This species I have found only in the vallies amongst the Circar mountains; it grows to a great size in standing sweet water.

Culms of great extent, creeping, or floating, with their extremities erected above the water; on dry ground I have also found it, where it is in general from one to four feet high.—*Leaves* as in other grasses; *sheaths* covered with stiff white hairs.—*Spikes* corymbed, numerous; from four to eight inches long, filiform, sub-erect, secund, level-toped. *Rachis, common* striated; *partial* three-sided.—*Flowers* oblong, in two rows, paired; one very short pedicelled, the other longer.—*Calyx* two-valved; the exterior valve small, the other as large as the corol, and striated, but no third or inner valve.—*Corol* as in the family.—*Seed* oblong, smooth white.

Sect. 4th. Spikes alternate on one side only of a common rachis.

19. P. *brizoides. Linn. Sp. Pl. ed. Willd.* i. 338.

Culms sub-erect. *Leaves* short. *Spikelets* remote, adpressed. *Flowers* oval, solitary, polygamous; inner two-valves of the calyx equal; nectarial petals long; *seed* ovate, pointed, rugose.

P. flavidum. *Linn. Sp. Pl. ed. Willd.* i. 339. *Retz. Obs.* iv. 15. appears to be the same plant growing on a barren soil; this was the opinion of König. P. *flaccidum* of the *Banksian Herbarium* seems the same, or a variety only.

Teling. Oda, or Woodoo-gaddi.

It is common in every soil and situation, even in deep water; in one that is rich and moist, it is often from two to four feet long, and again on one that is dry and barren only as many inches. It grows in tufts, various parts of it are often tinged purple.

Culms near the base resting on the ground and rooting above, regularly ascending, compressed, smooth.—*Leaves* bifarious, smooth, their length differs much; *mouths* of the sheaths bearded.—*Spikes* compound, secund, their length various according to the size of the plant. *Spikelets* secund, number very various, alternate, erect, pressing on the common rachis; below they are often two or three times their length from each other; above less remote. *Rachis, common,* somewhat three-sided, with a groove on the outside where the spikelets rest.—*Flowers* always disposed in two rows, sub-globular.—*Calyx* two-flowered. Inner valve the size of the corol, middle valve rather shorter; all are striated with green nerves. —*Corol* one *hermaphrodite,* and one *male,* as in some of the species already described, always present, except in very minute plants, and even then the stamens of the second floret are only sometimes wanting.—*Nectary,* in each two obovate scales.—*Seed* short, oval, pointed, slightly rugose.

20. P. *fluitans. Linn. Sp. Pl. ed. Willd.* i. 338.

Culms from two to three feet high, creeping at the base. *Leaves* long. *Spikelets* numerous, approximate, longer than the interstices. *Flowers* hermaphrodite, oblong, two exterior valvelets of the calyces minute ; *seed* oblong.

Beng. Peti-nar.

Teling. Doosa.

Delights in a moist rich soil, such as the banks of water-courses, borders of rice fields, &c.

Culms at the base creeping, above nearly erect, round, smooth, from one to four feet long, including the part that rests on the ground.—*Leaves* smooth, except about the mouths of the sheaths where they are bearded, and sub-ciliate.—*Spike* compound, secund, exceedingly like that of P. *brizoides,* only here it is generally longer, with often as many as twenty-five or thirty spikelets ; their distance from one another is not so regular and is generally less than their own length.—*Flowers* disposed in two rows on the outside of the spikelets, which are oblong.—*Calyx* one-flowered, the two exterior valves very small.—*Seed* oblong, pointed, rugose.

21. P. *lanceolatum. Linn. Sp. Pl. ed. Willd.* i. 337.

Culms, creeping. *Leaves* lanceolate, mouths of their sheaths elevated and bearded. *Flowers* hermaphrodite ; valvelets of the calyces equal, the exterior one awned. *Seed* smooth, oblong.

P. *aristatum,* of the same author, seems the same grass.

Compare with P. *compositum. Linn. Sp. Pl. ed. Willd.* i. 346. In the *Banksian Herbarium* the same species is labelled P. *unguinosum.* Grows under the shade of trees.

Culms creeping, ramous, with their extremities, from one to two feet high, sub-erect.—*Leaves* lanceolate, waved, often tinged with purple ; *sheaths* shorter than the joints, hairy ; *mouths* elevated, stipula-like, and hairy.—*Spikes* composed, secund ; from six to twelve inches long. *Spikelets* alternate, somewhat remote, secund, direction between expanding and adpressed. *Rachis,* common and partial three-sided.—*Flowers* generally paired, one-sessile, the other short-pedicelled ; insertions often surrounded with hairs, when single there is an awn, or the rudiments of a second floret accompanying it.—*Calyx*, the exterior glume as large as the next within it, and ending in a long waved, purple awn; exterior glume rather longer.—*Corol* no *neuter*, nor *male* floret.—*Seed* oblong, smooth, slightly three-nerved.

Obs. Cattle are not fond of it.

22. P. *Burmanni. Linn. Sp. Pl. ed. Willd.* i. 339. *Retz. Obs.* iii. *N.* 16.

Culms, creeping. *Leaves* lanceolate, hairy; *flowers* hermaphrodite; all the valvelets of the calyces hairy, and awned. *Seed* slender, smooth, pointed.

P. *hirtellum. Burm. Ind.* 24. *t.* 12. *f.* 1.

Hippogrostis amboinica. *Rumph. amb.* vi. *p.* 14. *t.* 5. *f.* 3.

This species is always found on pasture ground, under the shade of large trees.

Culms, creeping, branchy, with their extremities erect.—*Leaves* sheathing, lanceolate, waved, hairy; sheaths half the length of the joints; very hairy.—*Spikes* compound, secund, erect. *Spikelets* from four to eight, alternate, secund, adpressed. *Rachis*, common and partial three-sided.—*Flowers* generally paired; one sessile the other pedicelled.—*Calyx*, the two exterior glumes are hairy, and have long awns, that of the inner is shorter and smoother.

23. P. *stagninum. Linn. Sp. Pl. ed. Willd.* i. 337.

Culms, from one to four feet high. *Leaves* smooth. *Flowers* polygamous, three-fold, sub-sessile. *Calyces* hairy, and hispid; inner valvelets awned.

Beng. Dul.

Teling. Pedda-woondoo.

Delights in wet, cultivated, paddy lands.

Culms, towards the base resting on the ground, and striking root; above erect, which part is from one to four feet high.—*Leaves* smooth, soft, with only the mouths of the sheaths ciliate.—*Spikes* compound, secund, erect. *Spikelets* alternate, sessile, secund, expanding a little. *Rachis,* both common and partial three-sided.—*Flowers,* very numerous; three-fold below, above in pairs, all sessile, or nearly so; their insertion surrounded with small bristles, involucre-like.—*Calyx,* only the inner valve awned, all striated, hairy and hispid.—*Corol,* complete; *male* florets, besides the usual hermaphrodite one.

Obs. This is a coarse species. Cattle are not fond of it.

24. P. *crus corvi. Linn. Sp. Pl. ed. Willd.* i. 337.

Culms, creeping at the base. *Leaves* downy. *Spikes* from six to ten, three times longer than their interstices. *Calyces* downy. *Corol* three-valved. *Seed* oblong, obtuse, smooth.

A middle-sized, delicate, rare species, found on pasture ground.

Culms, filiform, creeping at the base, above nearly erect, ramous.—*Leaves* soft, downy; sheath hairy.—*Spikes* compound, secund, from one to three inches long. *Spikelets* from four to twelve, alternate, adpressed, secund, about three-fourths of an inch long; sometimes there is a very small ramification at the lower part of one or two. *Rachis,* common, two-furrowed; partial, three-sided, downy.—*Flowers* paired, or single, in two rows; both pedicelled; *pedicels* hairy, of unequal length, but both short.—*Calyx* striated, very downy.—*Corol,* the neuter valve is here present.—*Seed* oblong, smooth, shining white.

25. P. *colonum. Linn. Sp. Pl. ed. Willd.* i. 338.

Culms, creeping at the base. *Spikes* from six to ten, short. *Flowers* three-fold, sub-sessile. *Glumes* of the calyces three-nerved. *Corol* three-valved; *seed* roundish, smooth, three-nerved.

Beng. Shama. There is a reddish variety, called *Lal-shama.*

Teling. Woondoo-gaddi.

Delights in rich pasture ground.

Culms, below resting on the ground, and rooting; above sub-erect, that part is from one to two feet high, branchy, a little compressed, smooth.—*Leaves* tending to be bifarious, short, smooth, tapering from the base to a sharp point.—*Spikes* compound, secund. *Spikelets,* from six to twelve, alternate, somewhat remote, secund. *Rachis,* both common and partial three-sided; the two sides from whence the spikelets issue concave.—*Flowers* two or three from the same point, generally three; all sessile, or very nearly so; sometimes a few small bristles mixed with them.—*Calyx* as in the genus, all the glumes three-nerved, a little hairy, and scabrous.—*Corol,* the neuter valve is here present, but no stamens.—*Seed* broad, oval, smooth, a little pointed, three-nerved.

Obs. Cattle are very fond of it, yet it is not cultivated in these parts.

26. P. *grossarium. Linn. Sp. Pl. ed. Willd.* i. 349.

Culms, creeping. *Leaves* soft and downy. *Spikes* from six to ten; *flowers* solitary, sub-sessile. *Corol* three-valved. *Seed* oblong, acute, transversely rugose.

Loves fresh pasture ground.

Culms, in a good soil creeping to a great length, in a poor soil nearly erect.—*Leaves,* tapering from the base to a fine point, somewhat downy, and very soft.—*Spikes* compound, secund. *Spikelets* from four to twelve, alternate, secund, expanding on one side, most simple, in two rows.—*Calyx* pretty smooth.—*Corol,* the neuter valve is always present. *Nectary* a rugose gland between the germ and exterior valve of the corol.—*Seed* oblong, white, slightly rugose in a transverse direction.

Obs. The form and rugosity of the seed distinguish it from P. *colonum,* the flowers also are here single, there two or three from the same point.

27. P. *umbrosum. Linn. Sp. Pl. ed. Willd.* i. 343.

Culms, at the base creeping. *Leaves* soft. *Spikes,* from six to ten; *flowers* paired on a common pedicel. *Corol* three-valved. *Seed* ovate, longitudinally streaked, and transversely rugose.

This is also a native of pasture ground, but chiefly that which is shaded. It is exceedingly like the last described species.

Culms, creeping, &c. as in the last.—*Leaves* also the same.—*Spikes,* the same as in P. *grossarium. Spikelets* or rather *racemes,* the lower one or two sometimes branchy in luxuriant plants. *Rachis,* common striated; partial three-sided, and downy.—*Flowers,* the inferior ones in pairs, upon a short, common pedicel; besides, having their partial pedicels, one of which is nearly as long as its flowers; above they are often single, and somewhat remote; hairs surround the insertion of all.—*Calyx* hairy.—*Corol,* the neuter valve is here present.—*Seed* ovate, a little pointed, rugose, like that of the last species.

Obs. The chief character consists in the lower flowers of the racemes being paired on a common peduncle, with pedicels of very unequal lengths and hairy involucres.

28 P. *cuspidatum. R.*

Creeping near the base, wholly smooth. *Leaves* linear, without ligula. *Spikes* compound, secund; *spikelets* adpressed, secund; *flowers* from one to three, equally sub-sessile, all hermaphrodite; *valves* of the calyx three-nerved, and cuspidate. *Seed* cuspidate.

A native of Bengal. Found in a rich moist soil, in flower and seed in April.

Culms, numerous in luxuriant tufts, creeping near the original root, then ascending to about two feet in height, ramous, smooth.—*Leaves* ensiform, smooth, with coloured margins. *Sheaths* shorter than the joints, smooth, and completely destitute of ligula.—*Spikes* compound, secund, slightly recurvate, with the secund spikelets on the convex side pressing on the common, three-sided, hispid rachis; often of a dull purple colour.—*Flowers,* from one to three together, and equally sub-sessile.—*Calyx,* all the glumes three-nerved, and cuspidate, and generally armed with a few minute bristles.—*Corol* three-valved; the third or accessary one membranaceous, and neuter.—*Seeds* ovate, smooth, cuspidate.

Sect. 5th. Spikes alternate on all sides of a common rachis.

29. P. *setigerum. Linn. Sp. Pl. ed. Willd.* i. 338.

Culms, below creeping, sub-ligneous. *Leaves* with cordate base and ciliate margins. *Insertions* of the *spikelets* and *flowers* bristly. *Corol* three-valved. *Seed* oval, transversely rugose.

Beng. Bura-jal-ganti.

P. *setigerum. Retz. Obs.* iv. 15.

Grows amongst bushes or under the shade of trees.

Culms long, straggling, rooting when they rest on the ground, firm, almost woody towards the base, but small, and very smooth; joints downy.—*Leaves* sheathing, broad-cordate at the base, tapering from thence to the point; margins ciliate, and waved; mouths of the sheaths bearded.—*Spikes* compound. *Spikelets* from four to twenty; secund, spreading, disposed equally on all sides. *Rachis,* common four, five, or six-sided; partial three-sided.—*Flowers* paired; pairs alternate, in two rows, one sessile, one pedicelled, intermixed with many long strong hairs, like an involucre.—*Calyx,* the two inner glumes downy and striated lengthways with many green nerves. —*Corol,* the neuter membranaceous valve always present, but no stamens.—*Seed* oval, transversely rugose.

Obs. When exposed to the weather, the leaves are of a smooth shining deep green, when under trees, of a pale colour, soft, and somewhat downy, which made me long think they were different species. Cattle are fond of it.

30. P. *repens. Linn. Sp. Pl. ed. Willd.* i. 347.

Creeping. *Leaves* broad, waved, base cordate, stem-clasping, and ciliate. *Spikes* from four to ten, equally disposed on all sides. *Flowers* awnless; *involucre* hairy. *Corol* three-valved.

P. *repens. Burm. Ind. t.* 11. *f.* 1. is like this, only the partial spikes are rather too long, and the leaves too narrow.

Common about the beginning of the rains on cultivated land, in gardens, &c. near Calcutta. It is a pretty, cespitose species which cattle are fond of.

Culms, creeping to the extent of from one to two feet, geniculate, smooth, often coloured.—*Leaves* lanceolate, base cordate, stem-clasping, and ciliate. *Sheaths* shorter than the joints, somewhat hairy.—*Spikes* from four to ten, small, short, sessile, secund, equally inserted on the four or six-sided, villous rachis.—*Flowers* generally paired, and both unequally pedicelled, with an involucre of long soft hairs on the outside of the pedicel.—*Calyx, exterior* valve minute, and truncate, the *inner* two equal, five-nerved.—*Corol,* with a neuter valve.

31. P. *hirsutum. Kön.*

Leaves broad-lanceolate, with cordate, stem-clasping base, waved, ciliate, and hairy. *Flowers,* polygamous, solitary, sessile. *Seed* ovate, transversely rugose.

Hind. Jal-ganti.

Teling. Salla woodoo.

Is found on the borders of cultivated land, on pasture ground, &c.

Culms, spreading at the base ; resting on the ground and rooting ; above ascending ; *joints* downy.—*Leaves* broad, base cordate and embracing the culms, much waved, hairy, with the edges ciliate ; sheaths also hairy with a thick beard round the mouths.—*Spikes* composed of six or twelve, simple, alternate, secund, expanding spikelets, surrounding a common, four or five-sided, hairy rachis. —*Partial, rachis* three-sided, hairy.—*Flowers* in two rows, single, alternate, sessile, often intermixed with bristles.—*Calyx,* all the valves hairy and striated.—*Corol;* a complete, one-valved male floret in the same calyx with the hermaphrodite.—*Seeds* ovate, transversely rugose, white, daggered.

Obs. Cattle are very fond of it.

32. P. *tomentosum. R.*

Culms, sub-erect. *Leaves* hairy ; *spikes* from ten to twelve, oblong, from six to eight-flowered, intermixed with hispid bristly involucels. *Seed* ovate, transversely wrinkled.

This is a delicate, rare species, found growing in tufts, or dry pasture ground over various parts of India.

Culms, ascending, compressed towards the base, branchy, from twelve to eighteen inches high.—*Leaves* soft, covered with much, long, soft hair; *sheaths* half the length of the joints; *mouth* bearded.—*Spikes* compound, terminal, consisting of ten or twelve (generally alternate, though sometimes in pairs) *spikelets* of six or eight flowers, intermixed with hispid bristles, they are not placed on one side, but round the common rachis, which is generally three-sided.—*Calyx*, the two interior valves five-nerved; the second, half the length of the corol, as in P. *glaucum*.—*Corol* has a third neuter valve.—*Seed* transversely wrinkled, as in P. *glaucum*.

33. P. *verticillatum. Linn. Sp. Pl. ed. Willd.* i. 334.

Spikes cylindric. *Spikelets* quatern, sub-verticelled. *Involucels*, backwardly hispid bristles. *Seeds* oblong, three-nerved, and rugose.

Hind. Dora-byara.

Teling. Chicklenta.

Delights in a rich soil in out of the way corners, where there is rubbish, &c.

Culms, below resting on the ground and striking root, above weak, often leaning over to one side, smooth, length of the creeping part, from one to two feet, and that of the more erect culms from one to two feet.—*Leaves* sheathing, linear-lanceolate, a little downy; margins hispid when felt backwards; mouths of the sheaths hairy.—*Spikes* columnar, compound, from three to four inches long, generally matted together by means of the bristly involucels.—*Spikelets* or rather *racemelets*, tending to be verticelled, generally four in the verticel, composed of three or four short-pedicels, each bearing a few flowers; all intermixed with very stiff bristles (*Involucels*) armed with short, stiff, recurved points, by which they adhere firmly to every thing that touches them.—*Calyx* as in the family.—*Corol*, a third neuter valve, but no stamens to it.—*Seed* three-nerved, and slightly waved across.

Obs. Cattle will not eat it, so that it is reckoned a troublesome weed wherever it is found.

Plants reared from English seed, were not near so luxuriant as those of India, but evidently the same species. Thunberg's large cultivated variety is certainly *Panicum italicum*. Curtis's figure in his *Flora Londinensis* is represented with the leaves broader at the base than our Indian plant, or even than in plants reared in India from English seed.

34. P. *italicum*. *Linn. Sp. Pl. ed. Willd.* i. 336.

Culms erect. *Spikes* nodding; *spikelets* scattered, ovate; *pedicels* from two to three or four-flowered, with smooth bristles intermixed. *Seed* ovate, three-nerved.

Panicum. Rumph. Amb. v. *p.* 202. *t.* 175. *f.* 2.

Sans. कङ्गु, *Kungoo,* प्रियङ्गु, *Priyungoo.*

Beng. Kungoo, and Kungnee.

Hind. Kora.

Teling. Kora; Koraloo, the grain.

This is one of the plants called dry or small grain. It is cultivated in many parts of India, and requires an elevated, light soil. I never saw it wild.

Culms several from one grain of seed, erect, from three to five feet high, round, smooth; roots issuing from the lower joints.—*Leaves*, margins backwardly hispid; mouths of the sheaths bearded.—*Spikes* compound, &c. There is more or less of a third, or neuter valve on the corol.

Obs. Small plants have the spike more erect, and uniform, without vacancies between the racemelets.

The seed is an article of diet with the natives. It delights in a light, elevated, tolerably dry soil. Seed time for the first crop, about the month of June and July; harvest time in September; produce about fifty-fold in a favorable season. A second crop may be had from the same ground, between September and the end of January.

35. P. *strictum. R.*

Culms straight, from two to three feet high. *Spikes* compound cylindric, straight, with numerous expanding spikelets. *Glumes* of the calyx hairy, and acute. *Corol* three-valved, the innermost one awned.

Beng. Gang-bena.

A native of Bengal, where it is found, though rarely, on dry barren spots.

Culms straight, from one to three feet high, slender as a crow quill, hairy, particularly at and near the joints.—*Leaves* short, straight, broadest at the base, and from thence tapering to a fine point; a little hairy, particularly underneath.—*Sheaths* more hairy, their mouths bearded with much long, soft, white hair.—*Spikes,* (panicles) cylindric, straight, composed of numerous, one-ranked, expanding, short spikelets.—*Flowers* paired, on shorter and longer pedicels which unite before their insertions into the rachis of the spikelets.—*Calyx,* valvelets acute; two exterior with a few straight hairs, inner one smooth; corol-like.—*Corol* three-valved, the inner and smaller one awned.

36. P. *hispidulum. Retz. Obs.* v. *p.* 18. *Linn. Sp. Pl. ed. Willd.* i. 346.

Culms erect, from three to four feet high. *Spikes* panicled, erect, lanceolate; *spikelets* secund; *flowers* two-fold, both sub-sessile; valvelets of the calyx bristly, the middle one daggered, the inner one awned; *corol* three-valved; *seed* ovate, pointed, polished.

Beng. Bura-Shama.

Grows on the borders of rice lands, and rich moist places. Is never cultivated nor made any use of.

Culms erect with a few erect branches.—*Leaves* long, narrow, fine-pointed, smooth.—*Spikes* straight, panicled, composed of numerous, secund, erect spikelets, surrounding the common four, five, or six-sided rachis.—*Flowers* numerous, almost always two together and equally sub-sessile.—*Calyx,* all the three valves harsh with numerous sharp bristles issuing from the nerves of the glumes.

The two exterior ones with sharp subulate points; the inner one ends in a pretty long, strong, hispid arista.—*Corol,* with a third, membranaceous, neuter valve.—*Seed* ovate, pointed, polished.

Obs. It comes near my P. *frumentaceum,* and may probably be the same in its wild state.

37. P. *frumentaceum. R.*

Culms erect, from two to four feet high. *Panicle* erect; *spikes* secund, incurved; *flowers* three-fold, unequally pedicelled. *Valvelets* of the calyces daggered, or awned; *seed* ovate, smooth.

Sans. श्यामाक, Shyamaka.

Beng. Shama.

Teling. Bonta-shama; shamaloo, the grain.

This I have only found in a state of cultivation, it delights in a light, tolerably dry, rich soil; the same ground yields two crops between the first of the rains in June, July, and the end of January.

Culms erect, ramous, a little compressed, smooth, from two to four feet high.—*Leaves* large, margins hispid.—*Panicle* erect, oblong, rigid, composed of numerous, secund, condensed, incurved spikes; they entirely surround the common rachis and sometimes tend to be verticelled. *Rachis, common* five or six-sided, a little hairy; *partial* three-sided, waved.—*Flowers* almost always three from the same point, one sessile, the second short pedicelled, and the third a little longer pedicelled.—*Calyx,* with all the glumes three-nerved; the two large daggered.—*Corol,* with a third neuter valve.—*Seed* ovate, pointed, smooth.

Obs. There are several varieties of it known to the Hindoo farmers. The seed is wholesome and nourishing, it is an article of diet, particularly amongst the lower classes of the natives. It yields about fifty-fold in a good soil. Cattle are fond of it.

Sect. 6th. Panicled.

38. P. *trigonum. Linn. Sp. Pl. ed. Willd.* i. 352.

Culms creeping. *Panicle* halved. *Leaves* stipuled. *Pedicels* two-flowered. *Inner* valve of the calyces vaulted; *corol* three-valved. *Seed*, half round, smooth.

Grows on pasture ground, under the shade of trees, &c.

Culms creeping, filiform, from one to two feet long, with a few inches of the extremities erect.—*Leaves* lanceolate, soft, a little downy; *sheaths* ciliate, with a stipule-like process.—*Panicle* halved, oblong, smooth, composed of from three to ten ramous branchlets, rising on one side only when the plant is small; *pedicels* two-flowered. *Partial pedicels* unequal.—*Calyx*, inner valve cucullate, obtuse; all the valves striated and hairy.—*Corol*, with a neuter, but small valve.—*Seed* smooth, obtusely and obliquely three-sided.

39. P. *patens. Linn. Sp. Pl. ed. Willd.* i. 352.

Culms creeping. *Leaves* with a cordate base. *Panicle* oblong. *Pedicels* one-flowered. The three-valvelets of the calyces equal. *Flowers* polygamous. *Seed*, flat on one side, smooth.

A small, delicate, creeping species; it delights in shady pasture ground.

Culms like those of P. *trigonum.*—*Leaves* broad, with a cordate base, from thence tapering to a fine point, downy; *sheaths* hairy, their mouths want the stipule-like process.—*Panicle* small, about one and a half or two inches long, oval, ramous; *ramifications* smooth.—*Flowers* oval, solitary.—*Calyx*, two-flowered, the three valves are nearly equal, all smooth but striated; the inner one appearing more like a second valve to the male floret than one of the calyx, and the more so as it drops off with the seed, leaving the two exterior ones behind.—*Corol*, one male, or sometimes one neuter, and one hermaphrodite.—*Seed* oval, smooth.

40. P. *brevifolium. Linn. Sp. Pl. ed. Willd.* i. 353.

Culms creeping. *Leaves* lanceolate, cordate. *Panicle* ovate, very ramous, partly hid in the sheath of the exterior leaf, second valvelet of the calyces bearded; *corol* three-valved. *Seed* ovate, smooth.

Grows, with the two preceding species on shaded pasture ground and is very much like them.

Culms creeping, as in the two last described species, but rather more robust.—*Leaves* broad, and somewhat cordate at the base, waved; margins of the sheaths ciliate.—*Panicle*, the lower part always confined within the sheaths of the extreme leaf, oval, from three to five inches long, composed of numerous very fine capillary ramifications, the common rachis and larger branches very woolly. —*Flowers* obliquely-oval, numerous, on very fine, long pedicels. —*Calyx*, the *exterior* valve lanceolate, the second boat-shaped, with the outside covered with very long soft wool, the third oval.—*Corol*, with a neuter valve, but no stamen.—*Seed* oval, smooth, white.

41. P. *tenellum*. *R.*

Culms ascending. *Leaves* long. *Panicles* oblong, very ramous. *Flowers* solitary, remote; second valvelets of the calyces recurved, and pointed. *Corol*, three-valved. *Seed* oval, smooth, five-nerved.

Grows on pasture ground, borders of cultivated lands, &c.

Culms many, branchy, near the ground prostrate, rooting at the joints, the rest ascending; from one to two feet high, and as thick as a sparrow's quill.—*Leaves* soft with white hairs.—*Panicles* oblong, generally as long as the rest of the plant, thin, composed of expanding, capillary, ramous branches.—*Flowers* remote, oval, small.—*Calyx*, the *glumes* striated, otherwise smooth.—*Corol*, with a third neuter valve.—*Seed* oblong, smooth, light-brown, five-streaked.

42. P. *serrulatum*. *R.*

Culms erect, from two to four feet high. *Leaves* sword-shaped, with serrulate margins. *Panicle*, spiked valvelets of the calyces sharp-pointed; *Seed* smooth, oblong.

A large erect, coarse species; a native of moist valleys.

Culms erect, ramous, from two to four feet high. *Leaves* somewhat cordate at the base, from thence tapering to a point, margins ciliate; *sheaths* as long, or longer than the joints, smooth.—*Panicle* erect, linear, composed of erect, linear, ramous ramifications, which stand somewhat remote, particularly below, so that the panicle appears to be a compound, interrupted spike.—*Flowers* lanceolate.—*Calyx*, glumes lanceolate, very acute, the two inner ones somewhat hispid on the back.—*Corol* single, hermaphrodite.—*Seed* oblong, smooth, white.

43. P. *paludosum. R.*

Culms erect, from two to three feet high. *Panicle* erect, ovate, thin; *ramifications* few, three-sided, with sharp hispid angles; pedicel two-flowered; valvelets of the calyces cuspidate. *Corol* three-valved. *Seed* oblong, smooth.

Beng. Boru*ti*, and K*ulus*-nar.

Teling. Soda.

It grows generally in sweet water amongst the Circar mountains.

Culms erect, from two to three feet high, swelled under the water, and there emitting roots from the joints.—*Leaves* erect, acute, retrorso-hispid; *sheaths* longer than the joints, with a thick bearded mouth.—*Panicle* oval, diffuse, thin, composed of scattered, sub-alternate, ramous branches; all nearly triangular, the angles sharp, and armed with fine points.—*Flowers* oblong, pointed, in pairs on a common pedicel with unequal partial pedicels.—*Calyx*, exterior valve short, broad, involving the bottoms of the others.—*Corol*, with a small neuter glume, always without stamens.—*Seed* oblong, smooth, brown.

Obs. This grass is of a coarse nature; cattle are not therefore fond of it. When it grows on dry ground, the grass is much smaller; but the panicle larger.

44. P. *uliginosum. R.*

Culms erect, from two to four feet high. *Panicle* thin; *ramifications* four-sided, smooth; *flowers* polygamous, solitary; *seed* oblong, smooth.

Beng. Bur*unda*.

Teling. G*undru*.

A native of the moist borders of rice fields, in wet ground.

Root fibrous, white.—*Culms* many, ramous, below resting on the ground and rooting; above erect, a little compressed, smooth; from two to four feet high.—*Leaves* short and sharp, inside a little downy; *sheaths* shorter than the joints; mouths bearded. In *paludosum* they are surrounded on the inside with hair. Here the ramifications are compressed, and somewhat four-sided, there triangular, and hispid.—*Panicle* more contracted than in *paludosum.*—*Calyx* the same.—*Corol,* with a complete male floret.—*Seed* oblong, smooth.

Obs. To distinguish it from P. *paludosum,* attention must be paid to the leaves, ramifications of the panicle, and male florets.

45. P. *sarmentosum. R.*

Perennial, inane, hairy, sarmentose, with sub-erect, floriferous shoots from the joints. *Panicle* patent. *Flowers* solitary; *calyces* smooth and obtuse, containing one hermaphrodite and one neuter floret.

A native of Sumatra; and from thence introduced by Dr. Charles Campbell into the Botanic Garden in 1804, where it grows luxuriantly, and blossoms throughout the year.

Culms round, sarmentose, and of great length, inane, with nearly erect, frequently ramous, floriferous shoots of about two or three feet in height rising from the joints; every part clothed with much soft hairy pubescence, particularly the runners.—*Leaves* ensiform, soon becoming reflex, margins hispid, length from six to twelve inches, from half to three quarter of an inch in breadth at the broadest part. *Sheaths* rather longer than the joints.—*Panicle* terminal, ovate; composed of many, single, alternate, patent, compound, smooth branches.—*Flowers* very numerous, awnless, smooth, solitary, pedicelled.—*Valves* of the calyx three, obtuse, slightly marked with green nerves, the exterior one long, containing a two-valved hermaphrodite flower, and a small neuter one.

46. P. *miliare. Lamarck.*

Culms erect, ramous, from two to three feet high, smooth. *Pa-*

nicle thin ; *flowers* paired on a common pedicel, with unequal partial pedicels ; *corol* three-valved ; *seed* ovate, smooth, five-streaked.

Teling. Nella-shama. Nella-shamaloo, the grain.

This species I have found only in a cultivated state, it is one of the sorts of dry or small grain which is generally cultivated on an elevated, light, rich soil.

Culms, many from the same seed, erect, branchy, round, smooth, from two to three feet high.—*Leaves* smooth.—*Panicle* oblong, beautifully bowing with the weight of the grain ; branches alternate, ramous ; capillary, angular, hispid.—*Flowers* oval, paired upon a common pedicel, with unequal partial pedicels.—*Calyces*, *glumes* smooth, striated lengthways with nerves.—*Corol*, accompanied by a neuter valve.—*Seed* oval, striated, smooth, shining brown.

Obs. The seed is an article of diet with those Hindoos who inhabit the higher lands. Cattle are fond of the straw.

47. P. *miliaceum*. *Linn*. *Sp*. *Pl*. *ed*. *Willd*. i. 348.

Culms erect, ramous, from two to four feet high ; the whole plant very hairy. *Panicle* oblong ; glumes of the calyces cuspidate ; *corol* three-valved ; adventitious valve two-toothed ; *seed* oval, smooth, five-streaked.

Sans. व्रीहिभेद, Vreeh*i*bhed*a*, अनु, *U*n*oo*.

Hind and *Beng* Cheen*a*.

Teling. Worga ; Worglo, the grain.

This is another of the cultivated *cerealia ;* the same soil suits it as is proper for the last described species. Time of culture immediately after the rains.

Culms many from the same seed, erect, ramous, very hairy ; from two to four feet high.—*Leaves* large, with long sheaths, which involve most part of the culms, every part covered with much white hair.—*Panicle* as in the former, only here and there a few hairs.—*Calyx* all the three glumes are much pointed.—*Corol* the neuter valve is here emarginate. *Nectary* two, triangular, emarginate bodies embrace the germ laterally.—*Seed* oval, mouth brownish, with smooth, coloured streaks lengthways, as in the last.

48. P. *tenue*. *R.*

Culms erect, ramous ; from one to six feet high. *Panicle* thin, flowers paired on a common hairy pedicel, with unequal partial pedicels ; *Corol* three-valved ; *Seed* ovate, transversely waved.

This is a wild species, which grows amongst the Circar mountains.

Culms erect, sometimes resting on the ground, and striking root there, ramous, smooth, from one to six feet high.—*Leaves* large ; upper-side hairy ; edges armed with very sharp points. *Sheaths* shorter than the joints, hairy, with their mouths bearded.—*Panicle* erect, till the seeds are ripe, then, as in the cultivated sorts last described, bowing from the weight of the grain, it is composed of alternate, compound, nearly secund racemes. *Rachis, common* from four to five-sided ; *partial* three-sided, winding, and hairy.—*Calyx*, glumes many-nerved, otherwise smooth.—*Corol*, the neuter valve is here present, and in luxuriant plants, the stamens also.—*Seed* oval, transversely wrinkled, which distinguishes it from all the other panicled species.

49. P. *plicatum*. *R.*

Erect, herbaceous, harsh with stiff hairs. *Leaves* lanceolate, plicate. *Panicle* lanceolate, ramifications compound, flowers paired, polygamous.

Introduced into the Botanic Garden, from the Island of Sumatra. Flowering time the beginning of the cold season.

Root perennial.—*Culms* erect, or ascending, below as thick as a goose quill, and then somewhat compressed ; every part filled with pith.—*Leaves* sub-bifarious, expanding, lanceolate, plicate, a little hairy ; from six to twelve inches long, and from one to two broad. —*Sheaths* about the length of the joints of the culms, armed with sharp, rigid hairs ; mouths ciliate.—*Panicle* oblong, lax ; lower ramifications rather remote, above approximate, and composed of unilateral short spikelets.—*Flowers* solitary, paired, or threefold, when more than one they are on pedicels of unequal lengths; sometimes there is a single coloured bristle from the pedicels of the exterior flowers.—*Calyx* three-valved, two-flowered, one of the flowers male, with a corol of one membranaceous glume; the other hermaphrodite, and having the usual two-valved corol.—*Seed* oblong, rugose on the convex-side.

Obs. It is of too coarse a nature for cattle, but its foliage make it ornamental in the shrubbery, or flower-border.

50. P. *nervosum. R.*

Erect, from three to four feet high. *Leaves* lanceolate, plaited; *sheaths* longer than the joints, with the mouths bearded. *Panicle* very thin, composed of long, simple, filiform, remote branches. *Flowers* generally single, with a neuter valve to the corol.

From Nepal*a* Dr. Buchanan sent the seed to the Botanic Garden, where the plants blossomed in October, just one year from the time the seed was sown.

Root perennial.—*Culms* erect, slender, completely invested in the sheaths of the leaves, except from six to twelve inches below the panicle; height three or four feet.—*Leaves* lanceolate, plaited, a little hairy, particularly near the base, from six to twelve inches long, and one and a half broad.—*Sheaths* longer than the joints, a little hairy; mouths bearded.—*Panicle* very thin, and composed of long, simple, filiform, drooping branches. *Spikelets* remote near the base, secund, bearing a few alternate, generally single flowers.—*Calyx* three-valved, smooth, from three to seven-nerved.—*Corol*, with a third neuter valve, but no stamens.

Obs. In foliage this agrees very exactly with my *Panicum plicatum*, but differs widely in the panicle.

51. P. *costatum. R.*

Culms creeping. *Leaves* lanceolate, plaited, very hairy, as also their sheaths, which are shorter than the joints. *Panicle* composed of short expanding, secund, spiked-racemes. *Flowers* polygamous.

Introduced into the Botanic Garden from the Mauritius, by Captain Tennant, in 1802.

Culms creeping, throwing out numerous strong roots from the joints; length from three to six feet.—*Leaves* lanceolate, plaited, hairy; about six inches long, by one broad.—*Sheaths* shorter than the joints, very hairy.—*Panicle* oblong, smooth, composed of somewhat compound, secund, spiked-racemes, with here and there an arista-like coloured bristle —*Calyx* three-valved, smooth, from three to five-nerved.—*Corol* three-valved, two-flowered; the exterior floret one-valved and male. *Seed* transversely rugose.

Obs. In the foliage this is perfectly like my P. *nervosum* and *plicatum.*

52. P. *montanum. R.*

Culms erect, smooth, from three to four feet high. *Leaves* lanceolate, unequally divided by the nerve. *Panicle* oblong; *ramifications* capillary, straight. *Corol* two-valved; *Seed* smooth, oval, three-nerved on the back.

A native of the Circar mountains.

Root fibrous, from a ligneous perennial head.—*Culms* erect, smooth, of a firm woody texture, as thick as a crow quill, and three to four feet high.—*Leaves* lanceolate, large, beautifully striated lengthways, somewhat hairy, lower margins next the mouths of the sheath ciliate; the nerve divides the leaf unequally, which is an uncommon circumstance in grasses.—*Panicle* linear, oblong, from twelve to eighteen inches long, composed of straight, sub-erect, capillary ramifications.— *Flowers* small, oval, remote.—*Calyx* as in the genus.—*Corol* as in the family; this is one of the few instances of the want of a male or neuter floret.—*Seed* smooth, brown, with three stripes.

PHLEUM. Schreb. Gen. N. 128.

Calyx two-valved, sessile, linear, truncated, with a bi-cuspid tip. *Corol* inclosed.

1. P. *crinitum.*

Grows in tufts; *Culms* ascending. *Panicle* linear, crowded with innumerable soft, bearded ramifications; both glumes of the *calyx* awned and ciliate, and the large glume of the corol awned.

A native of Nepal*a,* and of the northern part of India.

From the former country it was introduced into the Botanic Garden by Dr. Buchanan; and from the latter by Col. Hardwicke.

Culms numerous from a single root, ascending, length from six to twelve inches, round, smooth, and involved in the sheaths of the leaves, which are longer than its joints.—*Leaves* ensiform, from three to six inches long, smooth.--*Panicle* linear, oblong, dense, exceedingly soft from the immense number of most delicate, pale-coloured awns.—*Flowers* minute and numerous.—*Calyx, glume* one-flowered, two-valved; *valves* nearly equal, with a most delicate, long, soft arista from the apex; hairy, chiefly on the back, and membranaceous ciliate margins.—*Corol* two-valved; valves obtuse, much shorter than the calyx, as in general a very minute awn issues from the apex of the larger valve.—*Seed* smooth, oblong, but obtuse.

Obs. In the cold season of 1802 -3, I reared the same grass from seed sent from England by A. B. Lambert, Esq.

MILIUM. Schreb. Gen. N. 10.

Calyx, glume one-flowered, two-valved. *Corol* two-valved, less than the calyx.

1. M. *filiforme. R.*

Creeping, smooth. *Leaves* short, with a stipulary process at the mouth of the sheath. *Racemes* two or three, terminal, erect; *flowers* alternate, solitary, awnless.

Hind. Kanka juriya.

Grows on pasture ground, is a small, delicate species, has much the appearance of *Agrostis linearis.*

Culms creeping, except a small flower-bearing portion, which is erect, round, smooth; and from six to twelve inches high.—*Leaves* small, smooth; mouths of the sheaths stipuled.—*Racemes* two or three, terminal, sessile, erect, secund. *Rachis* membranaceous, a little waved.—*Flowers* single, alternate, imbricated in two rows, awnless.—*Calyx*, glumes nearly equal, membranaceous, a little hairy.—*Corol*, valves rigid, smooth.

2. M. *sanguinale. R.*

Culms sub-erect, smooth. *Racemes* sub-digitate; *flowers* three-fold, on unequal pedicels.

Panicum sanguinale. *Burm. Flor. Ind. t.* 10. *f. t.*

Beng. Chiri-chira.

A native of pasture ground.

Culms sometimes creeping near the base, the rest erect, smooth, and about a foot and a half high.—*Leaves* long, narrow, smooth in every part.—*Panicle* racemed, sub-erect. *Racemes* from four to eight, secund, erect, rachis three-sided.—*Flowers* generally in threes, though sometimes in pairs or in fours, all have pedicels of unequal length.—*Calyx* and *corol* as in the last species.

3. M. *ramosum. Linn. Sp. Pl. ed. Willd.* i. 361.

Smooth; *culms* sub-erect, from two to four feet high. *Panicle* six or eight, remote, with somewhat ramous branches; *flowers* paired on common pedicels. *Calyces* pointed, and very hairy.

Found growing in large tufts, on moist, rich pasture ground.

Root creeping.—*Culms* from one to four feet long, spreading near the base, and striking root at the joints that rest on the ground.—*Leaves* smooth.—*Panicle* oval, very thin, composed of a few, from four to eight-compound, spreading, secund, racemes. *Rachis* of the racemes three-sided.—*Flowers* below, on a common-pedicel, besides which each has its proper pedicel, but of unequal length ; above single.—*Calyx,* valves equal, pointed, and very hairy.—*Corol,* the exterior valves ends in an awled point.

AGROSTIS. *Schreb. Gen. N.* 111.

Calyx, glume one flowered, two-valved. *Corol* two-valved, larger than the calyx.

1. A. *coromandeliana. Linn. Sp. Pl. ed. Willd.* i. 374. *Retz. Obs.* iv. *N.* 52.

Culms from four to eight inches high. *Panicle* verticelled; *ramification* simple, secund; *inner* valve of the calyx as long as the corol; *exterior* valve minute. *Seed* naked, obovate, rugose.

Teling. Yellika-tungoo-gadi.

A native of pasture ground.

Culms spreading near the base, then ascending, from four to eight inches long, mostly involved in the sheaths of the leaves.—*Leaves* rather broad, waved and slightly ciliated ; mouths of the sheath bearded.—*Panicle* oblong ; large for the size of the grass, composed of expanding, simple, verticelled, secund racemes.—*Flowers* awnless, generally in pairs, unequally pedicelled.—*Calyx, exterior* valve small, like that of the accessory valve in *Panicum. Inner* as large as the corol, and gaping with it when the seed ripens.—*Seed,* it drops uncovered by the corol, leaving it and the calyx in their places ; it is of an oblique oval form, brown, and transversely rugose.

2. A. *tenacissima. Linn. Sp. Pl. ed. Willd.* i. 374.

Smooth, creeping. *Leaves* minute. *Panicle* filiform ; exterior valvelet of the calyx shorter than the corol.

Teling. Toema-gerika.

A small, rigid species ; growing on old, poor, stiff, pasture ground, where it forms extensive plants of tenacious turf.

Culms ramous, creeping to a great extent, with from four to eight inches high of their flower-bearing extremities, erect, smooth, filiform, and very firm.—*Leaves* very small, and smooth.—*Panicle* erect, linear; *ramifications* and *flowers* adpressed, from one to two inches long.—*Calyx*, exterior valve smaller than the interior.—*Corol* as large and long as the inner valvelet of the calyx.

3. A. *Matrella. Linn. Sp. Pl. ed. Willd.* i. 366.

Creeping. *Leaves* pungent. *Spikes* terminal, solitary, few-flowered.

A native of the barren sandy lands, near the sea on the Coast of Coromandel.

4. A. *diandra. Linn. Sp. Pl. ed. Willd.* i. 371. *Retz. Obs.* v. *N.* 37.

Smooth; *culms* erect, from one to three feet high. *Panicle* linear, branchlets short, expanding; *flowers* paired, diandrous. *Calyx* scarcely half the length of the corol. *Seed* naked, obovate, rugose.

Beng. Bena-joni.

A native of moist pasture ground, in the vicinity of Calcutta.

Culms nearly erect, from one to three feet high, almost covered with the sheaths of the leaves.—*Leaves* narrow, tapering to a fine point, smooth; mouths of the sheaths slightly bearded.—*Panicle* linear-oblong, bowing a little, often a foot long, composed of expanding, alternate, compound, seemingly simple racemes.—*Flowers* in pairs on a common peduncle, but unequally pedicelled.—*Calyx*, valves equal, scarcely half the length of the corol, smooth, membranaceous.—*Stamens* two. *Seed* narrow-obcordate, dropping from the calyx and corol, brown, obtusely rugose.

5. A. *maxima. R.*

Culms erect, from four to twelve feet high. *Panicle* ovate, expanding. *Calyces* minute, exterior valvelets of the corol fringed with long hairs.—*Stamens* two.

This is the largest species I know; it is a native of hedges, &c. amongst the mountains.

Root ligneous, perennial.—*Culms* erect, from four to twelve feet high, often as thick as the little finger, round, and smooth, inside replete with a fibrous spongy substance.—*Leaves* smooth, from one to two feet long; and from one to two inches broad at the base, tapering from thence to a fine point.—*Panicles* erect, oval, or oblong, from one to two feet long, composed of alternate, ramous, filiform, expanding or bowing branches.—*Calyx*, valves equal, four times shorter than the corol.—*Corol*, exterior valve fringed round the margins, with long, soft, white hairs.—*Stamens* two.

Obs. The leaves of this plant are much more like those employed by the Chinese to pack up various articles with, than any other I have yet seen.

HOLCUS. Schreb. Gen. N. 1565.

Polygamous. *Calyx*, glume two-valved, one to three-flowered. *Corol*, glume two-valved, the hermaphrodite awned.—*Stamina* three. *Styles* two.

1. H. *nervosus. R.*

Erect, hairy. *Panicle* oblong. *Glumes* three, and five-nerved, cuspidate, two-flowered; *flowers* both sessile, one hermaphrodite and awned; the other male and awnless.

A native of Coromandel.

Culms erect, ramous; long, wiry roots issuing from the lower joints.—*Leaves* and their sheaths papillose, hairy.—*Panicles* oblong, smooth. *Ramifications* alternate, simple.—*Flowers* paired; one sessile, the other pedicelled.—*Calyces* two-flowered, three to five-nerved, smooth, straight, cuspidate, longer than the flowers.—*Corol* in both two-valved, the exterior floret male, large, and awnless, the inner minute, hermaphrodite, and awned.—*Stamens* three.—*Styles* two, in the hermaphrodite.

2. H. *ciliatus. R.*

Erect, hairy, panicle linear, hirsute. *Glumes* papillose, ciliate, two-flowered, both flowers sessile, one hermaphrodite and awned, the other male and awnless.

A native of Coromandel.

Whole plant about a foot high, erect. The *culms* with bearded joints.—*Leaves* rather large, hirsutely papillose, as are also their sheaths.—*Panicles* linear, shaggy with the numerous hairs of the *calyces.*—*Calyx* two-valved, two-flowered; *valves* nearly equal, considerably longer than the flower, backs covered with little warts, from each of which arises a very conspicuous hair; *apices* cuspidate, and recurvate.—*Corol* two-valved in both.—*Exterior* floret rather larger, male, or male hermaphrodite, and awnless; *inner* hermaphrodite and awned.—*Stamina* three.—*Styles* two in the hermaphrodite, in the male hermaphrodite the rudiments of two, but no germ.

ISCHÆMUM. Schreb. Gen. N. 1569.

Polygamous. *Calyx* two-valved, two-flowered. *Corols* two-valved.

1. I. *aristatum. Linn. Sp. Pl. ed. Willd.* iv. 939.

Annual. *Leaves* sword-shaped, ciliate and hairy, with coloured margins. *Spikes* paired, secund. All the calyces two-flowered; *valves* striated; *exterior* of the sessile flower with a bifid membranaceous border round the anterior margin; an awned hermaphrodite floret in each flower.

A native of newly formed pasture land, in various parts of India, appearing about the close of the rains. It is a scarce grass, and does not appear to be fit for cattle.

Culms at the base resting on the ground and there rooting, ramous; erect portions from one to two feet long, slender, with woolly joints.—*Leaves* length of the spikes, ensiform, broadest at the base and there cordate, hairy, ciliate, with hispid, coloured margins.—*Spikes* two, terminal, secund, from one to two inches long. *Rachis* flexuose, jointed, three-sided, with the angles very hairy.—*Flowers* in pairs, one sessile, the other pedicelled. *Pedicel* three-sided, and ciliate.—*Calyx* of the sessile flowers two-valved, two-flowered; one floret hermaphrodite, the other male. *Exterior valve* striated, with its anterior margins and apex enlarged with a broad, membranaceous, bifid border; *inner valve* boat-shaped, keeled, ciliate, and sub-aristate, one floret male, the other hermaphrodite.—*Corol, glumes* of each floret two-valved; the exterior one of the hermaphrodite floret bifid, with a long arista in the fork.—*Calyx* of the pedicelled flower two-valved, two-flowered; *valvelets* equal, boat-shaped, keeled, ciliate, and sub-aristate; one floret male, the other hermaphrodite.—*Corol,* as in the sessile flowers, and also awned in the same manner.

Obs. I am not certain that this is the original I. *aristatum* of Linneus and Retzius.

2. I. *rugosum. Linn. Sp. Pl. ed. Willd.* iv. 940.

Erect, ramous. *Leaves* lanceolate; *spikes* paired, both flowers awned; *calyces* oblique, obtuse, hard, and transversely rugose.

Is generally found growing among rice, and is so much like it, the inflorescence excepted, that they are not, till in flower, to be distinguished.—*Leaves* large, smooth; *mouths* of the sheaths crowned with a long, two-parted ligula.—*Spikes* terminal, and from the exterior axils paired, erect, from two to three inches long. *Rachis* jointed.—*Flowers* in pairs; one sessile, the other on a short, thick, angular clubbed pedicel.

SESSILE FLOWER. *Calyx* two-flowered, two-valved; *valvelets* in size nearly equal, the exterior one rugose, and very hard; one of the florets hermaphrodite, the other male, each with two membranaceous valves; the inner valve of the hermaphrodite flower has a long twisted awn issuing from the middle of its back.

PEDICELLED FLOWER smaller. *Calyx* as in the other, but containing two male flowers, one of which is also awned.

3. I. *semisagittatum. R.*

Annual. *Leaves* ovate-lanceolate, semisagittate. *Spikes* paired, secund, all the calyces two-flowered, exterior valvelets transversely rugose; hermaphrodite floret of the sessile flower awned.

A native of newly formed pasture land in Bengal. It is a rare grass, appearing about the close of the rains.

Culms ramous, resting on the ground near the base, and there striking root at the joints, the sub-erect portion from twelve to eighteen inches, very delicate, firm, and smooth.—*Leaves* short, broad-ensiform, or rather ovate-lanceolate, with a cordate base, and one of the lobes generally projecting considerably beyond the culm; margins slightly hispid, length one and a half or two inches.—*Spikes* two, terminal, sub-secund, an inch and a half long. *Rachis* jointed, scarcely flexuose, three-sided; *angles* hairy.—*Flowers* in pairs, one sessile, the other sub-sessile, both with their base bearded.—*Calyces* of both flowers almost alike, two-flowered, two-valved; *exterior* valvelets obliquely-lanceolate and rather obtuse, the lower gibbous half transversely rugose, as in I. *rugosum*, upper portions triated lengthways; margins of the lower rugose portion notched; of the upper striated part ciliate; *inner* boat-shaped in both. one floret is male, the other hermaphrodite.—*Corol, glumes* of each floret two-valved; the exterior valvelet of the hermaphrodite floret bifid, with a long arista in the fork.

Obs. It ought to be compared with I. *barbatum*.

4. I. *conjugatam*. *R.*

Partly creeping. *Leaves* short, acute, with broad cordate base. *Spikes* conjugate, seemingly united. *Flowers* paired, with the exterior valves of both calyces even, and very woolly; the sessile one hermaphrodite, awned.

A native of pasture land in the vicinity of Calcutta, appearing in the cool season.

Culms very ramous, creeping near the base, filiform, smooth, length from six to eighteen inches, generally purplish.—*Leaves* with broad cordate base, from thence tapering to a fine point, sheaths smooth.—*Spikes* conjugate on a slender, clavate, purple peduncle;

united at the base, at all times continuing close as if one. *Rachis* very hairy, and jointed.—*Flowers* from four to eight pairs on each spike, one sessile, the other sub-sessile.—*Calyces* nearly alike, two-valved, two-flowered; the exterior valves even, acute, and uncommonly woolly on the backs; the inner boat-shaped, acute, and smooth. In each, one of the florets is hermaphrodite, the other male.—*Corols* all two-valved, hyaline; the inner valve of the hermaphrodite floret in the sessile flower bifid and awned.—*Stamina* three.—*Styles* two.

5. I. *geniculatum*. *R.*

Stems geniculate, creeping, smooth. *Leaves* ensiform; *spikes* conjugate, apparently only one. *Calyces* two-flowered, smooth, and crested; with the hermaphrodite flowers awned.

A small species, a native of Bengal. Flowering time the cool season.

Culms polished, from one to two feet long, very ramous, diffuse, geniculate; *joints* large, emitting smooth, very brown roots.—*Leaves* small, ensiform, smooth; *sheath* smooth, and shorter than the polished joints of the culm.—*Spikes* terminal, conjugate, generally so close as to seem only one, smooth, about an inch long. *Rachis* jointed, flexuose; *angles* sharp, and sometimes hairy.—*Flowers* paired, one sessile, one peduncled.—*Calyces*, both two-flowered; one hermaphrodite and awned, the other nale and awnless. *Exterior* valve of the sessile flower enlarged with two, hard, serrulate crests, one on each side of the apex; *inner* mucronate, with a similar crest on the back, under the apex; of the *pedicelled flower* both are boat-shaped, mucronate, or sub-aristate, and the exterior one crested on the back under the apex; all are smooth and marked with green, longitudinal veins.—*Corols* two-valved; the exterior valve of the hermaphrodite flowers bifid and awned.

6. I. *tenellum*. *R.*

Creeping near the base. *Leaves* linear; *spikes* paired; *exterior* valve of the sessile flower truncate, the rest cuspidate, hermaphrodite flower alone awned.

A weak straggling species.

Culms weak, creeping and ramous towards the base; towards the heads erect; the thickness of a pigeon's quill; joints bearded. *Leaves* a little hairy near the base, mouths of the sheaths ligulate, and hairy.—*Spikes* paired, &c. as in the last species.—*Flowers* paired, &c. *Calyx* of the sessile flowers two-valved, two-flowered; *exterior* valve retuse and bordered; *inner* valve ending in an awled point.—*Flowers* two, one hermaphrodite with a two-valved corol, the exterior of which ends in an arista, the other is male with an awnless two-valved corol.—*Calyx*, of the pedicelled flower both valves alike, and ending in an awled point; it contains a single flower, which is sometimes hermaphrodite, and sometimes male, with the rudiment only of the second flower.

7. I. *repens. R.*

Creeping. *Leaves* short and acute. *Spikes* paired. *Flowers* all polygamous and awnless; *glumes* of the calyces smooth and cuspidate. *Rachis* smooth.

A native of Pulo-penang.

Culms creeping, flexuose, smooth, from one to two feet long.—*Leaves* bifarious, sword-shaped, short, rigid. *Sheaths* as long as the joints, a little hairy, with a stipulary process at the mouth.—*Spikes* terminal, paired, uniting a little above what seems to be their base. *Rachis* jointed, three-sided, smooth.—*Flowers* about six pair, one of each pair sessile, the other peduncled, both the pedicelled and sessile one consisting of a two-flowered, two-valved, rigid, acute, smooth calyx.—*Flowers* polygamous, one hermaphrodite, the other male, both awnless, each floret having a corol of two acute valves.

8. I. *cuspidatum. R.*

Floating on sweet water with a portion of the tops of the plant emerging. *Leaves* linear, smooth. *Spikes* three-fold. *Exterior* valve of the calyces with a long, waved, ensiform point; all the corollets two-valved, and awnless.

A native of Bengal, where it is found floating on pools of sweet water, and blossoming about the close of the rains in October.

Culms appear to be perennial, a fathom or two long, jointed, as thick as the little finger, floating.—*Leaves* linear, acute, from one to two feet long, margins backwardly hispid, in other respects smooth.—*Spikes* terminal, three in the only plant that has yet blossomed in this Garden, each about six or eight inches long, a little compressed; when the flowers expand about noon they appear like Chevaux de frise, at other times they appear smooth.—*Flowers* in pairs, one sessile, containing one hermaphrodite, and one male floret, the other pedicelled and neuter, or with two male florets.—*Calyx* two-valved; *exterior* valve rigid, with hispid margins and long, flat, waved, hispid-margined, ensiform points; *inner* valve boat-shaped with a flat hispid keel.—*Corols* of each floret two-valved.

APLUDA.

Polygamous. *Involucre* one-valved, two-flowered, one-sessile, the other pedicelled with a pedicelled neuter rudiment. *Calyces* two-flowered, the sessile one polygamous, the pedicelled one the same, or with male only. *Corols* two-valved.

1. A. *aristata. Linn. Sp. Pl. ed. Willd.* iv. 938.

Perennial, creeping, or scandent. *Leaves* lanceolate. Hermaphrodite flower awned.

Beng. Goroma.

Teling. Pootstrangali.

Is commonly found in hedges, under the shade of trees and such like places.

Culms creeping, or climbing, as thick as a crow-quill, jointed, coloured, the lower part naked, inwardly spongy; from the joints spring, erect, ramous, flower-bearing branches of from one to seven feet high.—*Leaves* very elegant, somewhat bifarious, sheathing, petioled, linear-lanceolate, backwardly hispid; from six to eighteen inches long; *sheaths* smooth, mouths short-stipuled.—*Panicles* terminal, and from the exterior axils, or, the whole may be considered, as one linear leafy panicle.—*Bractes* three-flowered, boat-shaped, with an awned point.—*Flowers*, one of the three sessile, or rather sitting on a globular receptacle, consisting of a common, two-flowered calyx, one of which is sessile, awned, and hermaphrodite; the other sessile, male and awnless; just without the calyx on each side, there is a bent, compressed pedicel, one of these supports a common calyx, with two male, two-valved, awnless flowers, the other only the rudiments of one or more florets.

2. A. *geniculata.* R.

Perennial, creeping, or scandent, intricately geniculate. *Leaves* lanceolar. In each calyx one hermaphrodite, and one male flower, all awnless.

Found on the banks of the Ganges in flower during the latter part of the rains and the cool season.

Culms creeping, or climbing amongst bushes, reeds, &c. to an extent of many feet, smooth and slender, bent at angles about a span distance at the swelled joints, emitting numerous wiry roots, filled with spongy pith, the thickest about the size of a crow-quill. —*Leaves* linear lanceolate, and smooth, but much smaller than in A. *aristata*. *Sheaths* shorter than the joints. *Ligula* lacerate.—*Panicles* terminal, composed of numerous slender branchlets of distinct fascicles of flowers, each fascicle containing many short pedicelled, three-flowered, boat-shaped, cuspidate, smooth bractes or involucres.—*Flowers* one of the three sessile, containing one hermaphrodite, and one male floret; the other two elevated or broad sulcate; proper pedicels, one minute and abortive, the other polygamous as in the sessile flower.—*Calyces* of both the sessile and pedicelled flowers two-valved, two-flowered.—*Corol* in all of them two-valved; *valves* unequal, exterior one thin, and transparent.

Obs. The accurate Gærtner, Carp. vol. ii. p. 466 t. 175 gives to the sessile flower of A. *mutica,* one hermaphrodite, and one female floret, and to the pedicelled one two male ones, consequently this must be different, and a second awnless species.

AIRA. Schreb. Gen. N. 112.

Calyx two-valved, two-flowered, without the rudiment of a third.

1. A. *filiformis. Kön. Mss.*

Smooth, erect, one foot high. *Leaves* ensiform. *Panicles* oblong, compound, of simple, linear, secund racemes.

A native of pasture ground.

Culms erect, mostly naked, round, smooth, about a foot high.—*Leaves* small, smooth; *mouths* of the sheaths enlarged by a membranaceous, torn process.—*Panicle* erect, thin, oblong, composed of alternate, simple, one-ranked, filiform, expanding racemes.—*Flowers* in two rows.—*Calyx* two-flowered, large.

MELICA. Schreb. Gen. N. 113.

Calyx two-valved, two-flowered, with the rudiment of a third.

1. M. *digitata. R.*

Culms from four to five feet high. *Spikes* terminal, expanding, mostly five-fold. *Calyx* one-flowered; outer valvelets of the corol and rudiment awned.

A large beautiful species, a native of hedges, thickets, &c.

Culms near the base procumbent, but when supported by bushes erect and often from four to five feet high.—*Leaves* upper side somewhat hairy; *mouths* of the sheaths bearded.—*Spikes* four or five, digitate, secund, spreading, filiform, from six to nine inches long, hairy at the base. *Rachis* three-sided.—*Flowers* in two rows, sessile, awned.—*Calyx* one-flowered; *glumes* unequal, the interior being three or four times longer than the exterior, and short-awned.—*Corol,* valves unequal, the largest has a long awn issuing just below its apex on the outside.—*Corpuscle*, or neuter floret pedicelled, and awned.

2. M. *refracta. R.*

Perennial, ascending. *Leaves* lanceolate. *Panicle* ovate, with the ramifications refracted. *Petals* bearded, with stiff refracted bristles.

A native of the Moluccas, and from thence introduced amongst other plants, into the Botanic Garden, where it blossoms during the cold season.

Root perennial.—*Culms* ascending, from one to three feet long, round and smooth.—*Leaves* lanceolate, smooth; *sheaths* smooth, scarcely bearded at the mouths.—*Panicle* terminal, ovate; *ramifications* somewhat compound, and refracted like the drooping branches of some species of Pinus.—*Calyx* smooth, two-valved, two-flowered, with an abortive third, or the rudiment of one.—*Corol*, exterior glumes, chiefly of the second flower, bearded with refracted, sharp, stiff bristles which fix to every thing that touches them.

3. M. *diandra. R.*

Smooth; *culms* from two to four feet high. *Leaves* lanceolate. *Panicle* of long, simple, expanding racemes; *flowers* awnless, diandrous.

Poa *malabarica. Linn. Sp. Pl. ed. Willd.* i. 394.

A native of mountains on the Coromandel Coast.

Culms ascending, from two to four feet high, firm, smooth, about as thick as a fine quill.—*Leaves* bifarious, lanceolate, striated, smooth; *margins* waved, unequally divided by the nerves, about an inch and half broad, and six inches long; *mouth* of the sheaths projecting above in a membranaceous process.—*Panicles* large, thin, oblong, bowing a little, composed of sub-alternate, simple, spreading racemes.—*Flowers* awnless.—*Calyx* two-flowered, with the characteristic corpuscle, which is also here pedicelled, but awnless.—*Corol*, exterior valve of the inner one has its margins and keel ciliate.—*Stamens* two.

4. M. *latifolia*. *R.*

Perennial; *culms* erect, simple, from four to eight feet high. *Leaves* lanceolate. *Panicles* large and open, with alternate, compound drooping branches.

A native of the Garrow hills, from thence it was brought to the Botanic Garden, by Mr. Robert Kyd, where it blossoms in March, at which period and indeed at all times it is one of the most elegant plants in the whole order.

Stems very numerous from the same root, erect, straight, and always without branches, more or less invested in the sheaths of the leaves; very smooth, texture remarkably hard, and filled with firm pith, generally about as thick as a goose quill, and when in flower from six to eight feet high.—*Leaves* sessile on their sheaths, lanceolate, smooth; from six to twenty inches long, and from one to four broad. They are much like the leaves employed by the Chinese to put between the boxes and lead canisters in which their Teas are packed.—*Seed* ventricose-oval, smooth, both ends rather pointed.

CHLORÌS.

Polygamous. *Calyx* two-valved, from two to six-flowered; hermaphrodite flowers sessile; male, or neuter ones pedicelled. *Corol* of the hermaphrodite flower two-valved; of the male, or neuter one-valved, all of them awned.

1. C. *tenella*. *R.*

Spikes solitary; *Spikelets* with three or four hermaphrodite awned florets, and a neuter rudiment.

A native of the Peninsula of India.

Culms delicate, erect, smooth, about a foot high.—*Leaves* rather large in proportion to the rest of the plant, smooth, and soft.—*Spikes* solitary, secund, scarcely two inches long.—*Spikelets* alternate, and alternately pointing two ways.—*Flowers* of the spikelets, or to each calyx, three, four, or five, all hermaphrodite, (at least in the rather imperfect state in which I have found this rare species) except the innermost one which is a one-valved corpuscle or rudiment, as in *Melica.*—*Calyx* from three to five- flowered, two-valved; *valvelets* unequal, broad lanceolate, smooth, acute, permanent.—*Corol* of the hermaphrodite two-valved; exterior *valvelets* obcordate, cucullate, awned; *margins* smooth, but with three hairy nerves on the back and sides; *interior* oblong, with edges ciliate and always awnless.

2. C. *montana. R.*

Spikes digitate, secund, united at the base; *hermaphrodite* flower ciliate, and awned; *neuter* of three awnless valves.

This a native of mountainous tracts only. It differs from *Chl. barbata* only in the following respects. 1*st.* The spikes are in number from four to six, and united into a common flower-bearing rachis near the base. 2*d.* The neuter floret has a third awned valve and the awns are all straw-coloured.

3. C. *barbata. Linn. Sp. Pl. ed. Willd.* iv. 926.

Spikes digitate, secund; *hermaphrodite* corol ciliate, and awned, neuter of two awned valves.

Andropogon *barbatum. Linn. Mant.* 302. 588.

Konda-pulla. *Rheed. mal.* 12. *p.* 95. *t.* 51.

Grows in large tufts on pasture ground, &c. very common every where.

Culms, below creeping, and ramous, the joints perfectly smooth.—*Leaves* near the base bifarious; *margins* near the base and mouths of the sheaths ciliate.—*Spikes* terminal, from six to twelve, sessile, secund, expanding, from one to two inches long. *Rachis* striated, not hairy, nor jointed.—*Flowers* pedicelled, alternate, in two rows on the outside of the spikes —*Calyx* two-flowered, two-valved, membranaceous and awnless. The lower flower is sessile and hermaphrodite, its corol consists of two unequal ciliate valves, the base of which is surrounded with hairs, the exterior one ending in a long, coloured awn. The other floret I have always found neuter;

it stands on a short pedicel and consists of two obcordate, awned valvelets; *awn* and *margin* of the valvelets coloured.

Obs. Cattle eat it till in flower; after which I never saw any animal touch it.

4. C. *polystachya. R.*

Spikes from ten to twenty, fascicled. Exterior glume of the hermaphrodite flower with ciliate margins; neuter florets two, with single smooth valves.

A native of the Peninsula of India.

Culms below procumbent for a little way, then erect, and about two feet high.—*Leaves* as in other grasses, and smooth.—*Spikes* about sixteen, in a terminal, sub-fastigiate umbelliform fascicle, secund with the alternate spikelets, or flowers pointing to opposite sides, length from two to three inches.—*Calyx* of two, unequal, lanceolate, boat-shaped, smooth, permanent awnless valves, which contain one, two-valved, hermaphrodite, awned flower; and two peduncled, one-valved, awned, neuter florets. The exterior valvelet of the hermaphrodite flower has its margins ciliate.

POMMEREULLA. Schreb. Gen. N. 97.

Calyx, glumes two-valved, from three to four-flowered; *valvelets* four-cleft, awned on the back.

1. P. *Cornucopiae. Linn. Sp. Pl. ed. Willd.* i. 314. *Vahl. En. Pl.* ii. 393. *R. Corom. Pl.* ii. *N.* 131.

A very singular, small, rare grass; growing under bushes on dry uncultivated ground.

Root, Culms, and *Leaves,* as described by König, but mine has the spike compound and secund, the rest as he says, alba base abvolluta folia sub-spatheformi.—*Spiculis* distichus, each having an involucre of two, lanceolate, acute, membranaceous valves. Within these two valves, is a short, clubbed, downy pedicel supporting the partial spikelets of four flowers, with their common calyx, it is joined to the involucre by an articulation and drops off most readily, particularly when dry, leaving the involucre attached to the rachis.—*Calyx, Corol,* and *Stamens,* are as König describes them; but the germ is obcordate, and there are always two styles with feathery recurved stigmas.—*Seed* as he describes it.

POA. Schreb. Gen. N. 114.

Calyx two-valved, containing many flowers of two unequal valvlets disposed alternately on distichous spikelets.

1. P. *bifaria. Vahl. symbol.* ii. 19. *Linn. Sp. Pl. ed. Willd.* i. 401.

Smooth; *culms* straight, from one to two feet high. *Spike* terminal, straight, secund. *Spikelets* sessile, alternate, bifarious, the inferior from four to six-flowered, above many-flowered.

Teling. Wooda-tallum.

I believe König called it *Poa malabarica.*

Grows on mountains and other dry situations.

Culms straight, simple, from twelve to twenty-four inches high, a little compressed, smooth.—*Leaves* small, smooth.—*Spikes* straight, secund, from four to eight inches long. *Spikelets* alternate, sessile, lanceolate, in two spreading rows, though issuing from one side of the common rachis.—*Calyx* below, from four to six-flowered; above many-flowered.

2. P. *chinensis. Linn. Sp. Pl. ed. Willd.* i. 395.

Smooth, from two to four feet high. *Panicle* of many alternate, simple, expanding, secund ramifications; *Spikelets* alternate, short-pedicelled, from four to six-flowered.

Beng. Bura-pini-nuti.

A large, beautiful species, growing on the borders of water courses, and places where there is much moisture.

Culms ramous, creeping near the base, above expanding, from two to four feet high, every part smooth, and a little compressed.—

Leaves as in other grasses, smooth ; *mouths* of the sheaths bearded.—*Panicle* large, oval, composed of long, alternate, filiform, simple, expanding, secund branches. *Spikelets* alternate, from four to six-flowered, short-pedicelled, expanding, in two rows from one side of the simple branches.

3. P. *procera*. *R.*

Smooth, erect, from three to five feet high. *Ligula* large ; *ramifications* of the panicle simple ; *spikelets* pedicelled, rather remote, linear, many-flowered.

Teling. Rewa.

Delights in a moist rich soil.

Culms nearly erect, branchy, from three to five feet high, round, smooth, much covered by the sheaths of the leaves.—*Leaves* long, slender and smooth. *Sheaths* longer than the joints, with their mouths crowned with a long ragged membranaceous process.—*Panicle* large, from nine to eighteen inches long, oblong, bowing a little, composed of numerous, long, filiform, expanding racemes, scattered round the common *rachis;* which is filiform, waved, three-sided, and hispid. *Spikelets* alternate, pedicelled, linear, remote, many-flowered.

4. P. *cynosuroides*. *Linn. Sp. Pl. ed. Willd.* i. 393.

Smooth, straight, from one to three feet high. *Leaves* long and acute. *Panicle* straight, sub-cylindrical; *ramifications* horizontal, *spikelets* depending, from six to twelve-flowered.

Uniola bipinnata. *Linn. Sp. Pl.* 104.

Sans. कुशः, Koosh*a*, कुथः, Kooth*a*, दर्भः, D*u*rbh*a*, पवित्रं, P*u*vitr*u*ng.

Cusa or *Cusha*. *Asiat. Res.* iii. 255. and 490. and iv. 249.

Beng. Koosh*a*.

Teling. D*u*rbha, D*u*bha, or D*u*rpa.

A strong coarse species, a native of dry barren ground.

Root creeping, perennial.—*Culms* straight, rigid, round, smooth, from one to three feet high.—*Leaves* numerous, very long, chiefly about the base of the culms, rigid margins hispid.—*Panicle* erect, linear-oblong, often tending to a conical form, composed of many somewhat three-fold, verticelled, horizontal, short, rigid, secund ramifications. *Spikelets* many-flowered, depending, in two rows, from the under side of the ramifications.—*Corol,* valves pointed, the inner one rather the largest.

Obs. It is employed by the brahmuns in their religious ceremonies. Cattle do not eat it. Can this be Gramen capillaceum, &c. *Pluk. Alm. p.* 176. *t.* 34. *Fig.* 2.—*Cusa,* or *Cusha,* the Sanscrit name of this much venerated grass, was given to it at a very early period, by the Hindoo Philosophers, and believed, by Sir William Jones, to have been consecrated to the memory of *Cush,* one of the sons of *Ram.* See *Asiatic Researches, vol.* iii. *p.* 490. If so, we have here a very ancient precedent for the present very convenient practice of giving such arbitrary names to plants, and to their families.

5. P. *ciliaris. Linn. Sp. Pl. ed. Willd.* i. 402.

Smooth, sub-erect, from one to two feet high. *Panicle* contracted, linear, with the lower ramifications remote. *Spikelets* from six to twelve-flowered; *inner* valvelets of the corol ciliate.

Delights in the same soil and situation with the last described species.

Culms near the base procumbent, above erect for one or two feet.—*Leaves* as in the last.—*Panicle* linear, from three to six inches long, the ramifications short, those of the lower part stand at some small distance, their insertions are not hairy.—*Calyx* from six to twelve-flowered, not ciliate.—*Corol,* only the inner valve ciliate. —*Seed* as in the last.

6. P. *cylindrica. R.*

Erect, slender, from twelve to eighteen inches high. *Leaves* slender. *Panicle* cylindric, crowded with short, adpressed, ramous, alternate branches; *spikelets* from six to twelve-flowered; *flowers* diandrous, inner valve of the corol ciliate.

From Canton in China the seed was received into the Botanic Garden, where the plants grow freely and blossom during the rains and the cool season.

7. P. *ciliata. R.*

Smooth, sub-erect, from one to two feet high. *Panicle* columnar; *spikelets* from six to twelve-flowered; *margins* of all the valvelets of the calyces and corols ciliate. *Seed* globular.

Is generally found on a poor dry soil.

Root perennial.—*Culms* erect, rigid, smooth; from one to three feet high.—*Leaves* smooth, mouth of the sheaths downy.—*Panicle* columnar, from two to four inches long, branches thereof crowded, insertions hairy, *spikelets* from six to twelve-flowered.—*Calyx,* margins ciliate.—*Corol,* the margins of both valves much ciliate, the exterior one has three nerves and is pointed, the inner one two-nerved —*Seed* obovate, smooth, brown, dropping from the corol when ripe.

8. P. *interrupta. Kön. Mss.*

Erect, smooth, from one to three feet high; *ramifications* of the linear panicle short, and collected into remote fascicles. *Spikelets* from six to twelve-flowered. *Seed* obcordate.

Teling. Nakurnaral.

Grows about at the borders of rice fields.

Clums generally erect, from one and a half to three feet high, round, smooth.—*Leaves* narrow, long, sharp, and smooth.—*Panicles* erect, from six to twelve inches long, linear; *ramifications* sessile, nearly erect, remote, particularly those of the lower part, where they are often twice their length asunder; one, two, or more from the same point, which makes it appear verticelled.—*Spikelets* minute, pedicelled, from six to twelve-flowered.—*Corol,* valves smooth.—*Seed* as in the last two species.

9. P. *nutans. Linn. Sp. Pl. ed. Willd.* i. 395. *Retz. Obs.* iv. 19.

Erect, smooth, from three to five feet high, panicles contracted;

ramifications adpressed, fascicled, and approximated. *Spikelets* from eight to fourteen-flowered ; *seed* oblong.

Teling. Urenka.

Delight in a rich, moist soil, such as the banks of water courses, borders of rice fields, &c.

Culms erect, from three to five feet high, generally simple, round, smooth.—*Leaves* narrow, long, sharp, and smooth.—*Panicle* linear, from one to two feet long ; *ramifications* filiform, peduncled, adpressed, one, two, or more from nearly the same place, but seldom so far asunder as their own length. *Spikelets* pedicelled, from eight to fourteen-flowered.—*Calyx* and *corol* smooth.—*Seed* oblong, smooth, brown.

Obs. The best mark to distinguish it from the last species is the form of the seed, which in that is obovate, in this oblong. Cattle are not fond of any of these tall, erect, coarse species.

10. P. *diandra. R.*

Erect, smooth. *Leaves* long, fine-pointed ; *panicle* linear, half the length of the whole plant; *ramifications* scattered, compound, expanding, smooth ; *spikelets* from four to eight-flowered. *Flowers* diandrous.

A native of Bengal, where it blossoms during the cold season.

Culms erect, growing in tufts, with few branches, smooth ; height of the whole plant in a good soil, from three to six feet.—*Leaves* from one to two feet long, smooth in every part, tapering to a long fine point.—*Panicle* linear, half the length of the whole plant; *ramifications* thereof scattered, expanding, smooth, compound, from two to three inches long.—*Spikelets* lanceolate, small, smooth, from four to eight-flowered.—*Corol,* with both valves smooth, and rather obtuse.—*Stamens* two. *Anthers* purple.

11. P. *viscosa. Linn. Sp. Pl. ed Willd.* i. 398. *Retz. Obs.* iv. *p.* 20.

Culm ascending, from nine to eighteen inches high, clammy. *Panicle* linear-oblong; *ramifications* verticelled, short, spreading;

spikelets from three to twelve-flowered; inner glumes of the corol ciliate.

This species grows in tufts on dry pasture ground.

Culms numerous, spreading, with their extremities ascending, ramous, from nine to eighteen inches long.—*Leaves* small, margins involute: *sheaths* shorter than the joints, with their mouth surrounded with long, slender, white hair.—*Panicles* linear-oblong, from two to four inches long, composed of short, sub-verticelled, sessile, expanding ramifications.—*Calyx* from three to four-flowered.—*Corol*, inner valve ciliate.

Obs. Every part of the plant is covered with tenacious gluten.

12. P. *tenella. Linn. Sp. Pl. ed. Willd.* i. 395.

Erect, smooth, from one to two feet high. *Panicle* linear-oblong, verticelled. *Spikelets* pedicelled, from four to six-flowered, smooth.

A native of pasture ground.

Culms erect, round, smooth, from one to two feet high.—*Leaves* smooth, *mouth* of the sheaths much bearded.—*Panicles* linear-oblong, composed of verticelled, filiform, expanding, ramous branches, having always at their insertions and sub-divisions a few fine, long hairs. *Spikelets* pedicelled, from four to six-flowered.—*Calyx* and *corol* smooth.—*Seed* oval, brown, smooth, falling from the corol when ripe, as do all the other species of this genus that I have yet seen.

Obs. The anthers shine through the valves, as in P. *punctata*, and indeed in most, if not all the *Poas*.

13. P. *plumosa. Linn. Sp. Pl. ed. Willd.* i. 403. *Retz. Obs.* iv. 20.

Sub-erect, smooth. *Panicle* ovate-oblong; *ramifications* scattered, horizontal; *insertions* hairy; ramuli depending. *Spikelets* pedicelled, from four to six-flowered; *inner* valvelets of the corols ciliate.

Tsjama-pullu. *Rheed. mal.* xii. *p.* 75. *t.* 41.

Gramen fumi. *Rumph. Amb.* vi. *p.* 10. *t.* 4. *f.* 3.

Is found growing in tufts on pasture ground.

Culms filiform, many, ramous, spreading a little at the base, above erect, from one to two feet high, round, smooth.—*Leaves, mouths* and *margins* of the sheaths fringed, with long delicate white hairs.—*Panicle* oblong, composed of many, capillary, alternate, horizontal, ramous branches, their insertions and sub-divisions are fringed with a few, fine, white hairs. *Spikelets* minute, pedicelled, depending, from four to six-flowered.—*Calyx* smooth.—*Corol,* inner valve fringed with hairs in the back.—*Seed* oblong, brown, smooth.

14. P. *punctata. Linn. Sp. Pl. ed. Willd.* i. 395.

Erect, smooth, from six to eighteen inches high. *Panicle* oblong; *ramifications* simple, expanding, lower ones verticelled. *Spikelets* from twelve to fourteen-flowered, smooth.—*Seed* obliquely oblong.

This species grows in tufts on pasture ground.

Culms erect, generally simple, round, smooth; from six to eighteen inches high.—*Leaves* small, smooth; mouths of the sheaths bearded.—*Panicle* oblong; *ramifications* simple, expanding, the lower ones collected into one verticel; above alternate.—*Spikelets* pedicelled, rather remote, linear, from twelve to fourteen-flowered.—*Calyx* and *corol* smooth, thin, showing the unexpanded anthers and stigmas through them.—*Seed* oblong, smooth, brown.

15. P. *elegans. R.*

Sub-erect, smooth, from one to three feet high. *Panicle* oblong, nodding ramifications solitary, ramous, adpressed. *Spikelets* from eight to twelve-flowered. *Seed* globular.

A very elegant, tall, delicate species, a native of moist pasture ground.

Culms simple, at the base spreading for a few inches, the remaining part is erect, round, smooth; from one to three feet high —*Leaves* few and very small, smooth in every part, except the mouth of the sheaths, which are bearded.—*Panicles* linear-oblong, nodding, composed of erect, adpressed, distinct, somewhat remote ramification, each of which forms a small, but similar panicle. *Partial rachis* with the angles hispid. *Spikelets* pedicelled, linear, from ight to twelve-flowered, purplish.—*Corol,* outer valve three-nerved.—*Seed* globular, smooth, and brown.

16. P. *multiflora*. *R.*

Sub-erect, smooth, from six to eighteen inches high. *Panicle* longer than the rest of the plant, nodding. *Spikelets* crowded, from fifty to seventy-flowered. *Seed* globular.

Is found on dry elevated places.

Culms nearly erect, naked, except at the base, from six to twelve inches high, round, and smooth.—*Leaves* one or two near the base of each culm, short; *mouths* of the sheaths a little hairy.—*Panicle* oblong, bowing; *ramifications* sub-sessile, short, filiform, simple, the lower collected into one or two remote branches; the upper ones alternate. *Spikelets* pedicelled, very long, linear, from fifty to seventy-flowered.—*Calyx* smooth, obtuse.—*Corol* smooth, outer valve three-nerved.—*Seed* round, smooth, brown.

17. P. *unioloides*. *Linn. Sp. Pl. ed. Willd.* i. 393. *Retz. Obs.* v. 19.

Erect, smooth, from one to two feet high. *Panicle* ovate; *ramifications* short, simple, below fascicled; above solitary, horizontal; *spikelets* long-pedicelled, from sixteen to twenty-flowered. *Seed* oblong.

Beng. Konee.

A most elegant species, a native of dry elevated situations.

Culms, below ramous, with a small portion resting on the ground, the rest erect, and from one to two feet high, round and smooth.—*Leaves* few, small, and smooth; mouths of the sheaths bearded.—*Panicles* half as long as the whole plant, erect, oblong, thin; *ramifications* peduncled, filiform, disposed as in *multiflora*. *Spikelets* ovate, from sixteen to twenty-flowered, generally of a bluish purple colour. *Corol*, outer valve three-nerved, as in P. *multiflora*.—*Seed* oblong, smooth, brown.

18. P. *flexuosa*. *R.*

Sub-erect, flexuose, nodding, smooth, from eighteen to twenty-four inches high. *Panicle* ovate, crowded; *ramifications* solitary, ramous, expanding, insertions hairy, *spikelets* twenty-flowered. *Seed* globular.

A pretty large species, growing in tufts on old walls, so exactly resembling P. *unioloides* as to be easily mistaken for it. It has much the habit of *Briza Eragrostis*.

Culms sub-erect, winding, round, smooth; from one and a half to two feet high.—*Leaves* much larger than in the last species, and the mouths of the sheaths more hairy.—*Panicle* oval, very large, generally more than half the length of the whole plant, branches alternate, nearly horizontal, ramous, with a brown, hairy gland in the axil of each division of the panicle.—*Spikelets* white, or pale green, lanceolate, large, about twenty-flowered.—*Seed* globular, brown, and smooth.

19. P. *paniculata*. *R.*

Erect, smooth, from two to four feet high. *Leaves* long; mouths of their sheaths bearded. *Panicle* oblong; *ramifications* most numerous, filiform; insertions hairy. *Spikelets* from four to sixteen-flowered; *valvelets* of the calyces ciliate. *Seed* globular.

Culm sub-erect, round, and smooth; length from one to two feet high to the panicle, which is about the same length, making the whole height from three to four feet.—*Leaves* large, mouths of the sheaths very hairy.—*Panicle* oblong, very large, composed of numerous, ramous, filiform, expanding branches, their insertions and subdivisions involved in fine, long, white hairs. *Spikelets* long-pedicelled, linear, from four to sixteen-flowered.—*Corol, outer* valve round at the apex, and three-nerved; *inner* one slightly ciliate on the back.

20. P. *gangetica*. *R.*

Grows in very dense tufts. *Panicle* thin; *spikelets* linear, from thirty to forty-flowered.

A native of the banks of the Ganges, but scarce.

Culms numerous, and ramous, growing in crowded tufts, smooth in every part, general length about a foot.—*Leaves* few, small, and smooth.—*Panicle* ovate, composed of a few, remote, alternate, sub-diverging branchlets, each supporting a number of heaped, linear, acuminate, smooth spikelets, of from twenty to fifty-flowers each; as they advance from the apex the lower-florets drop, leaving the zigzag rachis naked as they ripen.—*Calyx*, both valves smooth and obtuse.—*Seed* oval, smooth, brown.

DACTYLIS. Schreb. Gen. N. 117.

Calyx two-valved, compressed, one valvelet larger, keeled. *Seeds* few.

1. D. *lagopoides. Linn. Sp. Pl. ed. Willd.* i. 410. *Burm. Ind.* 28. *t.* 12. *f.* 2.

Creeping. *Leaves* rigid, with spinescent points. *Heads* globular. *Calyces* from six to eight-flowered, hairy, as are also the corols; *anthers* blue.

It is generally found growing on a salt sandy soil near the sea.

Culms creeping to a great extent, with from four to eight inches of the flower-bearing portions, erect, they are of a firm, ligneous texture, round and smooth, every part of the erect portions covered by the sheaths of the leaves. *Leaves* numerous, small, approximate, firm, acute, resembling thorns.—*Head*, or *spike* terminal, short-pedicelled, globular, composed of many, closely crowded, sessile, roundish spikelets.—*Calyx* from six to eight-flowered; *glumes* equal, shorter than the flowers, obtuse and hairy.—*Corol* two-valved, exterior striated, hairy, and obtuse; *inner* membranaceous, involving the stamens and pistillum.—*Stamens* three. *Anthers* blue.

2. D. *brevifolia. Linn. Sp. Pl. ed. Willd.* i. 410.

Creeping, ramous. *Leaves* short. *Heads* sub-globular, composed of three, secund spikes. *Calyces* acute, from six to twelve-flowered. *Corols* hairy, acute.

It is found on dry sandy ground at a small distance from the sea.

Culms spreading, ramous, creeping, with their flower-bearing extremities ascending, from six to twelve inches long. *Branches* generally opposite. *Leaves* short, with their margins ciliate.—*Head* or *spike* terminal, long-peduncled, globular, composed of two or three, sessile, secund, short spikes, which are again composed of two rows of alternate, from six to twelve-flowered, sessile, ovate-oblong spikelets.—*Calyx*, from six to twelve-flowered, equal, acute, striated.—*Corol*, outer valve keeled, acute, three-nerved, very hairy near the base, inner valve as in *Poa*, with its back ciliate.—*Seed* oval, compressed, smooth, and brown; dropping from the corol when ripe.

Obs. Both these species are of a coarse nature.

The Linnean definition of the genus does not accord well with the only two species I have met with in India. That of *Poa* agrees better with their essential character; and had not *Burman*, in his *Flora indica*, made them both belong to this genus, I should certainly have considered them as species of *Poa*, though in their general habit they by no means agree with the elegant plants of that genus. When I had an opportunity I forgot to examine the seeds in a sufficiently exact manner. I suspect they have the aril of the following genus, because they have much of its general habit.

ELEUSINE. Gært. Carp. i. *p.* 7.*

Calyx two-valved, containing many flowers of two equal valvelets. *Seed* with a complete membranaceous aril.

1. E. *Coracana. Gært. Carp.* i. 8. *t.* 1.

Culms erect, from two to four feet high, compressed. *Leaves* bifarious. *Spikes* digitate, incurved. *Calyces* from three to six-flowered. *Seed* round.

Cynosurus Coracanus. *Linn. Sp. Pl. ed. Willd.* i. 415.

Tsjetti-pullu. *Rheed. Hort. Mal.* xii. *p.* 149. *t.* 78.

Panicum gramineum seu Naatsjoni. *Rumph. amb.* v. *p.* 203. *t.* 76. *f.* 2.

It is called Nutchanee by Europeans on the Coromandel Coast.

* I adopt Gærtner's name on account of the aril, which envelops the seed, and the total want of the involucre of Cynosurus in all the Indian species that I have yet met with.

Ponassa, or early Soloo, is the Telinga name of the grain, and Sodee the name of the plant.

Beng. Murooa.

Raggee of the Coast Mahomedans.

This species is cultivated during the rains. I never saw it wild.

Culms erect, generally several from the same grain of seed; from two to four feet high, a little compressed, smooth.—*Leaves* bifarious, large, smooth; *mouths* of the sheaths bearded.—*Spikes*, from four to six, digitate, incurvate, secund, from one to three inches long, composed of two rows of sessile, from three to six-flowered spikelets. *Rachis* compressed, a little waved.—*Calyx* from three to six-flowered, exterior glumes twice as long as the interior; both are keeled, obtuse, and membranaceous-margined.—*Corol*, valves nearly equal.—*Seed* globular, dark brown, a little wrinkled, covered with a thin, pellucid, membranaceous aril.

2. E. *stricta*. *R.*

Culms erect, from two to five feet high, compressed. *Leaves* bifarious. *Spikes* digitate, straight. *Calyces* from three to six-flowered. *Seed* round.

Teling. Pedda, viz. great Soloo.

Hind. Raggee.

This is still more cultivated than the last, and differs from it only in having the spikes straight, being generally of a larger size, and more productive, the great weight of the seed, when full grown bends the spikes down into a horizontal direction.

There is a variety of this straight-spiked sort, which the Telinga farmers called *maddee rooba soloo;* it grows to still a larger size, with a greater number of straight spikes.

They all require a light rich soil, on which the water does not remain after heavy rains. The first or early sort does not require so rich a soil as the other two, it is sown earlier by which means the same ground yields two crops. From July to January inclusive they reckon on an increase of about one hundred and twenty fold if the soil and season are favourable.

The second sort requires a richer soil; it is sown later in the season; its increase is greater.

The third sort, *maddi ruba soloo,* requires a still richer soil; land fit for it is scarce, and the rent high. I am informed that it is nearly twice as much as that of land fit for the first sort, the increase of this kind is prodigious, five hundred fold, the farmers say, if the season and soil are favourable.

About twenty years ago, there came up accidentally amongst some rubbish in my garden at Samul-cotah, two tufts of this plant, each, upon examination, I found to be the produce of one seed, each had twenty-five culms, and each of these culms had on an average two lateral branches making in all seventy-five culms and branches, each produced upon an average, six spikes, (for they had from four to eight) in all four hundred and fifty spikes, each of these had at a medium sixty spikelets, and each spikelet ripened on an average three or four seeds, total produce eighty-one thousand. I was myself particularly attentive in counting the above produce because it was so astonishingly great.

In the Rajamundry Circar, where these observations were made, the soil is in general rich, and the season favourable.

3. E. *ægyptica. R.*

Culms with a creeping base, from one to two feet high. *Spikes* four-fold, cruciate. *Calyces* daggered, from three to four-flowered. *Seed* oval, somewhat three-sided, and transversely rugose.

Cynosurus ægypticus. *Linn. Sp. Pl. ed. Willd.* i. 416.

Gramen vaccinum. *Rumph. amb.* vi. *p.* 10. *t.* 4. *f.* 1.

Cavara-pullu. *Rheed. Mal.* xii. *p.* 131. *t.* 69.

Hind. Mak*ura*-jali.

Grows in pasture ground, and by road sides, &c.

Culms, creeping near the base; the remaining part nearly erect for a foot, or a foot and a half, ramous, a little compressed, and smooth.—*Leaves* fringed with hairs.—*Spikes* terminal, from three to five, horizontal, secund, &c. as in E. *Coracana.*—*Calyx* from three to four-flowered; *glumes* acute, particularly the exterior one which ends in a short awn.—*Corol*, outer valve very acute.—*Seed* oval, somewhat three-sided, wrinkled, arilled, but in this species the aril is so thin and delicate, as to be often removed by the time the seed is full grown.

Obs. Cattle are fond of it.

4. E. *indica. Gært. Carp.* i. *p.* 8. *t.* 1.

Erect, smooth, from one to two feet high. *Leaves* bifarious. *Spikes* from five to seven, digitate. *Calyces* from three to five-flowered, *seed* oblong, three-sided, rugose.

Cynosurus indicus. *Linn. Sp. Pl. ed. Willd.* i. 417.

Hind. Mal-ankuri.

Teling. Kuror, viz. wild soloo.

A common coarse species, found growing in large tufts on pasture ground, road sides, &c.

Culms ascending at the base, ramous, a little compressed, smooth, from one to two feet high.—*Leaves* bifarious, large, with a few long white hairs scattered over the insides and sheaths.—*Spikes* terminal, from three to six, expanding; one generally standing single a little below the rest, linear, secund, from two to four inches long; insertions glandular and hairy, spikelets in two rows on the underside.—*Calyx* from three to five-flowered; *glumes* equal, as long as the flowers, smooth and rather obtuse.—*Corol*, valves smooth, obtuse.—*Seed* obtusely three-sided, rugose, arilled.

Obs. Cattle are not fond of it.

5. E. *verticillata. R.*

Erect, smooth, from one to four feet high. *Leaves* bifarious. *Spikelets* numerous, panicled. *Calyces* from eight to twelve-flowered; *glumes* daggered; *seed* oblong, wrinkled.

It is a native of moist pasture ground.

Culms erect, round, smooth, from one to four feet high.—*Leaves* as in the last described.—*Panicle* erect, oblong, composed of many verticelled, expanding, round, sessile, linear spikes, much resembling those of E. *indica; spikelets* imbricated in two rows on the under side of the spikes.—*Calyx* from eight to twelve-flowered, corol-like.—*Corol, exterior valve* ending in a long dagger, inner valve membranaceous.—*Seed* oblong, wrinkled, arilled; in this species the aril is fully as delicate as in E. *ægyptica,* and is nearly found whole on the full grown or ripe seed.

6. E. *calycina. R.*

Erect, from two to three feet high. *Panicle* linear. *Spikes* numerous, alternate; glumes of the *calyces* equal, many times larger than those of the corol, daggered, three-flowered. *Seed* three-sided, wrinkled.

Teling. Wadat*a*-tok*a*-g*u*ddee.

Grows in small tufts on dry pasture ground, but generally amongst bushes.

Culms erect, generally about two feet high, much covered by the leaves.—*Leaves* very thinly scattered with hairs.—*Panicle* linear, erect, from eight to twelve inches long, composed of many sessile, expanding, alternate, secund spikes; *spikelets* occupying, as in the other species, the under side of the spike.— *Calyx* generally three-flowered, glumes equal, twice as long as the three flowers they embrace, each ending in a long, sharp dagger.—*Corol,* valves obtuse.—*Seed* three-sided. wrinkled. I cannot say that I have seen the aril in this species, though I have often looked for it; it may therefore be a species of *Dactylis.*

ARUNDO. Schreb. Gen. 124.

Calyx two-valved, containing one or more two-valved flowers, surrounded with long hairs.

1. A. *bifaria. Retz. Obs.* iv. 21. *Linn. Sp. Pl. ed. Willd.* i. 455.

Erect, smooth, leafy, from twelve to eighteen feet high. *Panicle* erect, linear, oblong, dense; *calyces*, three-flowered, much longer than the florets; glumes of the corol woolly.

Teling. Patoo-ederoo.

Is found chiefly among bushes on the banks of rivulets, lakes, &c. on the Circar mountains.

Culms sword-shaped, smooth, about ten to fifteen inches long, and from one to two broad.—*Panicle* erect, oblong, very much crowded with innumerable ramifications, which stand erect and press close on one another while in seed.—*Flowers* erect, on a common woolly pedicel within the calyx.—*Calyx* three-flowered; *glumes* equal, pointed, larger than those of the flowers.—*Corol*, exterior valve twice the size of the inner one, it is pointed and woolly on the outside.

Obs. The leaves seem to be one of those sorts with which the Chinese line the inside of their Sugar-candy tubs and other packages.

2. A. *Karka. Linn. Sp. Pl. ed. Willd.* i. 456. *Retz. Obs.* iv. 21. Erect, smooth, leafy, from eight to twelve feet high. *Panicle* loose, verticelled. *Calyces*, from three to four-flowered, exterior glume of the corols three-nerved, and daggered.

Sans. नड, *Nuda*, नल, *Nula*, धमन, *Dhumuna*, पेाटगल, *Potugula.*

Beng. Nal, Nur.

Teling. Naga-sara maitantos.

Is a native of similar places with the last described.

Culms erect, ramous, from eight to twelve feet high, and as thick as the last, round, smooth, entirely covered with the sheaths of the leaves. —*Leaves* approximate, sword-shaped, smooth, from twelve to twenty-four inches long, and one or one and a half broad; mouths of the sheaths bearded.—*Panicle* bowing with the wind, but if not disturbed erect, oblong, composed of many filiform, erect, (in luxuriant plants drooping,) sub-verticelled, ramous branches.—*Rachis* of the branches angular and hispid.—*Florets* alternate, on a longer common woolly pedicel than in A. *bifaria*, within the calyx.—*Calyx* from three to five-flowered; *glumes* unequal, the largest shorter than the exterior valve of the corol.—*Corol*, exterior valve three-nerved, tapering into long fine subulate points, inner valve scarcely half so long. —*Stamens* often only two, and the pistil is frequently wanting.

Obs. It does not agree with Retzius's description of A. *Karka* in the number of flowers in the calyx. Pipes are made of the culms, particularly those used by the people who carry about the dancing snakes.

In Bengal it is more luxuriant than on the Coast. The common Durma mats of that place are made of the stalks split open. Vessels from the port of Calcutta are generally dunnaged with them.

3. A. *bengalensis*. *Linn. Sp. Pl. ed. Willd.* i. 455. *Retz. Obs.* v. 20. *Culms* from six to ten feet high, fistulous. *Leaves* bifarious, sword-shaped, drooping. *Panicle* lax; *valvelets* of the calyx equal, from two to three-flowered; exterior valve of the corol woolly, daggered, and three-nerved.

Beng. Gaba-nul.

This elegant species is found growing in ditches, &c. low places in the vicinity of Calcutta, where it blossoms during the month of October and November.

Culms erect, from six to ten feet high, fistulous.—*Leaves* very numerous, approximate, bifarious, drooping, sword-shaped, from one to two inches broad at the base, from thence taper to a fine point; smooth on both sides, and in the margins.—*Sheaths* also smooth, with a long membranaceous margin round the mouth.—*Panicle* oval, lax, from one to two feet long, erect; *ramifications* alternate, ramous, drooping a little.—*Flowers* pedicelled, scattered.—*Calyx* two or three-flowered; *valves* nearly equal, and of the length of the flowers.—*Corol*, exterior valve three-nerved, and woolly on the back, with a very fine, long, taper point. *Inner* valve small and smooth. —*Nectary* two, cunei-form scales embrace the germ on the two sides.

Obs. The wool which girds the base of the corol, in most of the species, is in this species, found growing on the back of its exterior valve.

ARISTIDA. *Schreb. Gen. N.* 125.

Caylx two-valved, one-flowered. *Corol* one-valved, with three awns at the top.

1. A. *setacea. Retz. Obs.* iv. 22. *Linn. Sp. Pl. ed. Willd.* i. 460.

Erect, smooth, from two to four feet high. *Panicle* linear-oblong, composed of fifteen to twenty sub-alternate, erectish ramifications.

Teling. Shi*pur*-gad*i*.

Grows in a dry, barren, binding soil.

Root perennial.—*Culms* straight, generally simple, from two to four feet high, and about as thick as a crow quill at the base, solid, and of a very firm ligneous texture, round and smooth.—*Leaves* few, narrow, margins involute, nerveless, smooth.—*Panicle* bowing with the wind, linear, from six to twelve inches long, composed of subsessile, remote, adpressed ramifications.—*Calyx*, *corol*, &c. as in the family, except that the three awns are erect.

Obs. Cattle do not eat it, yet it is very useful. The Telinga paper-makers construct their frames of the culms; it also serves to make brooms and tooth-picks. It is employed in preference to other grasses for making the screens called *Tatties*, for this purpose it is spread thin on *bamboo frames*, and tied down, these placed on the weather side of the house, during the hot land winds and kept constantly watered during the heat of the day, renders the temperature of the air in the house exceeding pleasant, compared to what it is without. The Thermometer in the out side exposed to the wind, but not to the sun, will then be at one hundred or one hundred and fifteen degrees, or even more; and within, if the *Tatties* are properly disposed and well watered, they will keep it down to from eighty-five to ninety, with two or even three rows of *Tatties*, made very thin, and all kept well watered, the Thermometer, when it blows hard, may be brought down to eighty, but then it is absolutely chilling, and disagreeably cold. The difference between the open air and this refreshed air, is to the feeling inconceivably great. The Thermometer in the sun, at this season rises to from one hundred and thirty to one hundred and forty. This contrivance we received from Bengal, and by it the hot season which used to be so much dreaded, is now rendered to those who can keep within doors tolerable. I have never observed any bad effects from this cool moist air; however the hot season is the most healthy of the whole year.

2. A. *Hystrix, Linn. Sp. Pl. ed. Willd.* i. 459.

Culms diffuse, about two feet long. *Panicle* oval, large, thin; *ramifications* spreading, two-forked.

Teling. Shilp*uroo*-kall*i*.

This species is a native of the same soil with the last described.

Culms depressed, straggling, crooked, rigid, somewhat branchy, about two feet long.—*Leaves* as in the last species.—*Panicle* large, oval, thin; direction various, composed of diverging, generally two-cleft peduncles and pedicels.—*Calyx, corol,* &c. as in the genus.

Obs. This species is of no use that I know of.

3. A. *depressa. Retz. Obs.* iv. 22. *Linn. Sp. Pl. ed. Willd.* i. 461.

Ascending, ramous, from twelve to eighteen inches long. *Panicle* linear, of six to eight alternate ramifications; *glumes* of the calyces keeled.

Teling. Nall*i*-poot*i*k*i*.

Is found upon the same kind of land with the two last species.

Culms below depressed, ramous; above ascending, filiform, round, smooth, solid, rigid; from twelve to eighteen inches high.—*Leaves* as in the former.—*Panicle* smaller, and with fewer ramifications than in A. *setacea*.—*Calyx*, the glumes have a large green keel.—*Corol,* &c. as in the genus.

Obs. This is also so far as I know, perfectly useless.

MANISURIS. Schreb. Gen. N. 1570.

Polygamous. *Rachis* jointed. *Hermaphrodite* calyx two-valved, one-flowered. *Corol* two-valved, less than the calyx. *Male,* or neuter calyx two-valved, one or two-flowered. *Corol* two-valved.

1. M. *Myurus. Linn. Sp. Pl. ed. Willd.* iv. 945. *Corom. Pl.* 2. *N.* 117.

Culms creeping, smooth, from nine to eighteen inches high. *Spikes* solitary; exterior valve of the *hermaphrodite* calyx panduriform and emarginate; *male* calyx two-flowered.

Teling. Nalla-*punookoo.*

Is a native of dry elevated sandy ground.

Culms several, ramous, diffuse near the base, rooting at the joints, extremities ascending, from nine to eighteen inches long, as thick as a crow quill, compressed, smooth, not piped.—*Leaves* small and smooth; mouths of the sheaths increased by a lacerated membranaceous border or stipule.—*Spikes* solitary, terminal, and from the exterior arils peduncled, secund, jointed, smooth, a little compressed, from one to two inches long.—*Peduncles* articulated, and bracted at the middle.—*Flowers,* the hermaphrodite ones occupy a waved row of pits on the four sides of the spike, while the male ones possess two sides, the back is naked.

Hermaphrodite flowers. *Calyx* two-valved, one-flowered; the exterior valvelet has a broad coloured margin, which is emarginate, and deeply indented at the sides, like a fiddle; inner valvelet oblong, lodged in a pit of the rachis.—*Corol* two-valved, membranaceous. *Nectary* two obcordate, crenulated scales embrace the germ.—*Stamens* three. *Styles* two.

Male flowers. *Calyx* two-flowered, two-valved, valvelets nearly equal, boat-shaped.—*Corol* with two membranaceous valves. —*Stamens* three.—*Pistil* none.

2. M. *granularis. Linn. Sp. Pl. ed. Willd.* iv. 945. *Corom. Pl.* 2. *N.* 118.

Ramous, sub-erect, hairy, from one to two feet high; *spikes* fascicled. *Hermaphrodite* calyx oval, and rugose; *male* or *neuter* one-flowered.

Grows among bushes.

Culms very ramous, nearly erect, filiform, hairy, from one to two feet high.—*Leaves* numerous, very hairy, stiff and sharp.—*Spikes* terminal, and axillary, peduncled, several together, small, compressed, from half an inch to an inch long.—*Panicles* as in the former. *Rachis* jointed, much waved, and excavated as in Rottbœllia.—*Flowers* from four to ten of each sort, their situation exactly as in the last species, viz. the hermaphrodite occupy the forepart, while the male, or neuter are placed on the sides, the former are globular, and the latter lanceolate.

HERMAPHRODITE FLOWERS. *Calyx* one-flowered, two-valved, exterior valve round or rigid, and very rugose ; inner valve less rigid.—*Corol* two-valved, membranaceous.—*Stamens* three.—*Styles* two.

MALE OR NEUTER FLOWERS. *Calyx* two-valved, one-flowered. —*Corol* two-valved.—*Stamens* three or none.

ROTTBŒLLIA. *Schreb. Gen. N.* 1572.

Rachis (generally) jointed, variously excavated for the reception of one, or two flowers in a two-valved calyx, in each excavation.

1. R. *glabra. R.*

Near the base creeping, with extremities erect, smooth. *Leaves* bifarious, all hermaphrodite. *Calyx* one-flowered, *corol* three-valved.

Beng. Buksha.

A native of Bengal, where it grows on pasture lands, the borders of rice fields, and other moist places. Cattle are fond of it.

Culms jointed, smooth, compressed, filled with pith, creeping near the principal root for a foot or more, the remaining part from one to two, or even three feet, erect.—*Leaves* alternate, bifarious, approximate, short, spreading, ensiform, smooth.—*Sheaths* shorter than the joints of the culm, smooth.—*Spikes* axillary, solitary or in pairs, with a solitary terminal one, subulate, three-sided, smooth, short-peduncled, the peduncles with generally a part of the spikes hid in the sheaths of the leaves; general length from two to three inches.—*Flowers* opposite, in general all, or the greatest part hermaphrodite and as long as the joints of the rachis.—*Calyx* one-flowered, two-valved, smooth, cartilaginous. The exterior valve of one of the pair obliquely emarginate, the other acute.—*Corol*, the hermaphrodite and male three-valved, one valve on the inside, and two on the outside; all tinged with deep red at the base.

2. R. *compressa*. *Linn. Sp. Pl. ed. Willd.* i. 465. *Corom. Pl.* ii. *N.* 156. *Retz. Obs.* iii. 12.

Culms climbing, from five to twenty feet high, smooth, compressed; *spikes* axillary and terminal; *rachis* jointed; *flowers* opposite, hermaphrodite calyces one-flowered. *Corol* three-valved.

Beng. Pansheroo.

Teling. Shervoo, i. e. lake panookoo.

Is found on the borders of lakes, amongst other roots of long grass, and brush wood.

Culms several, creeping or climbing, ramous, compressed, from five to twenty feet long, piped, smooth, about as thick as a common quill, and very hard.—*Leaves* numerous, bifarious, small for the size of the plant, smooth and soft; *sheaths* short, compressed, and smooth.—*Spikes* peduncled, terminal, and from the exterior axills, generally from two to five together, compressed, a little smooth, from two to four inches long.—*Peduncles* articulated at the middle and there bracted, lanceolate, chaffy, bractes also surround their insertion.—*Flowers* all hermaphrodite, opposite, decussated, lodged in the excavations of the jointed rachis.—*Calyx* one-flowered, two-valved, valves nearly equal; the flowers have alternately their valves pointed and somewhat longer than the excavation in the rachis, both cartilaginous;—and obtuse, equal to the pits in the rachis, with the exterior one only cartilaginous.—*Corol* three-valved, valves membranaceous. *Nectary*, two triangular bodies embracing the base of the stamens and germ.—*Stamens* three.—*Styles* two; *stigmas* plumose.

3. R. *exaltata*. *Linn. Sp. Pl. ed. Willd.* i. 466. *Corom. Pl.* 2. *N.* 157.

Polygamous. *Culms* erect, hirsute, from six to ten feet high.

Spikes solitary, secund. Polygamous flowers on the front with neuter ones on the sides.

Beng. Bura-swooate.

Teling. Konda panookoo.

A native of mountains.

Root consisting of strong woody fibres.—*Culms* erect, ramous, a little compressed, inwardly of a spongy nature, from six to ten feet high, and as thick as the little finger near the base, were they are armed with strong and short, white hairs.—*Leaves* large and numerous, smooth on the outside, and hairy on the inside ; *margins* hispid, sheaths very hairy; hairs elevated on glandular points, very stiff and sharp.—*Spikes* terminal, and from the exterior axills, generally solitary, cylindric, &c.—*Flowers* of three sorts, hermaphrodite, male, and neuter, the first and second occupy alternate pits or one side of the spike, while the neuter stand on each side of them, so that the spike may be called secund.—*Calyx* of the hermaphrodite and male floret two-valved.—*Corol* as in the last.—*Stamens,* &c. as in the genus.—*Calyx* of the neuter flower two-valved, valvelets oblong. —*Corol,* glumes two, membranaceous.

4. R. *corymbosa. Linn. Sp. Pl. ed. Willd.* i. 446. *Corom. Pl.* ii. *N.* 181.

Polygamous, erect, smooth, from three to five feet high. *Spikes* fascicled, terminal and axillary; *rachis* jointed; *flowers* alternate, on opposite sides of the spike. *Calyces* generally two-flowered.

Teling. Pedda-panookoo.

R. punctata. *Retz. Obs.* iii. 12.

Aegilops eraltata. *Retz. Obs.* ii. 27.

A native of low rich pasture ground, grows in erect tufts.

Culms many, straight, mostly naked, the sheaths of the leaves being short and covering but a small part of them ; round and smooth, from three to five feet high, and like those of the last species, very firm, and not piped.—*Leaves* a few, from six to twelve inches long; *margins* sometimes fringed with a few hairs ; *sheaths* short and smooth.—*Spikes* terminal and from the exterior axills several, peduncled.—*Peduncles* filiform, curved, jointed near the base and there bracted, besides there are chaffy, filiform bractes at their insertions.—*Flowers* alternate in two rows; one row on each side of the spike.—*Calyx* two-flowered, two-valved.—*Corol*, one hermaphrodite within, and one male without, each has two valves, sometimes there is only one hermaphrodite floret, in which case it has three valves.

5. R. *perforata*. *Corom. Pl.* ii. *N.* 182.

Culms erect, smooth, from three to five feet high. *Spikes* solitary, below the flowers are opposite, and the rachis perforated. *Calyx* generally two-flowered; one flower male, the other hermaphrodite.

Beng. Kurki.

Teling. Panookoo.

This species is rather scarce, it grows on low rich pasture ground.

Culms many, erect, simple, round, smooth, jointed, but not piped; from three to five feet high, and about as thick as a crow quill.—*Leaves* small for the size of the grass, slender and smooth; *margins* hispid; *mouths* of the sheaths and a little way up the base of the leaves woolly.—*Spikes* terminal, and from the exterior axills, solitary, peduncled, round, smooth, as thick as a crow quill, and from three to six inches long; where the flowers are opposite, there is an oblong perforation of the rachis, so that the backs of the inner glumes of the calyx touch one another.—*Flowers* in pairs, nearly opposite on the lower half or more of the spike; above alternate, sometimes all hermaphrodite, sometimes hermaphrodite and male mixed.—*Calyx* one or two-flowered, two-valved; *valvelets* simple; the exterior one rigid, the interior one firm and white, both obliquely linear, oblong.—*Corol*, when single, hermaphrodite, and three-valved; when double the exterior one is male, in which case the two have four valves, viz. two valves each; all membranaceous, in shape conforming to that of the calyx, but smaller.

6. R. *setacea. Corom. Pl.* ii. *N.* 132.

Erect, setaceous, smooth, from two to six inches high. *Spikes* solitary, terminal, unilateral; *rachis* not joined. *Calyces* one-flowered. *Corols* two-valved, hermaphrodite.

Grows on old walls.

Culms many, erect, ramous, filiform, smooth, from two to six inches high.—*Leaves,* inside a little hairy, in size proportioned to the plant.—*Spikes* terminal, solitary, awled, secund, about an inch, or an inch and a half long. *Rachis* excavated on one side, but not joined.—*Flowers,* they are all hermaphrodite, imbricated transversly on the excavated side of the rachis.—*Calyx* one-flowered, two-valved; *valvelets* equal, membranaceous margined.—*Corol* two-valved, membranaceous, hairy.—*Stamens* three.—*Styles* two.

7. R. *Thomæa. Linn. Sp. Pl. ed. Willd.* i. 464. *Corom. Pl.* ii. *N.* 132.

Erect, compressed, about two inches high. *Spikes* solitary, terminal, two ranked. *Rachis* not jointed. *Calyces* one-flowered. *Corol* two-valved.

R. Thomæa and pilosa. *Linn. Sp. Pl. ed. Willd.* i. 464. 5. are, I suspect this very species.

Grows, with the last species, on old walls.

Culms minute, erect, compressed, about an inch high, exclusive of the spike, and two with it.—*Leaves* bifarious, numerous.—*Spikes* terminal, solitary, subulate, distichous, compressed. *Rachis* waved, excavated on the opposite side, not articulated.—*Flowers* all hermaphrodite, disposed alternately in the excavations of the rachis.—*Calyx* one-flowered, two-valved; *exterior* valvelet rigid, linear; *interior* boat-shaped, and less rigid.—*Corol* two-valved, membranaceous, hairy.—*Stamens* three.—*Styles* two.

8. R. *biflora. R.*

Setaceous. *Spikes* simple, subulate; *rachis* not articulate, but bifariously excavated for the reception of the two-flowered spikelets.

A native of the peninsula of India; and in size and appearance very like R. *Thomæa* and *setacea*; and were they not two-flowered in the same calyx I should no doubt have taken it for R. *incurvata, Linn.*

Culms scarcely more than may be called the scapes of the spikes, the whole plants being but two or three inches high.—*Leaves* many, filiform; *rachis* rigid, and with the sheaths somewhat pilose.—*Spikes* terminal, solitary, subulate, about two inches long. *Rachis* not jointed, but alternately excavated on the opposite sides for the reception of the flowers.—*Calyx* two-valved, two-flowered; *valves* very unequal, the exterior being many times larger than the inner one, which is a minute scale in the bottom of the excavations of the rachis, and not readily detected.—*Florets* two in each calyx, both hermaphrodite, one-sessile and the other short pedicelled, each with a corol of two equal membranaceous, smooth valves.

HORDEUM. *Schreb. Gen. N.* 129.

Calyx lateral, two-valved, valves narrow, accuminate, distant, altogether forming a six-leaved involucre, one-flowered, by threes at each toothlet of the rachis.

1. H. *hexastichon. Linn. Sp. Pl. ed. Willd.* i. 473.

All the flowers hermaphrodite and awned, placed regularly in six rows.

Sans. यवः, Y*uva*, सितशूकः, S*itu*shook*a*.

Beng. J*u*b*a*.

Arab. She-eer.

This species is much cultivated in most of the temperate parts of Hindoostan during the cool season.

Obs. I have had specimens and abundance of seeds of this and also of what I consider to be Siberian Barley (H. *coeleste*) both gahered in 1818 in Tartary, the latter in Dapa, by my highly esteemed friend Capt. W. S. Webb, surveyor at Kemaoon, who has favored me with the following observation on them: "These two sorts of Barley are called *Oo-a*, and deserve to be particularly examined. They are not known in the low lands of India, produce the hardiest of all grains, and are most nourishing, as the lusty Tartars almost exclusively live upon them and Tea. The grain cleared from the husk resembles no kind of Barley that I am acquainted with, and if they should prove new species might from their defiance of cold and many good qualities be a valuable acquisition in the Highlands of Scotland."—N. W.

TRITICUM. Schreb. Gen. N. 130.

Calyx two-valved, solitary, generally three-flowered.

1. T. *æstivum. Linn. Sp. Pl. ed. Willd.* i. 476.
Calyces four-flowered, ventricose, smooth, imbricate, awned.
Sans. गोधूमः, Godhoom*a*, सुमनः, Soom*u*n*a*.
Beng. Gom.
Hind. G*i*oon.
Pers. Gundum.
Arab. Burr.

Two varieties are very generally cultivated in the interior and northern parts of Hindoostan during the cool season.

2. T. *hybernum. Linn. Sp. Pl. ed. Willd.* i. 477.
Calyces four-flowered, ventricose, even, imbricated, with little or no awn.

Two varieties of this species are also cultivated in the northern parts of Hindoostan during the cool season.

TRIGYNIA.

MOLLUGO. Schreb. Gen. N. 139.

Calyx five-leaved. *Corol* none. *Capsule* superior, three-celled, three-valved, many seeded. *Embryo* annular, and furnished with a perisperm.

1. M. *pentaphylla*. *Linn*. *Sp*. *Pl*. *ed*. *Willd*. i. 492.

Annual, diffuse. *Leaves* verticelled cuneate-lanceolar; *flowers* panicled.

Hind. and *Beng*. Khet-papara.

Teling. Veri Shadrasi.

A common weed in gardens throughout India.

2. M. *triphylla*. *Lour*. *Cochinch*. 79.

Annual, diffuse; *branches* angular, smooth. *Leaves* sub-verticill-ed, lanceolate, unequal. *Panicles* of long dichotomous racemes of alternate flowers; *seeds* dotted.

Beng. Jul-papara.

Alsine multiflora. *Pluk*. *Phyt*. 21. *t*. 259. *f*. 2. good.

Mullugo stricta. *Linn*. *Syst*. 129. is probably the same plant, if so, the specific name is a very improper one. Is also a weed in gardens all over India.

3. M. *verticillata*. *Linn*. *Sp*. *Pl*. *ed*. *Willd*. i. 492.

Annual, diffuse, dichotomous, smooth. *Leaves* sub-verticelled, sessile, cuneiform peduncles inserted amongst the leaves, one-flowered. *Seed* furrowed along the back.

Obs. Found over most parts of India, and a common weed in our gardens during the dry season chiefly.

HOUTTUYNIA.

Spathe four-leaved. *Spadix* oblong, covered with naked flowers. *Capsule* one-celled, many-seeded.

1. H. *cordata*. *Thunb*. *Jap*. 12. *et*. 234. *t*. 26. Sir. J. E. Smith in Rees's Cyclop. in loco.

Polypara cochinchinensis. *Lour*. *Cochinch*. *ed*. *Willd*. i. *p*. 78.

A native of the valley of Nepala, where it grows in very great abundance, blossoming during the hot and rainy seasons, and producing ripe seed from August to the end of February., The Hon.

Mr. Gardner has likewise sent me specimens gathered in the vicinity of Gossein Than (also called Neel-*kuntha*) in Nepal*a*. I have been favored with specimens found by Dr. G. Gowan, the superintendent of the Botanic Garden at Saharunpore, on his tour to the Sewalic mountains.

Name in Nepal*a*, *Kala* or *Kullum-Soa*.

An erect, slender, pretty smooth, from two to three feet high, annual plant; rising from a long, thin, somewhat creeping root, with many short, capillary, verticilled fibres.—*Stem* round, slightly zigzag, about the thickness of a common goose quill at its base, with obscure, remote, pubescent, leaf-bearing joints: generally simple, though sometimes sending forth several short undivided branchlets toward its summit.—*Leaves* petioled, alternate, spreading, broad-cordate, with rounded distant lobes, acuminate, entire, from two to three inches long, dotted with copious semi-pellucid points, glaucous, and while young slightly furfuraceous below, reticulated, five-nerved, the three middle nerves generally uniting near the petiol.—*Petiol* stipulary slender, channelled, about half as long as the leaf, into the base of which it gradually widens.—*Stipules* membranaceous, flaccid, oblong, obtuse and rounded, slightly ciliated and sometimes marked with a few toothlets, scarcely stem-clasping at the base, having their petiol inserted about the middle of their back, persistent, the lowermost forming remote bractes on the base of the stem.—*Peduncles* leaf-opposed and terminal, naked, solitary, club-shaped, generally longer than the petiol, sometimes reaching to the middle of the leaf.—*Spathe* consisting of four equal, white, spreading, persistent, obovate or cuneate, obtuse, smooth, veined leaves, during the æstivation imbricating in the shape of a cone, about six lines long.—*Spadix* cylindric, much lengthened as the flowers decay, generally twice as long as the spathe and when fruit bearing sometimes measuring an inch and a half. *Flowers* very numerous and small, close together, covering entirely the spadix, which they render oblong and obtuse, sub-verticilled, destitute of both calyx and corol.—*Filaments* invariably three, subulate, somewhat longer than the pistils, inserted on the angles of their ovarium, a little above its base.—*Anthers* erect, oblong, two-celled, bursting lengthways along their margins.—*Ovarium* round, three-cornered, most slightly pubescent, one-celled, many-seeded; *ovula* attached to the sides of the cavity. *Styles* three, persistent, rising from the apex of the angles on the ovarium, above the insertion of the filaments, leaving its vertex naked and somewhat depressed; they are spreading awl-shaped, acute, marked along the upper part and the furrowed inside (the stigmas?) with numerous short papillæ.—*Capsule* about the size of a mustard seed, brownish green, of the same shape as the ovarium, membranaceous and unilocular, bursting at the top, with a triangular opening reaching half way up the styles.—*Seeds* about eight, brown, smooth, striated, oblong, acute at each end, somewhat darker coloured at their bases, which are attached to three parietal, oblong, elevated placentæ alternating with the angles of the cavity. *Integument* single, crustaceous. *Embryo* minute, lodged in a copious milk-white, mealy perisperm, towards the umbilicus, centrifugal.

Obs. In the numerous spadices which I have examined I have with Father Loureiro invariably found three stamina and as many styles attached to each ovarium, the former above the base, the latter at the apex of its angles; I have therefore not hesitated continuing this most interesting plant in the very class and order where it has been placed in the flora of Cochinchina. As there is no reason for considering it at all different from the original *Japan* plant I am at a loss to account for the difficulty which the celebrated Chevalier Thunberg experienced in determining its station in the sexual system; nor can there at present be any doubt of its neither belonging to Heptandria, Polyandria, or Monoecia.

The seeds are so small and their embryo proportionably minute that I have not been able to ascertain the structure of the latter; they appear to me however, to bear great resemblance to those of Tacca and Aristolochiæ, to which family the plant is still further allied by the epigynous insertion of the stamens. But its chief affinity seems to be to the *Aroideæ*, forming an additional connecting link between the two mentioned families. The leaves are much like those of *Aristolochia Clematitis*, as has been justly remarked by the illustrious author of the article *Houttuynia*, in Rees's Cyclopædia.

The plant is used while young as a potherb by the inhabitants of Nepal*a* as well as those of Cochinchina. Loureir. l. cit.—N. W.

Class 4th.

TETRANDRIA MONOGYNIA.

RHOPALA. Schreb. Gen. N. 144.

Calyx none. *Corol* four-petalled, bearing the stamina above the middle. *Nectary* four-scaled. *Germ* one-celled; *ovula* two, attached to one side of the cell.

1. R. *moluccana. Brown. Linn. Trans.* x. *p.* 191.

Leaves alternate, sub-sessile, oblong, entire, racemes axillary, and from the branchlets, longer than the leaves. *Nectarial* scales naked.

A native of Pulo-pinang.

As my description is taken from dry specimens I cannot say whether it be a tree or shrub, but the large woody specimens evidently show it to be perennial.—*Leaves* short-petioled, alternate, obovate, oblong, obtuse-pointed, entire and smooth on both sides, six or more inches long and less than three broad.—*Stipules* none. *Racemes* generally solitary, from the naked woody branchlets below the leaves, very long and completely clothed to the base with numerous beautiful, diverging, pretty large, delicate flowers.—*Bractes* minute.—*Calyx* none.—*Petals* four, very long and slender, near the apex they expand into a concave lanceolate shape for the reception of the stamens, they first open in the middle then at the apex, and soon after become revolute. *Nectary,* four scales embracing the lower part of the germ.—*Filaments* four, short, inserted into the petals above the middle. *Anthers* linear, curved in an opposite direction to the petals.—*Germ* above, ovate, style long and slender. *Stigma* clubed.

2. R. *serrata. Brown. Linn. Trans.* x. *p.* 192.

Leaves alternate, sub-sessile, oblong, serrate. *Racemes* axillary; *pedicels* two-flowered. *Nectarial* scales ciliate.

A native of the Malay Archipelago.

My description and drawing being from dry specimens, I cannot say whether it is a tree or shrub.—Young shoots clothed with ferruginous down.—*Leaves* alternate, short-petioled, oblong, acute, tapering at the base, serrate, firm and smooth on both sides; from four to six inches long, and from one to three broad.—*Spikelets* none. —*Racemes* axillary, solitary, shorter than the leaves, clothed with ferruginous down.—*Pedicels* two-flowered.—*Bractes* minute, cordate, one at the base of each common pedicel.—*Calyx* none.—*Petals* few, long, slender, clavate; from the middle downwards, their sides touch, and form a slender tube.—*Nectarial scales* four between the base of the petals and germ, obcordate, ciliate.—*Filaments* four, inserted on the petals, about one-third below their apices. *Anthers* ovate.—*Germ* superior, very hairy. *Style* as long as the corol. *Stigma* clavate.—*Seed vessel* not seen.

3. R. *excelsa. R.*

Leaves alternate, short-petioled, cuneate-oblong, obtuse, pointed, smooth, with a few large blunt serratures near the apex.—*Racemes* axillary and terminal, as long as the leaves, downy.—*Nectarial* scales four, distinct, and naked.

A large timber tree, a native of the forests of Chittagong, where it flowers in March and April.

Trunk of the full-grown trees about four feet in circumference, covered with dark-coloured, scabrous bark; branches numerous, spreading far, and dividing much; young shoots clothed with ferruginous down.—*Leaves* alternate, short-petioled, cuneate-oblong, obtuse-pointed; interior margins generally with one, two, or three large, blunt serrature on each side, texture firm and polished; length from four to six inches, by two or three broad.—*Racemes* axillary, and terminal, but chiefly the former, one, two, or three together, nearly as long as the leaves, straight and simple, densely cloathed with ferruginous down. *Pedicels* short, two-flowered.—*Bractes*, a small oval one at the base of each common peduncle and one about the middle of each pedicel.—*Flowers* numerous, crowded, small, yellowish green.—*Calyx* none. *Petals* four, linearly spatulate, revolute. *Nectarial* scales four, rounded, smooth and distinct.—*Filament* short, inserted on the petals. *Anthers* linear.—*Germ* oblong, very completely covered with long, reddish brown hairs, one-celled and containing two ovula attached a little to one side of the bottom of the cell. *Style* filiform. *Stigma* clavate, entire.

4. R. *robusta. R.*

Leaves alternate, sessile, cuneate-oblong, smooth, remotely serrulate. *Racemes* axillary, and below the leaves, smooth. *Nectary* a smooth four-toothed cup.

Joweea, the vernacular name in Silhet where it is indigenous, and grows to be a large, stout timber tree. Flowering in June.

Young shoots smooth.—*Leaves* alternate, sub-sessile, cuneate-oblong, obtuse-pointed, remotely and acutely serrulate, smooth on both sides, about twelve inches long, by five or six broad.—*Racemes* axillary, and below the leaves, one, two, or three together, about half the length of the leaves, smooth. *Pedicels* short, two-flowered.—*Bractes* smooth, one at the base of each common pedicel, and one about the middle of each of the proper ones.—*Flowers* numerous, pale, greenish yellow, and fragrant.—*Calyx* none. *Petals* four, linearly spatulate, smooth, revolute.—*Nectary* one-leaved, cup-shaped, obtusely four-toothed, with the margins somewhat ciliate.—*Germ* superior, oblong, smooth, one-celled; ovula two, attached to one side of the cell. *Style* filiform. *Stigma* clavate, entire.*

* Both this and the preceding species are distinct from R. *cochinchinens's*, Brown., Linn. Trans. x. p. 192.—N. W.

DIPSACUS.

Common calyx many-leaved ; *proper* superior, single. *Receptacle* chaffy.

1. D. *inermis. Wall.*

Leaves oblong, serrate, villous, petioled, sub-lobate ; cauline connate ; *heads* globular, villous ; *chaff* cuneate, subulate-acuminate, scarcely longer than the tube of the corolla.

β ? with all the leaves pinnatifid-lobate.

Grows wild at Katmandu and Gossain Than, from whence Mr. Gardner has sent me specimens and seeds. Dr. Gowan has likewise found it to the Northward of Saharunpore. Blossoms during the rainy season.

Nepal*a* name, *Barberi* and *Gooja.*

Root strong, fibrous, ash-coloured.—*Stem* erect, slender, fistulous, four or more feet high, many-sided, brachiate, the lower half densely clad with reflexed soft hairs, the upper sparingly beset with inoffensive bristles, chiefly along its ribbed corners. *Leaves* unequally and doubly serrated, soft and villous on both sides, with a few scattered bristles ; *radical* about a foot long, three inches broad, acute, tapering much downwards and sub-decurrent on their equally long slender petiol ; sometimes with a deep lobe on one or both sides of their base ; those of the stem consisting of two or three remote pairs, five or six inches long, acuminate, undivided, generally more bristly than the others, connate at their base and forming there a short cup.—*Peduncles* very long and slender, few, erect, mostly with a pair of narrow short leaves at their middle.—*Heads* nearly round, obtuse, an inch in diameter; *involucre* consisting of six or eight linear channelled recurved ciliate leaves, an inch long.—*Receptacle* ovate, marked with square impressions for the insertion of the flowers, beset with membranaceous pale, hairy, cuneate, concave spreading *scales,* with a rounded apex ending in a straight ciliate purple-coloured lanceolate acute point, which is as long as the lamina itself. *Corolla* yellowish white, pubescent, with four almost equal laciniæ.—*Stamina* spreading, long.—*Seed* precisely as in D. *laciniatus. Gært. Carp.* ii. *p.* 40. *t.* 86.

Obs. The plant is approaching to D. *pilosus,* but seems to differ sufficiently in the form and size of the leaves, and involucrum. I am doubtful whether the variety suggested above may not form a distinct species. It seems on the whole to be smoother, its leaves deeply pinnatifid or pinnate at the base, with acuminate lanceolate segments.—N. W.

HEDYOTIS. Schreb. Gen. N. 153.

Calyx four-parted. *Corol* one-petalled, funnel-shaped. *Capsule* inferior, two-celled, many-seeded. *Receptacles* adhering to the partition their whole length. *Embryo* centripetal, and furnished with a perisperm.

1. H. *hispida. Retz. Obs. Bot.* iv. *p.* 23. *Linn. Sp. Pl. ed. Willd.* i. 565.

Diffuse, four-sided, rooting at the joints. *Leaves* sessile, lanceolate, with hispid margins. *Flowers* axillary, crowded.

It sprung up and blossomed in December in the Botanic Garden, amongst plants introduced from the Moluccas, in 1798-9.

Stem none, but several square, creeping branches, with their floriferous extremities ascending.—*Leaves* opposite, sub-sessile, lanceolate; *margins* hispid, and entire, smooth, from two to three inches long; the connecting membrane divided into many subulate portions.—*Flowers* axillary, sessile, small, white.—*Capsule* roundish, two-celled.—*Seeds* numerous.

2. H. *geniculata. R.*

Weak, straggling, jointed, round, smooth. *Leaves* sub-sessile, lanceolate, smooth. *Flowers* sub-verticilled, short-peduncled.

A native of the Malay Islands.

3. H. *scandens. R.*

Perennial, scandent, round, smooth. *Leaves* broad lanceolate. *Panicles* corymbiform, terminal; mouth of the tube of the corol and apex of the style bearded.

Gujee, the vernacular name in Silhet, where it is indigenous; flowers and ripens its seed during the hot and rainy seasons.*

Stems and branches perennial, round, smooth, scandent.—*Leaves* short-petioled, broad-lanceolate, smooth, entire.—*Stipules* cup-shaped, connecting the leaves, with a short process on each side between the leaves.—*Panicle* corymbiform, terminal, sub-globular, throughout trichotomous, all the ramifications a little villous.—*Bractes* linear, spreading.—*Calyx* four-toothed.—*Corol* with the tube longer than the calyx, its mouth and the base of the four linear segments of the borders of the corol very woolly.—*Germ* oval, two-celled; *ovula* numerous, attached to the sub-globular receptacles, rising from the middle of the partition. *Style* longer than the corol, apex woolly; *stigma* two-cleft.—*Capsule* ovate, two-celled, many-seeded.

4. H. *lineata. R.*

Annual, diffuse, hairy. *Leaves* sessile, ventricose-lanceolar, longitudinally marked with simple, parallel veins. *Peduncles* axillary, two or three together, pretty long, many-flowered. *Capsula* round, hairy.

Native of Chittagong, where it blossoms during the rains.

5. H. *Auricularia. Linn.*

Sub-dichotomous diffuse, adscending, villous. *Leaves* lanceolate, sub-sessile, smooth above, many-nerved below. *Flowers* sub-sessile, densely whorled. *Capsules* villous.

Mariguti. Rheed. Hort. Mal. x. *p.* 63. *t.* 32.

A native of Silhet and Nepala, where the plant blossoms during the hot and rainy seasons. From the former place the plant was

* It is very common about Katmandu where it blossoms from December to March. Native name *Gookee.*—N. W.

introduced in 1815 in the Botanic Garden near Calcutta, by Mr. Smith.

Beng. Muttia-lata.

Root fibrous.—*Stem* short, woody, divided into many slender generally once or twice dichotomous roundish or slightly four-sided branches, measuring a foot in length, densely villous sometimes rooting at the base.—*Leaves* lanceolate, tapering and acute at both ends, an inch or an inch and a half long, even above, with several oblique villous nerves underneath, margins scabrous; pairs approximate and exceeding in length their interstices.—*Petiols* very short, the uppermost somewhat longer, hispid, uniting into a membranaceous, villous, stipulary membrane, which is terminated by long hispid ciliæ.—*Flowers* small, crowded into sessile, globose, axillary, copious, villous, verticils, concealing, but scarcely longer than the petiols and rendering their stipules reflexed.—*Laciniæ* of the calyx linear.—*Tube* of the corolla scarcely longer than the calyx.—*Laciniæ* lanceolar acute.—*Throat* pubescent.—*Stamina* exserted, erect, shorter than the limb of the corolla, with pubescent *filaments.*—*Style* barbate.—*Stigma* oblong.—*Capsule* small, round, slightly furrowed. —*Dissepiment* entire.—*Seeds* many, three-cornered, shining, brown.

Obs. I took this plant to be H. *nervosa*, Lam. or *hirsuta* of the same author, until I discovered that Sir J. E. Smith had proposed them as varieties only of H. *Auricularia;* Rees's Cyclop. in loco.—The branchy dichotomous habit of my plant, its hairy petiols and capsules are wanting in them all and may perhaps be deemed sufficient to form it into a new species, which I am unwilling at present to add to this difficult genus. It differs very little, perhaps only in the rounded branches from the species described by Roxburgh as Retzius's *hispida.*

I have not cited Burman's *Valerianella palustris,* Thes. Zeyl. 227. *p.* 108. *f.* 1. because it does not well agree with Rheede's or my plant.—N. W.

6. H. *ulmifolia. Wall.*

Decumbent, hispid with short spreading dense hairs. *Leaves* ovate-lanceolate, nerved and plaited. *Corymbs* peduncled, aggregate, axillary, opposite, shorter than the leaves, open. *Flowers* pedicelled, mostly ternate. *Calycine segments* very long.

Sent to me from Nepal*a* by the Hon. Mr. Gardner; time of blossoming the rainy season.

Root fibrous, strong.—*Stem* short, woody, divided into several simple, long, very hairy *branches*, rounded, smoother, and jointed towards the base, somewhat flattened at their upper-ends.—*Leaves* in approximate pairs, from two to three inches long, somewhat more than an inch broad, acute at the base, attenuate and acuminate at their end, rough and hairy on both sides, with many approximate, parellel, downy nerves, deeply furrowed and plaited above, acquiring on being dried, as well all the younger parts, a yellow appearance.—*Petiols* half an inch long, broad and slightly marginated, furrowed.—*Stipulæ* broad-lanceolate or triangular, as long as the petiols, adpressed, tapering into a very long, sometimes bifid, or trifid, hispid acumen.—*Peduncles* slender, an inch long, several in each axil, erect, once or twice trichotomous, with subulate bractes under each division.—*Flowers* rather large, pedicelled, mostly in threes.—*Calyx* round, hispid, with four lanceolate persistent laciniæ, almost as long as the smooth tubular *corolla*, the segments of which are linear-lanceolate, acute.—*Stamina* short. *Filaments* and *Style* pubescent. *Stigma* two-lobed.—*Capsule* globose, as large as a Coriander seed, crowned, hispid, many-seeded.

Obs. In general appearance and in the leaves this species agrees with H. *capitata*, Lam.; but it differs in having opposite aggregate corymbs, and in the flowers not being capitate.

7. H. *gracilis. Wall.*

Erect, slightly scabrous, with lengthened, sub-dichotomous, naked branches. *Leaves* linear. *Stipules* truncate, fimbriated, at last naked. *Peduncles* terminal, very long, tern, sub-panicled. *Stigmas* linear.

H. *stricta, Wall. in Asiat. Res.* xiii. *p.* 369.

A native of the Turraee and of Nepal*a*; sent to me from the former place by Mr. Assistant Surgeon William Jack, from the latter by Mr. Gardner. Flowering time the hot season.

Nepala name *Goshega-soa.*

Obs. I take this opportunity of correcting an error which I committed in the description referred to above by calling my plant H. *stricta:* a name which had already previously been given to a different species from Sierra Leone, by Sir J. E. Smith, in Rees's new Cyclopædia, and which therefore I have altered to *gracilis.*

SPERMACOCE. *Schreb. Gen. N.* 155.

Calyx four-leaved. *Corol* funnel-shaped. *Capsule* inferior, crowned, two-celled, two-valved. *Seed* solitary, longitudinally furrowed on the inside. *Embryo* erect, and furnished with a perisperm.

1. S. *sumatrensis. Linn. Sp. Pl. ed. Willd* i. 571. *Retz. Obs.* iv. 23.

Shrubby, erect, four-sided. *Leaves* lanceolate, remote, short-petioled. *Corymbs* terminal, decompound. *Capsule* opening from the base.

A native of moist places amongst the Circar mountains. Flowers during the rainy season.

Stem erect, woolly, obsoletely four-sided. *Branches* opposite, four-sided.—*Leaves* opposite, spreading, short-petioled, lanceolate, downy; size very various.—*Stipules* connecting, membranous, two or three-toothed.—*Umbel* compound, terminal.—*Flowers* white, very small.—*Calyx* four-leaved.—*Corol* with short, gibbous tube.—*Capsule* oblong, two-celled, two-valved, two-partible from the base.—*Seed* solitary.

Obs. This species ought to be carefully compared with *Hedyotis fruticosa, Retz. Obs.* ii. 8 ; and the capsule of the same species in *Linn. Flor. Zeyl. N.* 63. should be well examined.* The stipules seem the chief difference. In my plant they are always a lacerated connecting membrane.

* The reader is requested to consult the observation on this plant in Rees's Cyclopædia, Vol. xvii.—N. W.

Knoxia corymbosa, Linn. Sp. Pl. ed. Willd. i. *p.* 582, which he considers as *Gærtner's* K. *stricta,* appears to me to be this plant. Indeed the two genera *Spermacoce* and *Knoxia* differ so little from one another as to induce me to think one of them might be spared.

2. S. *teres. R.*

Shrubby, erect, round, villous. *Leaves* opposite, remote, lanceolate, entire, villous. *Corymbs* terminal, decompound, alternate; divisions becoming *spikes* when in seed. *Stamens* erect, but hid in hairs. *Style* twice the length of the corol.

Knoxia umbellata, Banks. Herb.

Found by Dr. Buchanan in the woods of Koorg, from whence he sent the seed to the Botanic Garden, where the plants thrive well, and blossom during the rainy season.*

Stem erect, perennial, round, while young villous. *Branches* opposite, a little inflected, round and villous, height of the whole plant about three feet.—*Leaves* opposite, short-petioled, lanceolate, downy, entire; length from two to four inches high.—*Connecting* membrane with several erect, filiform divisions; when they accompany the divisions of the inflorescence, they are more numerous, and may be called an involucre.—*Corymbs* terminal, brachiate, with the ultimate divisions lengthened into short spikes by the time the seeds are ripe.—*Corol,* mouth of the tube shut with much long white hair, which entirely conceals the large, erect anthers lodged amongst it.—*Style* erect. *Stigma* two-cleft with divisions involute.

Obs. This has much the habit of *Hedyotis fruticosa,* but differs widely in the stipules, (even if the pericarp of that plant should be that of a Spermacoce,) and in the form of the stem and branches. It is also very like S. *sumatrensis,* but there the branches are four-sided, here perfectly round; there the tube of the corol is smooth, short and gibbous, here cylindric, and shut with hairs; finally the capsule is there bipartible, opening from the base, with oblong brown seed, in

* I have received specimens of this and the preceding species from Nepala, varying, however, a little in their character from those described above.—N. W.

reality it is that of a *Knoxia ;* here it does not open spontaneously, and has black round seed.

3. S. *exserta. R.*

Shrubby, straight, tender parts villous. *Leaves* remote, opposite, petioled, lanceolate ; *corymbs* terminal; decompound. *Stamina* and the bifid *stigma* exsert. *Capsule* oval.

A native of the Circars, where it blossoms in October and November.

Obs. Is nearly allied to S. *teres,* farther examination may prove them the same.

4. S. *glabra. R.**

Shrubby, round, smooth, erect. *Leaves* linear-lanceolate, smooth. *Corymbs* axillary, compound ; *styles* hairy. *Stigma* clavate. *Capsules* globular, smooth.

A native of Pulo-pinang, has much the habit of a strong grass or slender bamboo, the situation of the leaves excepted.

Branches round, smooth, straight, jointed, in the dry specimens fistulous.—*Leaves* opposite, sub-sessile, linear-lanceolate, smooth, entire, length from four to six inches; connecting membrane with unequal awned processes from its mouth.—*Peduncles* axillary, bearing several cross-armed, minute umbellets or headlets of small cream-coloured flowers.—*Bractes* minute.—*Corol* bell-shaped, mouth of the tube hairy.—*Style* longer than the corol, hairy. *Stigma* clavate, grooved.—*Capsule* globular, smooth, size of a large pin's head.

5. S. *lævis. R.*

Biennial, straight, round, smooth. *Leaves* sub-sessile, lanceolar. *Corymbs* terminal. *Anthers* hid in the bearded mouth of the infundibuliform corol.

A native of the interior of Bengal; in the Botanic Garden it

* This name was previously given to an American distinct species by Michaux. —N. W.

blossoms and ripens its seed from September till February and is by far the prettiest Spermacoce I have yet met with.

Stems erect, perfectly straight, round, smooth. *Branches* opposite, and like the stems, whole height about three feet.—*Leaves* opposite, sub sessile, lanceolate, smooth, entire; from one to three inches long.—*Connecting* membrane divided into several subulate segments.—*Corymbs* terminal, compound and decompound.—*Flowers* numerous, large, pale pink.—*Bractes* small, and at the divisions of the corymbs chiefly.—*Calyx* one of the four segments large, all the segments permanent.—*Corol* funnel-shaped, before expansion the apex is large, with four-horns; after expansion these projections form a conical knob under the apex of each segment. *Mouth* of the tube bearded.—*Filaments* short; *anthers* linear, just within the tube.—*Germ* two-celled, with one seed in each, attached to the top of the partition; *style* longer than the tube of the corol; *stigma* bifid, *segments* revolute.—*Capsule* oblong, crowned, two-celled, two-valved; *valves* keeled on the back. *Seed* solitary, longitudinally grooved on the middle of the inside.

Obs. It may be readily distinguished from S. *teres* (the only species known to me that it can be mistaken for) by its smoothness, greater size of the flowers, and the segments of their border being enlarged with a conic knob on the outside of their points.

6. S. *lineata. R.*

Erect, ramous, hairy. *Leaves* short-petioled, lanceolate, acuminate, linear. *Stipules* triangular and shaggy. *Heads* axillary on trichotomous short peduncles; *stamina* protruded.

A native of the Moluccas. The lineate, hairy, very acute, tapering leaves, long shaggy triangular stipulæ; and small peduncled heads of axillary flowers readily distinguish this species from all the other East Indian species known to me.

7. S. *nana. R.*

Erect, simple, nearly round and smooth; *leaves* petioled, lanceolate, entire, smooth; *flowers* axillary and terminal, sessile, crowded.

A very small species, less than six inches high, a native of the Island of Honimoa.

8. S. *stricta. Linn. Sp. Pl. ed. Willd.* i. 573.

Annual, erect, four-sided. *Leaves* sessile, linear-lanceolate; *flowers* in dense, globular, axillary verticils; *capsules* scabrous, gaping at the apex.

Teling. Sookukada.

An annual, a native of wet rice fields, appears and flowers during the rainy season.

Stem erect, branchy, four-sided, about a foot high, angles acute and scabrous. *Branches* decussated, stem-like, but slenderer.—*Leaves* opposite, sessile, lanceolate, entire, a little scabrous.—*Stipule*, a connecting membrane, ciliate.—*Verticils* globular.—*Involucres* leaf-like.—*Flowers* most numerous, minute, white.—*Tube* cylindric.—*Stigma* globular.—*Capsule*, two-celled, opening at the apex.—*Seed* solitary.

9. S. *costata. R.*

Diffuse, very downy. *Leaves* broad-lanceolate, or oblong, ribbed. *Flowers* crowded on little axillary, proliferous umbellets. *Staminæ* protruded.

*Hedyotis Auricularia. Linn.**

Supposed to be a native of the Moluccas, as the plant sprung up in the Botanic Garden, from some earth brought with the spice plants from those Islands in 1800. It blossoms and ripens its seed during the greatest part of the year.

Root biennial, if not perennial.—*Stems* or rather branches many, diffuse, round, from one to three feet long, clothed with long, soft diverging hairs.—*Leaves* opposite, sub-sessile, broad-lanceolate, entire, ribed with numerous large, simple, parallel veins; downy on both sides, from two to three inches long, and about one broad;

* This is unquestionably a distinct plant from that described here, which in habit and leaves resembles my *Hedyotis ulmifolia.* See above, *p.* 370.—N. W.

connecting membrane with generally three, unequal, slender, subulate divisions.—*Flowers* axillary, numerous, pale pink, very small, crowded together into small round heads, on short peduncles, and pedicels the whole forming little, proliferous, villous umbels.—*Involucre* and *Involucels* annular, villous, from three to six-parted.—*Calyx*, divisions equal, acute, hairy.—*Corol* funnel-shaped, hairy on the outside and in the throat of the tube; divisions of the border recurved.—*Stamens* protruded beyond the mouth of the corol.—*Germ* beneath, broad-turbinate. *Style* about as long as the tube of the corol. *Stigma* two-cleft, villous.—*Pericarp* with two seeds.

10. S. *semierecta. R.*

Stem variously bent, but ascending, square. *Branches* brachiate, alternately smaller. *Leaves* ovate. *Flowers* in small, compact verticils. *Stamens* and *style* inclosed. *Capsule* turbinate, hairy.

A native, I suppose, of Sumatra, as the plant from which this description is taken sprung up in a bed, where earth from that Island had been thrown, and in one year was about a foot and a half high.

The *lower branches* brachiate; the *superior* often solitary, they are all very exactly four-sided with angles sharp and somewhat hispid; in other respects they are very smooth.—*Leaves* opposite, with smaller ones in their axills, short-petioled, ovate, accuminate, entire, very slightly scabrous, particularly the margins, with nerves and veins on the underside.—*Connecting* membrane divided into several filiform portions.—*Flowers* numerous, very small, white, sessile, collected into small compact, round verticills.—*Corol*, divisions of the border bearded in the centre.—*Anthers* within the tube.

11. S. *scabra. Linn. Sp. Pl. ed. Willd.* i. 572.

Annual, diffuse, round, hairy. *Leaves* opposite, sessile, round, obovate, the hairy margins much curved, and somewhat curled. *Flowers* axillary, sessile. *Stamens* and *style* erect.

Tardavel. *Rheed. Hort. Mal.* 9. *p.* 149. *t.* 76. The leaves however are a little too sharp for our Coromandel plant.

Stem short and generally annual. Lower *branches* opposite, superior alternate, round, coloured, very hairy, somewhat scabrous, weak, spreading and variously curved.—*Leaves* opposite, sessile, round, obovate, or nearly oval, somewhat acute; hairy on both sides, edges recurved and more or less curled.—*Connecting* stipule with three, four, or five filiform processes.—*Flowers* axillary, generally in pairs, though often several together, but expanding in succession, sessile, small, purple.—*Calyx*, the four divisions or in this species rather leaflets, lanceolate, nearly as long as the tube of the corol.—*Corol* sub campanulate, half four-cleft.—*Stamens* as long as the segments of the corol.—*Style* as long as the stamens. *Stigmas* two, two-lobed, segments revolute.—*Capsule* oval, two-celled. *Partition* membranaceous.—*Seed* solitary, oval, equally obtuse at each end.

12. S. *articularis. Linn. Sp. Pl. ed. Willd.* i. 572.

Annual, diffuse, four-sided. *Leaves* opposite, sessile, broad-lanceolate; connecting stipules from three to five-bristled. *Flowers* verticelled. *Corol* infundibuliform, four times longer than the calyx. *Stamens* and *style* erect.

A native of the sandy soil on the Coast of Coromandel near the sea, and in flower the greatest part of the year.

Root generally annual.—*Stem* scarcely any, but many diffuse straggling, four-sided, alternate, hispid branches from one to two feet in length.—*Leaves* opposite, sessile, lanceolate, a little hairy; veins simple, from one to two inches long, and half an inch broad.—*Connecting* stipules cup-shaped, with from three to six bristles from each side.—*Flowers* axillary, sessile, two, three, or four expanding in succession from the same axill.—*Calyx* with slender, linear divisions.—*Corol* infundibuliform. *Tube* slender, three or four times longer than the calyx.—*Stamens* erect.—*Stigma* erect, bifid; lobes broad, and revolute.—*Capsule* two-celled, partition membranaceous.—*Seed* solitary, oval, furrowed on the flat side.

13. S. *hispida. Linn. Mant.* 558. *Sp. Pl. ed. Willd.* i. 572.

Annual, with diffuse, obsoletely four-sided, hairy branches. *Leaves* obovate, cuneate, waved, scabrous. *Flowers* verticelled, two, three, or four in each axil. *Tube* of the corol twice the length of the calyx. *Stamens* and *style* erect.

Teling. Madana-bunta-kada.

Galeopsis zeylanica. *Burm. Zeyl. t.* 20. *f.* 3. agrees pretty well with my plant except in the numerous pedicelled flowers.

Common on the sandy lands near the sea on the Coast of Coromandel.

14. S. *pusilla. Wall.*

Erect, simple, four-sided, scabrous, with short hairs.—*Leaves* sessile, linear, smooth and silvery beneath, rather shorter than their interstices. *Stipulary ciliæ* very long. *Verticils* capitate. *Stamina* exserted. *Capsule* villous at the top and bursting there in two bifid valves.

Native of the Valley of Nepala; discovered by Mr. Gardner during the rains.

Root small, fibrous.—*Stem* single, undivided, four-sided, filiform, from four to six inches high, as well as the upper surface of the leaves, rough with short, whitish, subulate hairs; erect, somewhat adscending at the base.—*Leaves* about an inch long, spreading, tapering into a point, dark-green above, smooth and of a silvery white colour below, with an elevated hairy rib; nerveless; above the base they are a little contracted, thence they widen into an ample membranaceous truncated *stipula* which is crowned with five capillary long ciliæ. *Flowers* very small, pale, sub-sessile, crowded together into four or five dense axillary verticils, of which the uppermost are larger and capitate and supported by four reflexed floral leaves.—*Segments* of the *calyx* erect, linear, as long as the filiform, smooth tube, deciduous. *Throat* of the corolla wide; *laciniæ* lanceolate, equalling the spreading filaments. *Ovarium* oblong, slightly compressed, with a longitudinal furrow on each side, smooth at the base, villous at the top.—*Cap-*

sule ovate, chartaceous, bursting at the apex in two bifid slightly recurved valves. *Dissepiment* evanescent. *Seeds* two, linear, with a ventral furrow.

Obs. The description of *Crateogonum (amboinicum) minus* s. *verum,* Hort. Amb. vi. p. 25 agrees well with this plant; but it has been quoted under *Spermacoce stricta* of the younger Linneus, which differs in having lanceolate lineate leaves and a branchy stem. The same plant has been cited by Sir J. E. Smith under his *Hedyotis verticillata (Oldenlandia, Linn. Mant.)* in Rees's Cyclopædia. On these grounds I conclude that Rumphius's plant is different from mine, which approaches to S. *linifolia,* Vahl, and *remota,* Lam.—S. *nana* of Roxburgh differs in being smooth on all its parts, and in having petioled leaves.—N. W.

HYDROPHYLAX. *Schreb. Gen. N.* 159.

Calyx four-parted; *corol* companulate. *Germ* two-celled, one-seeded, inferior. *Capsule* inferior, angular, two-celled. *Seed* solitary; *embryo* erect, and furnished with a perisperm.

1. H. *maritima. Linn. Sp. Pl. ed. Willd.* i. 581.

Sarissus anceps. *Gært. carp.* i. *p.* 118. *t.* 25. *f.* 4.

Diodia orientalis. *Kön. Mss.*

A native of the sand hills along the shores of Coromandel and Malabar, where it is in blossom most part of the year.

Root perennial.—*Stem* none, but many long, slender, variously bent, stragling, round, coloured branches running over the sand and often striking root from the joints.—*Leaves* opposite, approximate, short-petioled, from a deep, stem-clasping, dentate cup.—*Flowers* axillary, sub-sessile, one or two together, of a pale pink colour.—*Calyx* above, four-parted; divisions ensiform, permanent.—*Corol* companulate, tube hairy on the inside near the bottom, the four divisions of the border ovate, spreading, with their apices somewhat revolute.—*Filaments* four, erect, shorter than the corol, and inserted on the mouth of its tube, opposite to its fissures. *Anthers* purple.—*Germ* oblong.

Style length of the corol; *stigma* of two round lobes.—*Pericarp* when ripe a dry berry of a lanceolate shape, crowned with the remaining calyx, two-celled with a single linear oblong seed in each, as detailed and figured by that very eminent and most accurate botanist Gærtner.

GALIUM.

Corolla of one petal, flat, superior. *Seeds* two, roundish.

1. G. *asperifolium. Wall.*

Leaves in sixes, cuneate-oblong, retuse with a subulate point, very spreading, short-petioled, scabrous above, with hispid rib and margins; *stem* villous, four-sided, branchy, flaccid; *panicles* slender; *corolla* awned; *ovarium* smooth.

A native of Nepala from whence this and the following species were introduced into the Botanic Garden at Calcutta in 1818, by the Hon. Mr. Gardner.

Stem four-sided, strong, clad with copious soft recurved hair; *branches* numerous, weak and leaning; *joints* slightly swelled.—*Leaves* mostly six in each whorl; those below remote, oblong-cuneate, an inch and more long, those above smaller and closer together, almost reflexed, cuneate; all are tapering at the base into a very short petiol, retuse or rounded at the end and invariably terminated by a short triangular withering greyish point; above dark-green, shining, scabrous, with short stiff hair, below pale and soft, except along the recurved, entire margins, and elevated rib, which are beset with stiff hair. *Panicles* lengthened, with ternate or opposite, copious, small floral leaves. *Flowers* fascicled. *Corolla* large, expanding; each lanceolate segment ending in a capillary point. *Style* half-cloven. *Stigma* capitate. *Ovarium* perfectly smooth.

Obs. This species has much of the habit of G. *Mollugo.* The stem and branches are very soft, while the leaves possess a considerable harshness on account of their stiff hair; the epidermis separates partly from their upper surface when dry, assumes a greyish shining colour and has the peculiar roughness of some of the species of Cordia.—N. W.

2. G. *Aparine.*

Leaves in eights, lanceolate ; keels scabrous, with prickles pointing backwards. *Joints* villous. *Fruit* hispid. *Linn. Sp. Pl. ed. Willd.* i. *p.* 597.

In the Botanic Garden at Calcutta it blossomed freely in March, 1819.

The description of this well-known species, which Miller and Lamarck have given, agrees well with the Nepal*a* plant. It measured when in flower about two feet in length, with abundance of weak opposite or alternate branches, by which it supported itself on the nearest plants. The *stem-leaves* in whorls of six or seven, from an inch to two long, about two lines broad at their middle, and from thence much tapering. The upper ones smaller, generally in sixes or fives, but all equalling their interstices, upper surface scabrous; the margins and keel as well as the sharp angles of the stem and branches beset with pellucid recurved prickles.—*Peduncles* axillary, mostly opposite and three-flowered.—*Calyx* imperceptible.—*Corolla* exceedingly small, of a yellowish tint.—*Styles* two.—*Fruit* thickly beset with hooked bristles.—N. W.

3. G. *elegans. Wall.*

Leaves in fours, ovate or round-ovate, obtuse, three-nerved, soft, hairy with villous nerves ; *stems* diffuse, hairy; *panicles* trichotomous; *seeds* beset with hooked bristles.

A native of Nepal*a*, from whence I received specimens in 1818 from Mr. Gardner.

Root slender, creeping, red, with verticilled fibres.—*Stems* weak, elongated, four-cornered, two or more feet long, very villous while young, afterwards smoother, having only some roughish slightly recurved hair along the corners ; with a few alternate branches.—*Leaves* about half an inch, sometimes an inch long, sessile, entire, mostly shorter than their interstices, slightly tapering towards the base, dotted and hairy above, smoother below; the recurved margins and elevated nerves very villous.—*Panicles* divaricate, terminal or axillary, generally opposite, with some pairs of opposite small leaves. *Ramifications* divaricate, trichotomous, filiform, with minute bractes at their base.—*Divisions* of the *corolla* lanceolate, acute, one-nerved.—*Seeds* round, covered with copious soft, hooked bristles.—N. W.

RUBIA. Schreb. Gen. 164.

Calyx scarcely any. *Corol* one-petalled, from four to five parted. *Berries* inferior, twin. *Seeds* solitary. *Embryo* erect, and furnished with a perisperm.

1. R. *Munjista. R.*

Perennial, scandent. *Leaves* four-fold, long-petioled, cordate, acute, from five to seven-nerved, hispid. *Corol* flat, five-parted, pentandrous. Fleming in Asiat. Res. xi. 177.

Sans. मंजिष्ठा, M*u*nj*i*stha, विकसा or विकषा V*i*k*u*sa, जिंगी, J*i*ngee, समंगा, S*u*m*u*nja, कालमेषिका or कालमेशिका Kal*u*mesh*i*ka, मंडूकपर्णी, M*u*ndook*u*p*u*rnee, भण्डीरी or भण्डिरी, Bh*u*n-deeree or Bh*u*nd*i*ree, भण्डी Bh*u*ndee, योजनवल्ली, Yoz*u*n*u*v*u*llee.

Beng. M*u*nj*i*t or M*u*nj*i*stha.

A native of Nepal*a*, and other mountainous countries, north and north east of Bengal. It grows in the Botanic Garden at Calcutta, but requires uncommon care to keep it alive during the rainy season and has never blossomed there.

Root perennial.—*Stems* woody, scandent, climbing over trees and bushes. *Branches* cross-armed, from the axill of the large pair of leaves, jointed, swelling at the joints, hairy for a little distance below them, four-sided; angles sharp and armed with small, recurved prickles; the exterior cortex, which forms these angles, drops off by age, leaving the longer branches and stems round, and covered with a soft reddish somewhat downy bark.—*Leaves* four-fold, petioled, one of the pairs always much larger, viz. about double the size of the other, and with longer petioles, all are beautifully cordate, entire, acute, pointed, generally five or seven-nerved, though in the largest there are sometimes nine, and in the smallest three, or even sometimes obliterated; these nerves are armed with small recurved prickles, like those on the angles of the branchlets; both sides channelled.—*Panicles* terminal, oblong, composed of numerous, opposite, small, subglobular, trichotomous panicles from the axill of the floral leaves. —*Peduncles* and *pedicels* four-sided, smooth.—*Bractes* opposite, sessile, long, cordate.—*Flowers* numerous, minute.—*Calyx*, if any, so minute as not to be visible.—*Corol* flat, five-parted; divisions lanceolate, acute, with points inflected.—*Stamens* five, shorter than the corol.—*Germ* beneath, twin. *Style* scarcely any, stigma large, two-cleft.—*Berries* two, or one; the second not always coming to maturity, globular, size of a small grain of pepper, smooth.—*Seed* single, round, smooth, with a deep pit on one side.

Obs. It differs from R. *cordifolia* in being pentandrous.* The roots, stems, and larger branches are used to dye red with.

2. R. *alata*. *Wall.*

Leaves lanceolate, three-nerved, petioled, quatern, very scabrous. *Stem* and *branches* four-winged prickly. *Flowers* pentandrous, hispid in trichotomous, elongated, straight, axillary, and terminal corymbs.

A native of Nepal*a*, from whence Mr. Gardner sent me specimens. It blossoms in June.

Root red.—*Stem* several feet long, apparently leaning, strong, together with the opposite or verticilled *branches* four-cornered, and four-winged, the wings beset with short, recurved prickles, otherwise pretty smooth; joints remote.—*Leaves* spreading, quatern, unequal, from an inch to an inch and a half long, much shorter than their interstices, rough on both sides, especially along their margins, fur-

* I strongly suspect that R. *cordifolia*, Linn. Mant. is this very plant, especially as *Lamarck* observes in his description that the flowers are four and five-cleft; *Encycl. Bot.* ii. 606. The plants agree in every respect. In the numerous specimens which have been sent to me by Mr. Gardner from Gossain-Than and Katmandu, and by Sir Robert Colquhoun from Almora I have always found the flowers pentandrous. —N. W.

rowed above, tapering towards the apex and pointed, rounded at the base, the lowermost small whorls ovate, sub-sessile.—*Petiols* about one-third the length of their leaves, deeply furrowed, smooth above, the sides margined, almost winged and with the keeled base priekly, apex contracted and jointed.—*Peduncles* lengthened, straight, four-winged, twice or thrice trichotomous, hispid. *Bractes* mostly ternate, lanceolate, sessile, hispid, the uppermost opposite, very small. —*Flowers* in peduncled fascicles. *Calyx* imperceptible. *Corolla* flat, white, hairy on the outside; *segments* five, lanceolate, acuminate. *Stamina* and *style* very short. *Stigma* capitate.—*Ovaria* slightly hispid.—N. W.

*IXORA.** *Schreb. Gen. N.* 167.

Calyx four-toothed. *Corol* one-petalled, funnel shaped. *Tube* long, with the stamina in its mouth. *Germ* two-celled, one-seeded, lateral. *Berry* inferior, two-seeded. *Embryo* erect, curved, and furnished with a perisperm.

1. I. *coccinea. Linn. Sp. Pl. ed. Willd.* i. 609.

Shubby. *Leaves* oblong, cordate, sessile, acute, smooth, entire. *Segments* of the corol ovate-lanceolate, acute. *Berry* crowned with the shut segments of the calyx.

Schetti. *Rheed. mal.* ii. *p.* 17. *t.* 12. is a pretty good figure of it, but the description of the colour of the flowers does not so well agree.

Jasminum flore tetrapetalo. *Burm. Zeyl.* 125. *t.* 57, good.

In the Tanjore country and in China I have found this most beautiful shrub in great abundance in its wild state. In a cultivated state it flowers all the year round. I have had it eight years in my garden, and it is not more than three or four feet high, with many erect branches, but scarcely any thing like a trunk.

* Sir William Jones observes (*Asiat. Res.* iv. 251), that no Indian god was ever named IXORA; he adds at the same time that ISWARA, which is indeed a title of *Shiva* would be a very improper appellation of a plant which has already a classical name.

Leaves opposite, sessile, approximated, decussate, oblong, pointed, entire, smooth, shining, from three to four inches long, and one and a half broad.—*Stipules* within the leaves with an awled process on each side.—*Corymb* terminal, three-forked.—*Bractes* small, rigid. —*Peduncles* and *pedicels* short, coloured.—*Flowers* numerous, pretty large, bright scarlet, inodorous.—*Calyx* four-toothed; divisions acute.—*Corol*, stamens and pistil as in the genus.—*Berry* size of a small cherry, red, juicy, two-celled.—*Seed* solitary, convex on one side, and flat on the other.

Obs. Flamma sylvarum. *Rumph. amb.* iv. *p.* 105. *t.* 46. is a very bad representation of this species, and much more like my *Ixora fulgens.* I have therefore doubtfully quoted it for that plant.

2. I. *Bandhuca. R.*

Shrubby, spreading. *Leaves* oval, stem-clasping. *Corymbs* crowded; *segments* of the corol ovate, obtuse. *Berries* crowned with the open calyx.

Asiat. Res. iv. *p.* 250.

Sans. रक्तकः, Ruk*tu*k*a*, बंधूकः, B*u*ndhook*a*, or बंधुकः, B*u*ndhoo-k*a*, बंधुजीवकः, B*u*ndhoojeev*u*k*a*.

Beng. R*u*ng*u*n, R*u*j*u*n*a*.

A very elegant, uncommonly ramous, bushy shrub, a native of Hindoostan. Sir William Jones observes that the flower is often mentioned by the best Indian Poets. It is in flower the whole year but chiefly during the rains, when it is highly ornamental.

Stem none, but branches innumerable, which divide much and spread close on the ground, forming a large hemispherical bush.—*Bark* of the large branches dark brown, and rather scabrous; that of the young shoots, smooth, and green.—*Leaves* opposite, stem-clasping, oblong, obtuse, with a small point, entire, firm, smooth on both sides.—*Stipules* interfoliaceous, annular, with a subulate process on each side.—*Corymbs* terminal, crowded, divisions many, but short. —*Flowers* numerous, when they first open scarlet, changing afterwards to crimson.—*Bractes* opposite, conic, acute.—*Calyx* superior, four-toothed, coloured, permanent.—*Corol* with a very long slender tube; *segments* of the border four, ovate, rather obtuse, spreading.—*Filaments* four, short, inserted without the mouth of the tube, spreading, anthers linear, spreading.—*Germ* two-celled, with one seed in each, attached to the middle of the partition. *Style* length of the tube of the corol. *Stigma* bifid.—*Berry* spherical, size of a large pea, smooth, fleshy, when ripe purple, two-celled.—*Seed* solitary, oval, convex on the outside, flat, with a deep pit on the inside. *Integuments* two; *exterior* nuciform, *inner* membranaceous. —*Perisperm* conform to the seed, cartilaginous.—*Embryo* erect, curved, nearly as long as the perisperm. *Cotyledons* reniform, cordate. *Radicle* cylindric, longer than the cotyledons, inferior.

Obs. This species is nearly allied to I. *coccinea*, but differs in the following respects:

1*st*. This is a very ramous spreading bush, that much less so, and nearly erect.

2*d*. In this the leaves are stem-clasping, and obtuse; in that sessile and acute.

3*d*. In this the segments of the border of the corol are ovate, and rather obtuse; in that ovate-lanceolate, and acute; and lastly,

4*th*. The calyx on the fruit is open in this species, in that shut into a conical point.

3. I. *fulgens. R.**

Shrubby. *Leaves* sub-sessile, lanceolate, acute. *Corymbs* terminal, highly coloured. *Segments* of the calyx cordate. *Tube* of the corol long and filiform, with smooth contracted mouth; *segments* lanceolate, and larger than the erect stamens. *Berries* two-lobed.

Flamma sylvarum. *Rumph. amb.* iv. 105. *t.* 46? pretty good.

This elegant, highly ornamental shrub is a native of the Moluccas, and from thence was introduced into the Botanic Garden, where it blossoms most part of the year.

* I. *longifolia*, Sir J. E. Smith in Rees's Cyclopædia.—I. *lanceolata*, Lam. Encycl. Bot. iii. 343, exclus. omn. synon. præter Rumphium.—N. W.

Trunk short, soon dividing into many straggling weak branches, covered with smooth dark-brown bark. *Leaves* opposite, short-petioled, lanceolate, entire, acute, smooth on both sides; from six to eight inches long, and from one to three broad. *Stipules* connecting, with an acute point on each side. *Corymbs* terminal, composed of short, decussate, highly coloured branches, and branchlets, supporting numerous, short-peduncled, long-tubed, pretty large, scarlet flowers, with the four segments of their border acute, and broad-lanceolate, stamens erect. *Berry* two-lobed, size of a marrowfat pea, smooth, when ripe of a deep purple, two-celled. *Seeds* solitary, nearly round, convex on the back with a deep cavity on the inside, and from thence attached by a thick short cord to the centre of the partition. *Integuments* two, the *exterior* corresponding with what Gærtner calls the *arillus* in *Coffea*, ash-coloured, and firm like parchment; the *inner* membranaceous, and darker coloured.—*Perisperm* conform to the seed, pale blue, and of a soft cartilaginous consistence. *Embryo* erect, one-third shorter than the perisperm, curved. *Cotyledons* two, cordate. *Radicle* sub-clavate, inferior.

4. I. *stricta. R.**

Shrubby, straight. *Leaves* sub-sessile, oblong. *Corymbs* dense, compound, hemispheric. *Laciniæ* of the corol round, spreading. *Anthers* bristle-pointed.

Ixora *coccinea*. *Loureir, Cochinch.* 95. *Curtis's Botanical Magazine, No.* 169.

Flamma sylvarum peregrina. *Rumph. amb.* iv. 107. *t.* 47.

This beautiful plant was brought to the Botanic Garden from the Moluccas in 1798, where it is in constant blossom the whole year round; but rarely ripens its fruits. The plant, when in flower is highly ornamental, though by no means so gaudy as I. *coccinea* and *Bandhuca*, which are certainly two of our most showy Indian shrubs.

Trunk scarcely any, but a few perfectly straight branches, covered

* I. *flammea*, Salis. Hort. 62, according to Sir J. E. Smith, l. cit. I. *chinensis*, Lam. l. cit. 344.—N. W.

with dark-brown smooth bark; height from three to four feet. —*Leaves* opposite, sub-sessile, oblong, entire, smooth on both sides. —*Stipules* interfoliaceous, long, taper, acute-pointed.—*Corymbs* terminal, very dense, sub-hemispheric, primary divisions brachiate, and short, extreme divisions trichotomous; all smooth and of a bright, deep red colour.—*Flowers* numerous, colour at first a lively orange approaching to scarlet, becoming deeper and deeper from the time of expansion.—*Calyx* bright red, fleshy; *divisions* short and obtuse.—*Corol,* tube cylindric. *Border* of four round spreading segments.—*Filaments* without the tube, short, spreading, flat, with their linear acute anthers over the divisions of the border of the corol.—*Stigma* two-cleft, elevated a little above the mouth of the tube. *Berry* spherical, smooth, succulent, red, two-celled, with a single rugose seed in each

Obs. In the Botanic Garden at Calcutta there is a more ramous variety of this charming plant introduced from China where it is called Hong-mou-tang, with pale pink flowers which it bears in abundance all the year.*

5. I. *alba. Linn. Sp. Pl. ed. Willd.* i. 906.†

Leaves sessile, broad lanceolar. *Corymbs* decompound, dense, sub-hemispheric; *laciniæ* of the corol obovate and reflexed.

Until I came to Bengal I had not an opportunity of seeing this elegant plant. I have found it only in gardens. Flowering time the hot and rainy seasons. The shrub itself is much like *Ixora coccinea.*

Ramous, erect, and smooth in every part. *Leaves* opposite, sessile, broad-lanceolar, smooth, a little waved, from three to six inches long.—*Stipules* within the leaves.—*Corymbs* terminal, compound, or decompound, dense, often hemispheric; *divisions* or *sub-divisions* by threes.—*Flowers* very numerous, inodorous, white, with a straight, rather longer and more slender tube than in *Ixora coccinea*; divisions of the border reflexed, obovate.

Obs. This is probably nothing more than a variety of I. *stricta,*

* I. *incarnata, Roxb. Mss.* according to Sir J. E. Smith, l. cit.—N. W.

† This species is widely different from I. *alba,* Linn.—N. W.

or that of this. It was originally brought from China where it is called Ta-mou-tang.

6. I. *cuneifolia. R.*

Shrubby. *Leaves* broad-cuneate, lanceolate, pointed. *Corymbs* terminal, long-peduncled. *Flowers* crowded; *segments* of the calyx oblong, conic.

A native of the country about Dacca, from thence the late Colonel Peter Murray sent seeds to the Botanic Garden, where the plants grow freely, blossom in March, and ripen their seed in August and September.

Trunk short. *Branches* opposite, nearly erect, round, and covered with smooth brown bark; young shoots smooth, and green.—*Leaves* opposite, short-petioled, broad, cuneate, lanceolate, recurved, entire, taper, obtuse-pointed, firm and polished, somewhat bullate; length from four to six inches, by one and a half or two broad.—*Stipules* tapering, subulate, pointed.—*Corymbs* terminal, long-peduncled, trichotomous, alternate divisions ending in fascicles of sub-sessile, crowded, pure white flowers, with a slight tinge of pink on the outside.—*Bractes* in opposite pairs, at the various divisions of the corymb and under the calyx subulate.—*Flowers* very numerous, pure white, fragrant.—*Calyx* four-cleft to the base, divisions long, narrow and acute.—*Tube* of the corol cylindric, very slender, three quarters of an inch long; the four segments of the border oblong and obtuse.—*Filaments* from the inner edge of the fissures of the border of the corol.—*Anthers* narrow, sagittate.—*Germ* oval, two-celled, with one seed in each, attached to the middle of the partition. *Stigma* of two linear lobes, elevated considerably above the mouth of the tube.—*Berry* round, turbinate, size of a small cherry, when ripe bright red, and smooth, two-celled.—*Seed* solitary, round, oval, convex on the outside, with a large deep pit on the inner. *Integuments* two; *exterior* nuciform; *inner* membranaceous, and greenish.—*Perisperm* conform to the seed.—*Embryo* erect, curved like the convex seed.—*Cotyledons* reniform. *Radicle* cylindric, length of the cotyledons, inferior.

Obs. Vahl's figure of I. *parviflora, Symbol. Bot.* iii. *t.* 52, is much like this plant; his description however agrees better with what I have always considered to be his *parviflora* and which is so labelled in Sir Joseph Banks's *herbarium.* They differ in the following respects. That (I *parviflora*) is a tree, this a shrub; in that the leaves are linear oblong with a cordate base, of a much firmer texture, and more polished; in this they taper for two-thirds of their length to the base (hence the specific name *cuneifolia.*) In that the segments of the calyx are short and obtuse, in this long and acute. In that the berries are round and black; in this short, turbinate and red. The corymbs also differ somewhat. In both species the flowers are white, but in this much larger.

7. I. *brachiata. R.*

Shrubby. *Leaves* short-petioled, lanceolar. *Panicles* brachiate, with remote, diverging branches.

A stout shrub, a native of the forests of Bengal. In the Botanic Garden it flowers in March, and the seeds ripen in May.

Trunk erect. *Branches* opposite, spreading much, but they are not so numerous as in I. *parviflora. Bark* brown, and pretty smooth.—*Leaves* opposite, short-petioled, lanceolar, obtuse, smooth, of a hard texture, from four to six inches long, by less than two broad.—*Stipules* triangular, acute.—*Panicles* terminal, brachiate; *ramifications* few, remote on long, round, smooth peduncles.—*Flowers* numerous, minute, much smaller than in I. *parviflora.*—*Calyx* small, deeply four-toothed, coloured.—*Corol* with a filiform tube; *segments* of the border obovate, emarginate.—*Filaments* none. *Anthers* linear, attached by their backs to the bottom of the fissures of the border of the corol.—*Germ* turbinate, two-celled, with one seed in each, attached to the partition. *Style* scarcely longer than the tube. *Stigma* large, two-lobed.—*Berry* the size of a large pea, smooth, succulent, dark brown purple, two-celled.—*Seeds* solitary, round, convex, on the outside concave, with a pit on the inner. *Integuments* two, *exterior* white, hard, thin, and elastic; *inner* tender and thin.—

Perisperm acetabuliform, amygdaline.—*Embryo* erect, curved. *Cotyledons* cordate. *Radicle* cylindric, curved, inferior.

Obs. A handsome shrub, and nearly allied to Vahl's I. *parviflora*, though very different from the plant so labelled in the *Banksian herbarium*, which grows in abundance in this garden.

1*st.* In the shape of the leaves, which in that species are sub-sessile, and generally have a broad stem-clasping base; in this the petiole is about half an inch long, and the leaves taper more towards the base than at the exterior end.

2*d.* In the thinness of the panicle, (though of the same conformation), and the size of the flowers, which in this species are much smaller, and every way more delicate, and also in their shape, which in that species is oblong before expansion, but in this round.

3*d.* In this the anthers are sessile, in that they have short filaments.

8. I. *villosa. R.*

Shrubby. *Leaves* broad-cuneate, lanceolar, villous underneath. *Corymbs* super-decompound to the sixth or seventh degree, divisions trichotomous, villous. *Calycine* segments oblong and ciliate.

Beng. Choonaree.

This species is a native of Silhet, where it grows to be an elegant middling sized shrub with large crowded corymbs of fragrant white flowers, which appear in May and June.

Young shoots straight and villous.—*Leaves* opposite, short-petioled, broad-cuneate, lanceolar, entire, acuminate, underneath soft with very short down; from six to ten inches long, by from two to four broad.—*Stipules* downy, with the ends subulate.—*Corymbs* terminal, from four to eight times trichotomous, all the divisions downy, and the last three-flowered.—*Bractes* opposite, downy; two slender small ones pressing the base of each calyx.—*Calyx* four-parted, segments oblong, ciliate.—*Corol* with a very long, smooth, filiform tube; segments of the border four, oblong, spreading; in the oblong obtuse bud they are imbricated.

9. I. *acuminata. R.*

Shrubby. *Leaves* petioled, lanceolar, acuminate, smooth; *floral* pair stem-clasping, and broader. *Corymbs* super-decompound, much crowded, and smooth. *Calycine* segments ensiform.

A very charming shrubby species, a native of the forests near Silhet, where it blossoms during the hot season, and perfumes the air with the fragrance of its flowers.

Addition by N. W.

The shrubs in the Botanic Garden have acquired a height of five to six feet.—*Bark* scabrous, ash-coloured.—*Branchlets* roundish, smooth, pale.—*Leaves* lanceolate-oblong, acute at their base, from six to eight inches long, from two to three broad, of rather a thin but firm texture, pale and glaucous, opaque, slightly bullate above, whitish and scabrous beneath, with copious, opposite, parallel nerves uniting into sub-marginal arches; veins reticulated.—*Petiol* short, roundish. —*Floral* pair of leaves smaller than the rest, ovate-oblong, rounded at the base.—*Stipules* broad, triangular, acute, purplish.—*Corymbs* large, sub-sessile, round, with snow-white, most fragrant, crowded flowers.—*Peduncles* twice or thrice trichotomous, very short, pubescent, with opposite linear-lanceolate *bractes* under each division. —*Calyx* ovate, with white lanceolate straight laciniæ supported by a pair of bractes.—*Tube* of the corol slender, an inch and a half long. *Laciniæ* spreading, ovate-oblong, obtuse, with recurved sides; *mouth* surrounded with a series of silky long hairs.—*Filaments* reflected between the laciniæ.—*Anthers* linear.—*Stigma* elevated above the mouth, club-shaped, bifid.

The opaque remarkably pale and glaucous leaves, the sub-sessile crowded corymbs of large white blossoms, with white calyces, sufficiently distinguish this elegant shrub from I. *barbata,* R. and all the other species.

10. I. *parviflora. Vahl. Symbol.* iii. *p.* 11. *t.* 52. *Linn. Sp. Pl. ed. Willd.* i. 609.

Arboreous. *Leaves* sub-sessile, oblong, smooth. *Panicles* ovate oblong, decussated. *Laciniæ* of the corol oval. *Style* hairy. *Berries* round.

Beng. Gundhul-rungun.

Teling. Tadda-pullu.

Among the mountains of the Coast of Coromandel this species grows to be a pretty large tree; in the low lands it is much smaller, seldom exceeding fifteen or twenty feet in height. Flowering time February and March.

Bark dark-coloured, scabrous.—*Leaves* opposite, short-petioled, linear-oblong, pointed at the base, often cordate; smooth, shining, firm, entire, from three to four inches long, and from one and a half to two broad.—*Stipules* as in I. *coccinea.*—*Panicles* ovate, terminal, erect, cross-armed; *divisions* always three-forked.—*Bractes* at the principal divisions stipulaceous, with four subulate processes; the rest are small, one below each ramification.—*Flowers* small, white, fragrant. The style is hairy.—*Berry* generally two-seeded, somewhat two-lobed, size of a pea, black.

11. I. *barbata. R.**

Tube of the corol long; mouth bearded. *Leaves* opposite, short-petioled, oblong, entire, smooth, shining; *floral* leaves round, cordate, sessile. *Panicles* open.

This elegant, densely ramous, large shrub, or small tree, I have found only in the Botanic Garden. Flowering time the hot season.

Trunk scarcely any, branches numerous, opposite.—*Leaves* opposite, short-petioled, oblong, entire, smooth, and shining on both sides, from six to nine inches long.—*Stipules* within the leaves as in the other species.—*Corymbs,* or rather panicles terminal, decompound, large, diffuse, always trichotomous, smooth in every part.—*Bractes,* the lowermost pair embracing the base of the common peduncle, large, and cordate, (they may be called floral leaves;) the

* This plant must not be confounded with *Pavetta barbata* of Sir J. E. Smith in Rees's Cyclopædia, in loco.—N. W.

rest gradually decrease in size till they become very minute at the ultimate divisions.—*Calyx* as in the genus, with acute divisions.—*Tube* of the corol long and slender, its mouths crowned with long white hairs.—*Style* the length of the corol. *Stigma* clubbed, just without the mouth of the tube, apex slightly bifid.—*Berry* red, smooth, size of a pea, two-seeded.

Obs. Bem-Schetti. *Rheed. Mal.* ii. *p.* 19. *t.* 13. is a tolerable representation of this plant, but a bad one of what I consider to be *Ixora alba.*

12. I. *undulata. R.*

Leaves opposite, broad-lanceolar, acute, much waved. *Panicles* terminal, open; *divisions* of the calyx acute. *Filaments* as long as the anthers. *Stigma* two-cleft. *Berries* transversely oval.

Beng. Paluka-Jooi.

A ramous shrub, a native of forests in Bengal, flowering time the hot season.

Leaves opposite, sub-sessile, spreading, broad-lanceolar or oblong, pointed, margins much waved, smooth on both sides but not shining. —*Stipules* within the leaves, with a long awled process on each side. —*Corymbs* or rather panicles terminal, decompound, &c. as in I. *barbata,* but much smaller.—*Bractes* as in the other species.—*Flowers* numerous, small, white.—*Calyx,* divisions acute.—*Corol,* divisions of the border reflexed, rather shorter than the tube.—*Filaments* from the fissures of the corol, pretty long, spreading and variously bent. *Anthers* as in the genus.—*Style* rather longer than the tube. *Stigma* two cleft; *divisions* recurved.

13. I. *Pavetta. R.*

Shrubby. *Leaves* short-petioled, smooth, entire. *Panicles* fastigiate, axillary, and terminal. *Style* twice the length of the corol. *Stigma* entire. *Berries* globular.

Pavetta *indica. Linn. Sp. Pl. ed. Willd.* i. 610.

Pavetta, s. Malleamothe. *Rheed. Hort. Mal.* v. 19. *t.* 10, is certainly this plant.

Beng. Kookoora-choora.

Teling. Nooni-papoota.

A middle sized shrub, common on road sides, hedges, uncultivated lands, &c. where there is a good soil.

Branches cross-armed, ascending. *Bark* ash-coloured.—*Leaves* opposite, petioled, oblong, entire, pointed, always smooth on both sides, which forms the chief difference between this and the next species; from four to six inches long, and two, or two and a half broad.— *Stipules* as in the former species.—*Corymbs* compound, terminal, large, nearly level-topped, cross-armed.—*Peduncles* and *pedicels* round, and smooth.—*Bractes* as in the former species.—*Flowers* white, somewhat fragrant.—*Calyx, divisions* obtuse.—*Style* twice the length of the tube. *Stigma* clubbed.—*Berry* globular, size of a pea, one or two-seeded.

14. I. *tomentosa. R.**

Shrubby. *Leaves* oblong, ventricose, entire, tomentose. *Panicles* lax, sub-globular, tomentose; *style* twice the length of the corol. *Stigma* entire. *Berries* globular.

Beng. Jooi.

Teling. Nam-papoota.

A large shrub, native of the same places, but not so common as the last described. Flowers the beginning of the hot season.

Branches cross-armed, ascending. *Bark* ash-coloured.—*Leaves* opposite, petioled, ovate, pointed, entire, both sides very downy, from four to eight inches long, and from three to five broad.—*Stipules* as in the former species.—*Panicles* terminal, globular, cross-armed, divisions always three-forked.—*Peduncles* and *pedicels* round, downy. —*Bractes* as in the former species.—*Flowers* numerous, white,

* Consult Rees's Cyclopædia under *Pavetta.*—I have had the species from Katmandu and Almora.—N. W.

faintly fragrant.—*Style* twice the length of the tube. *Stigma* club-bed.—*Berry* generally two-seeded.

Obs. In all the foregoing species the berries are two-seeded, though it often happen that but one comes to perfection. I never met with any that had three or four.—I do not find that the inhabitants of these parts make use of any of the trees, except for fuel.

15. I. *tenuiflora. R.*

Leaves short-petioled, oblong, obtuse, smooth; *panicles* terminal, corymbiform, contracted. *Segments* of the calyx cordate, and obtuse, tube of the corol long, and most slender; *segments* of its border oblong, obtuse, and longer than the erect style, or stamina.

A native of the Moluccas, and with flowers infinitely more delicate than I. *fulgens* which it most resembles.

16. I. *congesta. R.*

Shrubby. *Leaves* short-petioled, oblong, strongly veined. *Corymbs* terminal, short, dense. *Flowers* sub-sessile. *Segments* of the calyx obscurely reniform; those of the border of the corol oblong, and obtuse.

A native of the Moluccas, and differs from I. *fulgens* in the firm somewhat coriaceous leaves, with stronger parallel veins.

17. I. *lanceolaria. Colebr.*

Shrubby, smooth; *branches* slender, pendulous, sub-dichotomous. *Leaves* in approximate pairs, very long, linear-lanceolate, acuminate short-petioled. *Corymb* terminal, sub-sessile, small, with coloured scaly, crowded *peduncles. Tube* of the corol filiform; *laciniæ* linear-oblong. *Anthers* linear with deeply bifid base.

A native of Travancore. Introduced into the Botanic Garden by Dr. A. Berry in 1808. Time of blossoming the hot season; the fruit ripens towards the close of the rains.

An erect shrub, five, six, or even more feet high, with trichotomous, very little spreading, ash-coloured branches. *Branchlets* very slender and twiggy, hanging, green, with numerous compressed widened joints, like all the other parts of the plant smooth.—*Leaves* spreading, from five to seven inches long and almost as many times exceeding their interstices, an inch or less broad a little above their acute base, from thence tapering into a sharp acumen, opaque and somewhat glaucous above, pale beneath, with a slender white rib and very fine sub-opposite nearly transversal nerves which communicate with each other in sub-marginal arches.—*Petiols* about two lines long, channeled. *Stipules* adpressed, lanceolate, as broad as the interstices between the insertion of the two opposite leaves, tapering into a subulate point which generally exceeds the petiols in length.—*Corymb* small, supported by a pair of very short leaves, consisting of twice or thrice trichotomous, reddish, pubescent peduncles, with opposite linear, subulate bractes under each division, having a stipuliform, fimbriated process between their bases. *Pedicels* ternate or fascicled, two or three lines long, with three pairs of reddish fleshy subulate scales, the uppermost adpressed to the calyx.—*Calyx* very small, oblong, reddish, with erect subulate teeth; at the bottom, within, there is a series of fleshy, subulate, withering and sphacelated *ciliæ* as in the family of *Asclepiadeæ* and *Apocyneæ*. *Corolla* white; *tube* slender, half an inch long; *limb* spreading, equalling the tube, with linear oblong-fatcate, slightly pubescent *laciniæ*.—*Anthers* linear and long, the base bifid and ending in two whitish processes; *filaments* short, exserted. *Style* clavate; *stigma* two-lobed, subulate, spreading, elevated above the mouth of the corolla. *Berry* as large as a marrow-fat pea, red, smooth, crowned with the four subulate, erect teeth of the calyx.—*Seeds* cup-shaped, hemisphærical.

Obs. This species is so distinct from all the others as to be easily known. Its slender hanging branches, pallid and glaucous leaves, and the small corymbs of crowded flowers contribute to render it a very ornamental plant in the shrubbery.—N. W.

18. I. *rosea*. *Wall.*

Shrubby. *Leaves* oblong, acute with contracted sub emarginate base, pubescent beneath; sub sessile. *Corymbs* super-decompound, open and large, terminal and from the exterior axils. *Laciniæ* of the corol oblong-cuneate, acute.

Khasee name, *Mattiā-Chāndā.*

A native of the hilly parts on the borders of Bengal about Silhet, from thence introduced, in 1815, into the Botanic Garden at Calcutta, by Mr. M. R. Smith. It blossoms most freely during the hot and rainy seasons.

An erect, strong, branchy shrub, about four feet high, with scabrous, brown bark. The younger *branches* pubescent, slightly compressed towards the insertion of the leaves, ash-coloured.—*Leaves* sub-decussate, spreading, leathery and firm, five or six inches long, about two broad, tapering towards the base and there rounded or slightly emarginate, dark green, bullate, smooth and shining above, pale and pubescent beneath, especially along the elevated rib and copious nerves.—*Petiols* thick and woody, scarcely more than a line long.—*Stipules* broad ovate, pubescent ending in a long subulate adpressed point, connected and stem-clasping by means of a narrow membrane, continuing within the base of the petioles, and concealing a series of small rust-coloured *ciliæ.*—*Corymbs* round, as large as the leaves, mostly elevated on a peduncle, with trichotomous brachiate spreading pubescent reddish peduncles, the lower ones an inch long; the whole forming an ample terminal inflorescence of innumerable, most beautiful, pale, rose-coloured, ternate, sessile flowers.—*Bractes* opposite, lanceolate, connected by a ciliated margin.—*Calyx* very small, oblong with four reddish lanceolate ciliated teeth, supported by two adpressed bractes.—*Ciliæ* as in the preceding species.—*Corolla* pubescent; *tube* slender, a little more than an inch long; *laciniæ* smooth, spreading, at last reflected, equalling one-fourth of the tube in length.—*Filaments* reflected. *Anthers* linear-sagittate.—*Stigma* reddish.

Obs. This shrub is exceedingly elegant on account of its large round corymbs which for eight months of the year are produced in a constant succession. The colour of the flowers is a pale pink, gradually becoming reddish as they grow old, beautifully contrasting with the shining dark green leaves, which are not unlike those of I. *Bandhuca.* The colour of the flower resembles that of the pink variety of I. *cuneifolia;* it is paler than the variety of I. *stricta,* which Roxburgh formerly used to call I. *incarnata.*—N. W.

GARDNERIA.

Calyx inferior. *Corol* flat. *Stamina* partly cohering into a tube. *Berry* two-celled. *Seeds* solitary, cup-shaped.

1. G. *ovata. Wall.*

A native of the mountains in the district of Silhet in Bengal, and of Nepal*a.* It blossoms in the hot and rainy season, and ripens its fruit in the cold.

Khasee name, *Huldi.* In Nepal*a*, *Oocha-Soa.*

A strong climbing smooth shrub, with long slender round branches; the new twigs have a green bark and are obscurely four-sided.—*Leaves* opposite, approximated, spreading, petioled, ovate or lanceolate, acuminate, pointed at the base, entire, smooth on both sides, very dark and shining above, pale underneath, with elevated rib and transversal sub-opposite very fine nerves, which unite with each other and with the reticulated veins in sub-marginal arches. They measure from three to four inches, and are generally twice as long as their interstices.—*Petiols* round, channeled above, slightly curved, half an inch long, widening at the base and surrounding the branch with an elevated, fleshy, narrow, erect margin, concealing in their axils a series of short subulate fleshy withering *ciliæ.*—*Stipules* no others than the appendices just described.—*Corymbs* axillary, opposite, equalling or somewhat exceeding the leaves in length, slender, with one or two pairs of small lanceolar leaves under their primary divisions. *Peduncles* slender, opposite, an inch or an inch and a half long, three-flowered, pedicels clavate with two lanceolate minute bractes at the base, and generally a pair above their middle; the intermediate ones shorter.—*Calyx* very small, cup-shaped, persistent, divided into four concave, orbicular, ciliated, alternately imbricating *laciniæ*.—*Corol* thick and leathery, yellow, deeply four-parted; *laciniæ* ovate, acute, slighty concave, very spreading, almost recurved, about two lines long, alternating with those of the calyx and with the insertion of the stamens, covered above, particularly towards the thickened apex with a sericeous down, margins broad and flat; during the *aestivation* they are valvate.—*Tube* none.—*Stamens* four, erect, somewhat shorter than the corol. *Filaments* scarcely perceptible, inserted at the bottom of the corol just below the fissures. *Anthers* ovate, acute, and free at both ends, fleshy in the middle; the sides uniting into a ventricose, pervious, four-toothed tube; within, towards their margins, they are two-celled; the cells separated by a longitudinal partition, which becomes free and keel-formed at their bursting.—*Ovarium above,* and perfectly distinct from the calyx, ovate, minute, two-celled, two-seeded: ovula attached to the centre of the partition.—*Style* filiform, shorter than the tube of the anthers.—*Stigma* simple? acute.—*Berry* superior, round, somewhat depressed, smooth, crowned with the persistent base of the style, supported by the calyx, scarlet, about the size of a large Currant, fleshy, two-celled, two-seeded. *Pulp* pale red.—*Dissepiment* very thin, membranaceous, marked on each side with some vessels disposed in the form of a cross.—*Placenta* fleshy, short, originating from the centre of the dissepiment, entering the cavity of the seed and filling it entirely with a soft pulp.—*Seed* solitary in each cell, very thin, bullate and excavated in the shape of a hemispherical cup with a flat somewhat widened, exactly circular margin, gibbous at the back, smooth on both sides, of a grayish brown colour.—*Integument* single, coriaceous, closely adhering to the horny, pearl-coloured *Perisperm,* which has exactly the form of the seed.—*Embryo* dicotyledoneous, erect, cylindrical, adscending, from above the inferior portion of the margin, about one-third the length of the seed, slightly curved, milk-white.—*Cotyledons* leafy, lanceolate, flat, parellel to the sides of the seed. *Radicle* cylindric, inferior and slightly centripetal, twice as long as the cotyledons, its apex situated immediately within the margin of the seed at its lower part.—*Plumula* inconspicuous.

Obs. The general habit and structure of the fruit of this interesting shrub are those of the family of *Rubiaceæ*, especially its primary sub-divisions, with a single two-celled, two seeded fruit; while its *superior free ovarium* is irreconciliable with the character of that most extensive tribe. In this point, to which may be added the connate anthers and stipulation, the genus agrees with *Apocineæ.* But it differs from them both in the æstivation of its flowers, forming with the axillary subulate bodies a more remote affinity with *Asclepiadeæ*, among which Mr. R. Brown has recorded a solitary instance of a valvate corolla in his *Leptadenia.**

The difficulty thus encountered in assigning my plant to its proper station in the natural arrangement has fortunately been lessened by M. Savigny, who concludes his description of *Pagamea* of Aublet, inserted in Lamarck's Encyclop. Bot.† by suggesting that it might be united with *Gaertnera* of Lamarck‡ under a distinct order; and finally removed by Mr. R. Brown who, having on a previous occasion§ pointed at the existence of an intermediate family between *Rubiaceæ* and *Apocineæ*, afterwards confirmed its existence in his interesting observations on the plants collected in the vicinity of the River Congo by my lamented friend and fellow student the late Professor Christian Smith.‖ To this connecting group my plant necessarily belongs, requiring a separate section in addition to those proposed by Mr. Brown. The fruit appears to agree with *Gaertnera* of Lamarck, with *Pagamea*, especially if the doubts of Jussieu respecting the latter¶ should be verified, and with *Baconia* of Decandolle. Its general structure, as well as the horny and copious perisperm which, during germination, is raised up with the young plant and remains a considerable time in that state, and the relative posi-

* Memoirs of the Wern. Soc. i. p. 19 et 34. † Vol. iv. p. 694.

‡ Illustr. tab. 167 et Poiret in Encycl. bot. suppl. append. p. 684. § Prodr. flor. Nov. Holl. i. p. 455.

‖ Narrative of Capt. Tuckey's Expedition to the River Zaire; Append. p. 448.

¶ Gen. plant. ed. Usteri. p. 2$_{\circ}$2.

tion of the Embryo are perfectly like those of *Psychotria, Coffea* and *Ixora.* It has the excavation of the seed in common with *Pavetta indica** and several species of *Ixora;* the ventral aperture is however much larger here and circumscribed by a somewhat widened margin.—The younger parts of the plant as well as the dark-coloured shining leaves contain a yellowish juice, from which probably the Silhet name is derived; indeed they are not unlike those of *Guttiferæ.*

My Nepala specimens differ in some degree from the others; their leaves are smaller, lanceolate and more narrow; the segments of the corolla lanceolate and more villous, the berries large and crowned with the persistent long style; stigma bifid.

I have dedicated this new genus to the Hon. Edward Gardner, late resident at the court of the Raja of Nepal*a*, by whose indefatigable exertions and liberality the Botanic Garden at Calcutta has been enriched by the most splendid additions of growing plants, seeds, and specimens of the interesting and hitherto mostly unknown vegetable productions of that country, of which ample and frequent collections have been transmitted to Great Britain.—N. W

PYROSTRIA. Juss. Gen. ed. Ust. 229.

Calyx four-toothed. *Corol* with cylindric tube, and four-cleft border. *Drupe* inferior, with as far as eight, one-seeded nuts.

1. P. *hexasperma. R.*

Leaves opposite, short-petioled, oblong, acute, entire. *Corymbs* axillary, and also within the cuspidate, hairy, stipules, dichotomous. *Drupe* six-seeded.

A native of the Island of Honimoa.

Tender-shoots clothed with brown hairs—*Leaves* opposite, approximate, short-petioled, oblong, acute, entire, a little hairy; from three to four inches long.—*Stipules* within the leaves cuspidate, hairy.—*Peduncles* opposite; with the stipules many times longer than the petioles, each ending in a small twice dichotomous, corymbiform,

* Gaertn. Carp. i. p. 116. t. 25.

panicle of recurved secund spikes.—*Bractes* subulate, one-flowered. —*Flowers* rather small, hairy.—*Calyx* four-toothed.—*Corol* with a slender, cylindric, villous tube, and contracted, four-parted border. —*Filaments* none. *Anthers* linear in the mouth of the tube of the corol, and affixed to it by their backs.—*Germ* beneath. *Style* shorter than the corol. *Stigma* simple.—*Drupe* round, six-ribbed; size of a pea.—*Seeds* or nuts regularly six, sub-cylindric.

SALVADORA. Schreb. Gen. N. 220.

Calyx four-toothed. *Corol* one-petalled, four-cleft. *Style* none or short. *Berry* inferior, one seeded.

1. S. *persica. Linn. Sp. Pl. ed. Willd.* i. 695. *R. Corom. Pl.* i. *N.* 26.
Rivina *paniculata. Syst. nat.* x. *p.* 899.
Teling. Pedda-warago-wenki.

A middle sized tree, a native of most parts of the Circars though by no means common; it seems to grow equally well in every soil. Produces flowers and ripe fruit all the year round.

Trunk generally crooked, from eight to ten feet high, to the branches, and one in diameter. *Bark* very scabrous, and deeply cracked. *Branches* exceedingly numerous, spreading, with their extremities perfectly pendulous, like the *weeping Willow.*—*Leaves* opposite, petioled, oval or oblong, entire, very smooth, and shining on both sides, without veins; from one to two inches long, and about one broad.—*Stipules* none.—*Panicles* terminal, and from the exterior axills.—*Flowers* minute, very numerous, greenish-yellow. —*Bractes* minute.—*Calyx* inferior, four-toothed, permanent.—*Corol* one-petalled. *Tube* short, border four-cleft; *segments* oblong, revolute.—*Filaments* four, inserted into the tube of the corol under its fissures, and rather shorter than the border. *Anthers* oval.—*Germ* globular, style none. *Stigma* scabrous.—*Berry* very minute, much smaller than a grain of black pepper, smooth, red, juicy.—*Seed* one.

Obs. The berries have a strong aromatic smell, and taste much like garden cresses. The bark of the root is remarkably acrid, bruised and applied to the skin, it soon raises blisters; for which purpose the natives often use it. As a stimulant it promises to be a medicine possessed of very considerable powers.

CALLICARPA. *Schreb. Gen. N.* 175.

Calyx four-toothed. *Corol* one-petalled, tubular, four-parted. *Germ* four-celled, cells one-seeded; *ovula* attached to the middle of the axis. *Berry* four-seeded. *Embryo* erect, and furnished with a perisperm.

1. C. *arborea. R.*

Arboreous. *Leaves* long-petioled, oblong, ventricose, tapering much toward both ends, entire, hairy underneath. *Panicles* axillary, sub-globular, throughout dichotomous.

A stout tall tree: a native of Chittagong.* Flowers at various times through the year.

Trunk nearly erect, and in trees nine years old, thirty-three inches in circumference four feet above the surface of the soil. *Branches* spreading. *Branchlets* obtusely four-cornered, and mealy.—*Leaves* opposite, long-petioled, oblong, ventricose, tapering equally at each end, entire, smooth above, hairy underneath, while young and unexpanded very mealy on both sides; length from six to twelve inches, petiole included, which may be reckoned one-fourth of the whole.—*Panicles* axillary, solitary, or paired, the largest on a peduncle as long as the petioles, sub-rotund, from six or eight times dichotomous, all the ramifications somewhat four-cornered and mealy.—*Flowers* small, very numerous, lively purple, or lilac, small, rather offensive.—*Bractes* minute.—*Calyx* companulate, obscurely four-toothed, mealy on the outside.—*Corol* with the tube much longer than the calyx, one of the segments of its border smaller.—*Filaments* twice the length of the corol. *Anthers* brown.—*Germ* superior, round, four-celled, with one ovula in each, attached to the middle of the axis. *Style* as long as the stamens, and bent in an opposite direction. *Stigma* headed.—*Berry* superior, round, polish-

* It is very common also about Katmandu in Nepala.—N. W.

ed, lively purple, succulent, size of a small grain of black pepper, one-celled.—*Seeds* four, convex on the outside, with two flat faces on the inner, rough; *interior* integument hard like the shell of a nut.—*Perisperm* oval, soft, amygdaline.—*Embryo* straight, erect, nearly as long as the perisperm. *Cotyledons* oblong. *Radicle* oblong, the length of the cotyledons, inferior.

2. C. *lanata*. *Linn. Sp. Pl. ed. Willd.* i. 620.

Arboreous, all the young parts mealy. *Leaves* opposite, oval, entire, downy. *Corymbs* axillary, dichotomous throughout. *Corols* oblique

Tomex tomentosa. *Flor. Zeyl. N.* 59.

A pretty large tree, a native of the Circar mountains, and of the vallies between them.

Trunk perfectly straight, and of a considerable height.—*Bark* smooth, ash-coloured. *Branches* few, and near the top, forming but a small head for so large a trunk. *Young-shoots* somewhat four-sided, and covered with grey farina.—*Leaves* opposite, petioled, ovate, tapering to a point, entire, downy, below whitish with much down; from nine to ten inches long, and five or six broad.—*Petiols* round, two or three inches long, covered with down.—*Corymbs* axillary, erect, two-forked throughout; every part covered with grey farina.—*Bractes* small.—*Flowers* very numerous, small, purple.—*Calyx* short; border four-sulcated, permanent.—*Corol, tube* bent to one side. The other parts as in the genus.—*Style* descending.

Obs. The wood of this tree is white, spongy, and of course not fit for much use.

3. C. *cana*. *Linn. Sp. Pl. ed. Willd.* i. 620.

Shrubby, downy. *Leaves* ovate, glandular-dentate-serrate, downy underneath. *Panicles* axillary, sub-globular. *Stigma* bifid. *Berries* purple.

Callicarpa americana. *Loureiro. Cochinch. ed. Willd.* i. *p.* 88.

This shrub was introduced into the Botanic Garden from the

Moluccas in 1798, and in three years had attained to the height of from four to eight feet high, considerably ramous, with the young parts downy, and the ligneous ones covered with smooth, pale ash-coloured bark. They are in blossom in March and April chiefly, though more or less the whole year.

Leaves opposite, short-petioled, from broad cordate to oblong, reticulated with margins glandular-dentate-serrate; upper side soft and of a deep green colour, but very downy, and pale underneath.—*Cymes* axillary, dichotomous, scarcely longer than the petioles, villous.—*Flowers* numerous, small, pale red colour.—*Bractes* subulate, caducous.—*Calyx* woolly, four-toothed.—*Corol* companulate; divisions sub-cordate.—*Filaments* longer than the corols, and inserted into the base of its tube, ascending.—*Style* declined, length of the filaments. *Stigma* two-cleft.—*Berry* small, round, smooth, deep purple *seeds* four.

Obs. It differs from C. *americana, Willd.* in having a woolly, toothed calyx, two-lobed stigma and in the form of the leaves.

4. C. *incana*. *R.*

Shrubby, young shoots hoary. *Leaves* lanceolate, obtusely serrulate, fine- and entire-pointed, hoary underneath.

Mashandari *Asiat. Res.* iv. 233.

Beng. M*uttura*, M*uttrunja*.

A stout shrub, with all the tender parts and the under surface of the leaves densely clothed with long, soft, white, stellate pubescence: common in the vicinity of Calcutta, where it is in flower and seed nearly the whole year.

I long considered this to be Vahl's *macrophylla*, but on rearing what I also took for the same species from Silhet and Chittagong, in the Botanic Garden, I could plainly observe a striking difference when growing near each other, and as the Chittagong and Silhet sort agrees much better with Vahl's figure and description, I must consider it to be his *macrophylla*. In the Calcutta plant, which I now call *incana*, the leaves are never so broad in proportion to their length, more round at the base; much more pointed, with the long taper points entire; all the rest of the margin, except what may be called the base, obtusely-serrulate. In *macrophylla,* the leaves are crenate, more obtuse, and the margins cut to the very apex: the two are however very nearly allied, though I think sufficiently distinct to authorize their being considered as different.

5. C. *macrophylla. Vahl. Symbol.* iii. 13. *t.* 53.

Shrubby, downy. *Leaves* opposite, ovate lanceolate, serrulate, reticulate, hoary underneath. *Corymbs* axillary, dichotomous, rather longer than the petiols. *Berry* minute, white.

Native of Silhet and Chittagong.

A shrub, from four to eight feet in length. *Trunk* scarcely any, but several, round, erect branches, covered with white down.—*Leaves* opposite, petioled, lanceolate, or oblong lanceolate, fine-pointed, finely serrate, wrinkled, above soft, and a little downy, below covered with much whitish soft down, from six to nine inches long, and two or three broad.—*Stipules* none.—*Petiols* about an inch long, downy.—*Corymbs* axillary, peduncled, two-forked, nearly globular, downy, many times shorter than the leaves.—*Peduncles* as long as the petioles, round, downy.—*Bractes* lanceolate, one under each division of the corymb.—*Flowers* very numerous, small, rose-coloured.—*Calyx* woolly, the four divisions distinct and acute.

6. C. *acuminata. R.*

Shrubby, tender parts hoary with a stellate pubescence. *Leaves* broad-lanceolar, acuminate, remotely repand, denticulate. *Panicles* axillary, long-peduncled, dichotomous, shorter than the leaves.

A native of Silhet, flowers in May. In this species the panicles are elevated on longer peduncles than in the other species, the leaves and young parts very hairy, except the upper surface of the former when fully expanded, which is then naked and reticulate; from four to five inches long by nearly two-broad.

7. C. *longifolia, Linn. Sp. Pl. ed. Willd.* 1. 621.

Shrubby, with erect weak branches. *Leaves* rather long-petioled, broad lanceolate, serrulate, smooth above, downy underneath. *Panicles* axillary, dichotomous, length of the petiols. *Berries* white.

A native of Prince of Wales Island, where it blossoms in June, July, and August.

8. C. *cuspidata, R.*

Shrubby, all the tender parts, and the under surface of the short-petioled, elliptic, dentate, cuspidate. *Leaves* woolly. *Corymbs* axillary, their division and the calyces clothed with minute grains under the wool.

A native of the Moluccas; the leaves are always acutely dentate, and end in a long taper, acute point. The Berries are very small, smooth, and purple.

9. C. *lanceolaria, R.*

Shrubby, hairy. *Leaves* lanceolar, serrulate, acuminate. *Panicles* axillary, short-peduncled, sub-globular. *Berries* white.

H. *Koamoora.*

A pretty, shrubby species, with narrower leaves than any of the other species I have yet met with in India, they taper most toward the base, are nearly smooth on the upper surface, but very hoary underneath; as are all the other tender parts. *Flowers* numerous, minute, purple.

A native of the forests of Silhet, where it is in flower most part of the year.

10. C. *pentandra, R.*

Shrubby, tender parts meally. *Leaves* opposite, with an alternate one between, oblong, entire, cuspidate. *Corymbs* axillary. *Flowers* pentandrous. *Stigma* from three to four-lobed.

A native of the Moluccas.

11. C. *purpurea, Juss.*

Shrubby; *branchlets* and younger parts slightly scaly. *Leaves* sub-sessile, lanceolate, acuminate, serrate, with sub-entire cuneate base, smooth above, glandular-dotted below. *Corymbs* axillary, very small, dichotomous; *flowers* glandular.

Porphyra dichotoma. Lour. Cochinch. ed. Willd. i. 87.

A native of China. Introduced in 1812, into the Botanic Garden, at Calcutta, where it thrives extremely well, has attained a height of from four to six feet, and blossoms from April to September. The berries ripen at the close of the rainy season.

Shrubby, spreading, with roundish branches, covered with scabrous gray bark. *Branchlets* very slender, slightly scabrous, and beset with ferruginous stellate and somewhat scaly pubescence; new shoots almost mealy.—*Leaves* from lanceolar to oval, two inches long, tapering at both ends, but more so at the base, and only there entire: the rest of the margin serrulate; without any hair or down, but scabrous below, from a great number of minute brown shining glandular dots which impart on being rubbed, a faintly aromatic smell; nerves and ribs elevated and slightly scaly.—*Petiols* very short.—*Corymbs* axillary, or rather supra-axillary, small, round, two or three times dichotomous, many times shorter than the leaves; *peduncles* short, divaricate, beset with mealy pubescence, with minute lanceolate scales under each sub-division.—*Flowers* fragrant, purple, marked with glandular dots.—*Calyx* turbinate with very small obtuse teeth.—*Corol* about three times longer than the calyx, infundibuliform; *laciniæ* oval, obtuse.—*Filaments* almost twice as long as the corolla, purple; *anthers* large, orange-coloured, covered with many glandular dots.—*Stigma* sub-capitate, elevated above the anthers.—*Berry* very small, purple, at last red, with one or two fertile seeds.

Obs. I cannot but consider this as Loureiro's plant quoted above, which Jussieu has justly placed among the genus *Callicarpa*. It seems to be nearly allied to C. *japonica*, but differs chiefly in wanting the smoothness, the short stamina and pistillum, and the acute stigma of the latter. Its flowers are besides of a beautiful purple colour, especially within, while those of Thunberg's plant are white. —N. W.

BUDDLEIA. *Schreb. Gen.* N. 184.

Calyx four-cleft. *Corol* four-cleft. *Capsule* superior, bisulcate, bilocular. *Seeds* numerous.

1. B. *Neemda,* Buchanan.

Arborescent. *Leaves* lanceolate, sub-serrate, hoary underneath. *Spikes* terminal, lengthening, with *flowers* three-fold.

Neemda the vernacular name at Chittagong, where the plant is indigenous, and from thence introduced into the Botanic Garden by Dr. Buchanan, where it begins to blossom about the close of the cold season. Seeds ripen in March and May.*

Trunk erect; branches numerous, opposite. *Bark* of the longer woolly parts ash-coloured, young shoots covered with white down. Height of the tree, in two years, about fifteen feet.—*Leaves* opposite, short-petioled, lanceolate, slightly hairy; from four to eight inches long.—*Petioles* short, their bases joined by a connecting membrane.—*Spikes* terminal, generally three-fold, there being a larger, and longer in the centre, with a smaller on each side; long, slender, hairy, lengthening, and blossoming for three months.—*Flowers* generally three-fold, sub-opposite, sessile, very numerous, small, pure white, and of a pleasant smell.—*Bractes* lanceolate, one under each flower.—*Calyx* inferior, four, rarely five-toothed, villous.—*Corol* one-petalled, hypocrateriform. *Tube* the length of the calyx; mouth villous, border of four, rarely five, rounded, equal segments.—*Filaments* short, inserted near the base of the tube of the corol. *Anthers* broad, sagittate, hid in the tube.—*Germ* conical. *Style* short. *Stigma* large, oblong, bisulcate, even with the tops of the anthers.—*Capsule* ovate-oblong, two-celled, two-valved. *Seeds* minute, numerous, imbricated, compressed.

* I have received specimens of this charming plant collected by my friend Dr. Govan at Saharunpore.—N. W.

Obs. The whiteness of the leaves, and young shoots of this plant, independently of its numerous, beautiful small, fragrant, pure white flowers, makes it highly ornamental, particularly amongst plants with darker foliage.

2. B. *paniculata, Wall.*

Leaves lanceolate, acuminate, entire, acute at the base, villous above, densely hoary beneath; *racemes* short, hoary, sessile, crowded into terminal panicles; *flowers* fascicled.

A native of the valley of Nepal*a*, from whence the Honorable E. Gardner sent me specimens in 1818. Flowering time February to April.

A small ramous tree with round spreading branches, covered with smooth fibrous separating bark.—*Branchlets* slightly compressed, and, as well as the leaves and inflorescence, covered with dense white stellate easily separating tomentum; especially while young.—*Leaves* lanceolate, tapering into an acumen, quite entire, acute at the base, from four to five inches long, rugose above, soft and hoary below, with conspicuous sub-opposite nerves and reticulated veins.—*Petioles* half an inch long, obscurely furrowed above, connected by an elevated line running across the branch.—*Racemes* short and stout, obtuse, the lowermost from the terminal axils about two or three inches long, the others shorter and more simple, until they become a terminal elongated raceme; the whole forming an elegant crowded attenuated sessile *panicle* or *thyrsus*. *Common peduncle* about six inches long, and, as well as the outside of the flowers, hoary; *partial* one very short, and supported by opposite narrow linear acuminate *floral leaves* or *bractes*.—*Flowers* fragrant, white, numerous, sessile, crowded into approximate sub-sessile fascicles, each with a pair of subulate elongated *bractes*, besides some smaller ones under the exterior calyces.—*Calyx* with ovate teeth.—*Tube* of the corolla cylindric, three times longer than the calyx, slightly hairy within. *Laciniæ* short, oval, obtuse.—*Anthers* linear, sub-sessile below the mouth, enclosed.—*Ovarium* and *base* of the short style villous. *Stigma* clavate.—N. W.

EXACUM. Schreb. Gen. N. 185.

Calyx four-leaved. *Corol* salver-shaped. *Capsule* superior, two-celled, two-valve'd. *Seeds* many.

1. E. *bicolor, R.*

Annual, erect, four-sided. *Leaves* sessile, ovate, five-nerved. *Flowers* axillary, *apices of the segments* of the corol blue, the rest white. *Anthers* erect.

A native of Cuttack, where it appears during the cool season.

Root almost simple.—*Stem* erect, branchy, four-sided; angles somewhat winged, from one to two feet high.—*Leaves* sub-sessile, opposite, oval, five-nerved, smooth, size very various. *Floral leaves* small and narrow.—*Flowers* axillary, solitary, short peduncled, large, of a beautiful white and blue colour.

2. E. *tetragonum, R.*

Annual, simple, erect, four-sided. *Leaves* sessile, lanceolate, five-nerved. *Flowers* numerous, terminal. *Stamens* and *style* declinate.

Beng. Koochuri.

A native of Bengal, where it grows among long grass; and flowers during the rains.

Stem erect, simple, exactly four-sided, smooth; from two to three or even four feet high.—*Leaves* opposite, decussate, sessile, lanceolate, five-nerved, smooth and shining, from two to four inches long.—*Flowers* terminal, numerous, large, of a most beautiful blue colour, inodorous.—*Calyx* divided nearly to the base into four, daggered segments.—*Corol*; segments oblong, acute, spreading.—*Filaments* on the tube of the corol, short. *Anthers* linear, bending to one side, with two pores at the top, for the discharge of the pollen, as in Cassia.—*Germ* oval. *Style* projecting obliquely with the stamina; stigma enlarged.—*Capsule* two-celled, two-valved.—*Seeds* numerous.

Obs. This is one of the most elegant plants I have met with, the flowers being not only numerous, large, and beautiful, but retaining their beauty and expansion for many days, which is a rare circumstance in an Indian plant.

3. E. *teres*, *Wall.*

Stem round. *Leaves* linear-lanceolate, acuminate, tapering at the base, and sub-sessile, three-nerved. *Peduncles* opposite, three-flowered, forming a terminal corymb. *Segments* of the *Corolla* lanceolate, acuminate.

Communicated by Mr. M. R. Smith from Silhet, where it blossoms during the rainy season.

This plant is nearly allied to the preceding species (which I have also received from the Hon. E. Gardner from Katmandu, where it is called *Kashwa Soa*, and from Mr. William Jack, who discovered it in the Turrayee, bordering on the Nepal*a* mountains). Both agree in habit and beauty; but my species differs in the following point. It is smaller and more slender, the whole height not exceeding two feet; the *stem* perfectly cylindric, as thick as a small goose-quill, with a few branches at the base. The *leaves* are three or four inches long, and about half an inch broad at the middle, from thence tapering towards both ends, narrow at the base and almost petioled, finely acuminated; like the whole plant, smooth. *Peduncles* almost as long as the leaves. *Pedicels* slender, about half an inch to an inch long, terminated by a dark blue large flower. *Laciniæ* of the *Corolla* narrower and more acuminated than in E. *tetragonum*. In other respects the plants are very much like each other.—N. W.

4. E. *zeylanicum*, *R.*

Annual, pentandrous. *Leaves* opposite, subsessile, lanceolate, three-nerved. *Flowers* five-cleft. *Calyx* with five, broad membranaceous wings with subulate points.

Chironia trinervia, *Linn. Sp. Pl. ed. Willd.* i. 1065.

Lysimachia folio sinuato, acuminato trinervio, flore cœruleo, amplo calyce carinato, alato, *Burm. Zeyl. p.* 145. *t.* 67.

A native of Ceylon.

Stem erect, somewhat four-sided, smooth, annual.—*Leaves* opposite, short-petioled, lanceolate, entire, smooth on the sides, three-nerved, from one to three inches long.—*Racemes* terminal, and opposite from the exterior axils, the whole forming an expanded corymbiform panicle.—*Peduncles* and *pedicels* four-sided, smooth.—*Flowers* numerous, large, of a violet colour.—*Calyx* five-cleft; divisions with a broad membranaceous wing on the back, and a long subulate point.—*Corol; tube* gibbous, length of the calyx, some short filaments amongst the insertions of the stamina crown the mouth of the tube. *Border* of five, obovate, beautiful, violet-coloured divisions.—*Filaments* on the mouth of the tube, erect. *Anthers* linear, with two pores at the top, and also opening on the sides.—*Style* declining, rather longer than the filaments.—*Stigma* headed.—*Capsule* oblong, two-celled.

Obs. The anthers remain straight, and hard, with two pores at the top, and also bursting by two longitudinal openings on the sides, I have for that reason removed this plant from the genus. *Chironia,* the essential mark of which consists in its spiral anthers.

5. E. *carinatum, R.*

Annual, erect. *Leaves* sub-sessile, lanceolate, three-nerved. *Corols* four-cleft. *Calyces* broad, membranaceo-carinated. *Stamens* protruded. *Style* declined.

A native of Ceylon.

Root ramous, annual.—*Stem* erect, with one or two opposite pairs of branches about the middle; above these dichotomous, somewhat four-sided, smooth, from six to twelve inches high.—*Leaves* opposite, short-petioled, lanceolate, entire, three-nerved.—*Flowers* terminal, numerous, pretty large, yellow, collected on dichotomous corymbs, with a long peduncled one in each of the divisions.—*Calyx* four-parted, each division augmented by a broad membranaceous keel on the back.—*Corol* four-parted.—*Filaments* short, inserted on a membranaceous rim round the inside of the mouth of the tube. *Anthers* erect, yellow, opening on the sides.—*Style* longer than the stamens, declined. *Stigma* ascending, simple.

6. E. *sulcatum, R.*

Annual, erect, ramous, sulcated on the opposite sides of the stem and branches. *Leaves* sub-sessile, lanceolate, fleshy, three-nerved. *Calyx* with four ridges; *stamens* erect; *capsules* globular.

A native of Bengal; blossoms during the cool season; the seeds ripen in March.

Root fibrous, annual.—*Stem* erect, from six to twelve inches high, smooth, sulcated on opposite sides. *Branches* opposite, their flower-bearing extremities two, and three times two-cleft.—*Leaves* opposite, sub-sessile, lanceolate, fleshy, entire, three-nerved; from one to two inches long.—*Flowers* terminal, numerous, of a bright blue colour, those in the divisions of the branchlets long-peduncled.—*Peduncles* four-sided.—*Bractes* ovate, recurved.—*Calyx* deeply four-parted; divisions with an ample keel.—*Filaments* short, inserted in the mouth of the tube of the corol. *Anthers* ovate, oblong, touching at the sides, apices perforated.—*Style* longer than the stamens, declined; stigma two-lobed.—*Capsule* globular, two-celled, two-valved.—*Seeds* numerous.

Obs. This is a very beautiful small plant.

*PLADERA. Solander's Mss.**

Calyx cylindric, four-toothed. *Corol* funnel-shaped, border irregular. One of the *stamina* larger. *Stigma* two-lobed. *Capsule* superior, one-celled, two-valved. *Seeds* numerous.

1. P. *sessiliflora, R.*

Annual, erect, four-sided; *angles* winged. *Leaves* opposite, sessile, ovate, entire. *Flowers* sessile.

Exacum heteroclitum, *Linn. Sp. Pl. ed. Willd.* i. *p.* 639.

Gentiana heteroclita, *Mant.* 560.

A native of Coromandel and Bengal, appears on moist pasture ground during the rains and first part of the cool season.

Stem erect, with two or three pairs of opposite branches, which are again once or oftener dichotomous, all are four-sided, with a thin

* *Canscora*, Lamarck.—N. W.

wing to each angle; height of the whole plant about a span.—*Leaves* opposite, sessile, ovate, entire, three-nerved; differing much in size, the largest rarely an inch long.—*Flowers* solitary, in the divisions of the plant, sessile, small, pale red.—*Calyx* sub-cylindric, four-toothed; teeth acute.—*Corol* with the tube as long as the calyx. *Border* three-cleft, two of the divisions equal, oblong, and crowded at the apex, the third divided half way down.—*Filaments* four; three longer, and inserted in the mouth of the tube; the fourth just below the fissure of the bifid segments of the corol. *Anthers* equal, oblong.—*Germ* linear-oblong. *Style* as long as the stamina. *Stigma* two-lobed; *lobes* large, obovate, thin.—*Capsule* sub-cylindric, nearly as long as the permanent *calyx*, and hid in it, one-celled, two-valved. *Receptacles* consist in the incurved margins of the valves, (parietal, and two-lobed.) *Seeds* numerous, angular, very small.

2. P. *virgata, R.**

Annual, erect, four-sided, branches opposite, alternate. *Leaves* sessile, lanceolar, three-nerved. *Flowers* terminal, sub-panicled.

Exacum diffusum, *Linn. Sp. Pl. ed. Willd.* i. 637.

Gentiana diffusa, *Vahl. Symbol.* iii. 47.

A beautiful erect, ramous annual; growing on the banks of water courses, and on low fallow lands. Flowering time the wet season.

Stem about a foot, or sometimes two feet high, erect, four-sided, ramous, smooth; inferior branches opposite, above alternate, very slender and twiggy.—*Leaves* opposite essile, the lower lanceolate, the upper broader, all are pointed, and three-nerved, entire, smooth; size various.—*Flowers* numerous, terminal, peduncled, small, rose-coloured.—*Calyx* four-toothed, permanent.—*Corol* funnel-shaped. *Border* irregular, three-parted; the two upper segments equal, and orbicular, the lower one two-parted; with a deep groove; in the groove is lodged the fourth or large stamen.—*Filaments* four, inserted into

* *Canscora diffusa*, Brown, Prodr. Fl. Nov. Holl. i. 451.—Linn. Syst. Veg. ed. Roem. et Schult. iii. 301.—N. W.

the mouth of the tube, the lowermost longer than the other three, and bearing an orange-coloured, and longer *anther*, which when shedding its pollen, is bent over the fissure of the *stigma*; anthers of the other three filaments equal, pale yellow.—*Style* single, length of the short filaments. *Stigma* two-cleft, segments recurved.—*Capsule* one-celled, many-seeded.

3. P. *decussata, R.**

Annual, erect, four-winged, branches cross-armed. *Leaves* ovate-lanceolate, three-nerved. *Flowers* terminal, three-fold, peduncled.

Beng. Dunkoni.

This species resembles P. *virgata*, but is a little more robust, and seldom so high. It is a native of the same places. Flowering time the same.

Stem about a foot high, perfectly erect, four-sided; *angles* very sharp, or rather membrane-winged, smooth, ramous, branches always opposite, cross-armed, in other respects like the stem.—*Leaves* opposite, spreading, sessile, lanceolate, sharp-pointed, entire, smooth, three-nerved, size various.—*Flowers* terminal, and axillary, peduncled, the terminal flowers three-fold; the axillary single, larger than those of the last species, and white.—*Peduncles* four-sided, like the stem and branches.—*Calyx* large, four-toothed, four-sided, four-winged.—*Corol, stamens,* &c. exactly as in the last species.

4. P. *perfoliata, R.*

Annual, erect, four-winged. *Cauline* leaves lanceolate, three-nerved; *floral* leaves (or bractes) perfoliate.

Canscora, *Lamarck's Encyclop. Bot.* i. 601.

Kansjan-Cora, *Rheed. Mal.* x. *p.* 103. *t.* 52.

A native of India, where it delights in a low moist soil. Flowering time the rainy season.

Root ramous, annual.—*Stem* erect, somewhat four-sided, with

* Exacum alatum, Roth. Linn. Syst. Veg. ed. Roem. et Schult. iii. 159.—N. W.

four broad membranaceous wings. *Branches* opposite, and like the stem; all are smooth. Height of the whole plant about eighteen inches.—*Leaves* of the stem opposite, sessile, lanceolate, three-nerved. *Floral* leaves perfoliate, all smooth and entire.—*Flowers* terminal, peduncled, numerous, pretty large, pale rose colour.—*Calyx* four-toothed, striated, permanent.—*Corol* irregular, somewhat two-lipped; the upper one larger and deeply two-parted, the lower one consisting also of two lobes, but smaller.—*Filaments* four, three of them rather shorter, and inserted on the upper side of the tube; the third larger, and inserted below the fissure of the lower lip. *Anthers* oblong.—*Style* as long as the stamens. *Stigma* two-parted. *Segments* recurved.—*Capsule* oblong, one-celled, two-valved.

5. P. *pusilla, R.*

Annual, from one to two inches high. *Leaves* sessile, sub-cordate. *Flowers* terminal, sessile, three-fold, monandrous. *Corol* hid in the calyx. *Capsule* globular.

Compare with Hopea dichotoma.* *Vahl. Enumer.* i. *p.* 3; also with *Exacum sessile, Willd.* i. 635.

A beautiful minute plant, a native of Coromandel.

Stem erect, four-sided, branchy; from one and a half to two inches high.—*Branches* cross-armed.—*Leaves* opposite, sessile, cordate, minute, smooth.—*Flowers* terminal, sessile, three-fold, small, yellow.—*Calyx* one-leaved, sub-cylindric, as long as the corol, four-toothed, permanent.—*Corol* one-petalled. *Tube* a little gibbous, mouth four-toothed.—*Filaments* four within the tube of the corol, and inserted into it near the base. *Anthers,* three minute, and abortive; one large and fertile.—*Germ* globular. *Style* short. *Stigmas* two, globular.—*Capsule* oval, covered by the calyx, one-celled, two-valved. *Seeds* numerous, black.

* Sir J. E. Smith has confirmed the identity of this plant (which Dr. Roxburgh formerly referred to Exacum) with Hopea of Vahl. See that Article in Rees's Cyclop. —N. W.

MITRASACME, Labillard.

Calyx angular, four, seldom two-cleft. *Corolla* deciduous, with angular tube, and four-cleft equal border. *Style* single, bifid at the base. *Capsule* bursting between the divisions of the style.

1. M. *capillaris, Wall.*

Stem erect, hairy; *leaves* lanceolate, hairy above, connected at the base by means of a narrow ciliated membrane, the uppermost approximated, sub-quatern; *umbel* peduncled, mostly compound; *calyx* half the length of the corolla, its segments villous at their apex; *stigma* almost entire; apex of the *capsule* exserted.

A native of the valley of Nepal*a*, where it blossoms in September. Specimens were communicated by the Hon. E. Gardner.

Root fibrous. *Stem* erect, capillary, from three to six inches high, mostly undivided, slightly compressed.—*Leaves* spreading, acute, ciliate, three-nerved, sessile, smooth below, four or five lines long, exceeding their interstices; the uppermost two pairs so close together as to appear almost verticilled.—*Umbel* terminal, divided into three or four rays, which are again sub-divided into several slender pedicels. Two or three lanceolate, minute, connate bractes under the primary divisions.—*Calyx* four cleft; its segments lanceolate, acute, united at the base by a small membrane. *Corolla* sub-campanulate, slightly pubescent, white, with ovate laciniæ.—*Stamina* inclosed; *anthers* oblong, attached to the back of the filaments.—*Ovarium* ovate, bi-cuspidate. *Style* capillary, persistent; when the capsule ripens, the division at the base extends almost up to the slightly two-lobed, sub-capitate stigma.—*Capsule* ovate, smooth, bursting at the top with a roundish aperture between the divisions of the style; *valves* very short, scarcely equalling one-third of the capsule, acute, incurved. *Seeds* minute, whitish.

Obs. This little plant comes near to M. *connata*, Brown, Prodr. Flor. Nov. Holl. i. 454; it differs in the shorter limb and the ovate segments of the corolla, and the style not being entire when the flower opens.—N. W.

PLANTAGO. Schreb. Gen. N. 186.

Calyx four-cleft. *Corol* four-cleft, with border reflexed. *Stamina* very long. *Capsule* superior, two-celled, cut transversely.

1. P. *Ispaghula, R.—Fleming in Asiat. Res.* xi. 174.

Annual, caulescent. *Leaves* linear-lanceolar, three-nerved, slightly woolly. *Peduncles* axillary, naked, simple, length of the leaves; heads cylindric. *Capsules* two-celled.

Ispagool, the Hindee and Persian name, and that by which it is most generally known in Bengal and on the coast of Coromandel.

Native place uncertain, but it is cultivated during the cool season.

Root ramous, annual.—*Stem,* if any, very short, soon dividing into three or four ascending, round, villous branches, which are a few inches in length.—*Leaves* alternate, linear-lanceolate, three-nerved, luxuriant, in young plants remotely denticulate, somewhat woolly, toward the base channelled, and stem-clasping; length from six to eight inches by a quarter or half an inch broad.—*Peduncles* axillary, solitary, naked, erect, round, a little villous, the length of the leaves.—*Spikes* solitary, terminal, before the flowers expand ovate, afterwards cylindric; when in seed about an inch or an inch and a half long and erect. —*Flowers* numerous, imbricated, small, dull, white, withering.—*Bractes* one-flowered, ovate, concave, with the keel green, and the sides membranaceous.—*Calyx* four-leaved. *Leaflets* oblong, ovate, with broad membranaceous margins.—*Corol. Tube* gibbous. *Segments* ovate, acute.—*Filaments* oblong, two-celled, four, much larger than the segments of the corol.—*Germ* oblong, two-celled, with one seed in each, attached to the middle of the partition. *Style* rather longer than the corol. *Stigma* subulate, villous.—*Capsule* circumscissa, ovate, two-celled.—*Seeds* solitary, ovate-elliptic, convex on the outside; concave on the inner. *Integument* single, rather thick, smooth, tough, and brown.—*Perisperm* conform to the seed.—*Embryo* sub-cylindric, erect. *Cotyledon* long, linear. *Radicle* inferior, one-fourth the length of the cotyledon.

Obs. From the seeds a mucilaginous drink is prepared, and often prescribed by our medical gentlemen in India, where emollients are wanted. They are also used by the native practitioners in medicine and are to be met with for sale in the Bazars of India under the Persian name *Uspagool.*

2. P. *attenuata, Wall.*

Leaves lanceolate, ending in a bluntish acumen, tapering at both ends, with a few remote toothlets, smooth, five-nerved, decurrent on their short petiole, the base of which is woolly. *Scape* five-sided, sulcated, much longer than the leaves, with adpressed hairs. *Spike* cylindric, dense; *bractes* acuminate; *calyx* four-leaved, ciliate.

Specimens collected along the fields about Katmandu were communicated to me by the Hon. E. Gardner. Flowering time the hot season.

Root thick, sending forth a number of long ash-coloured fibres; immediately within the insertion of the leaves and on the inner surface of the base of their petiols there is a quantity of long whitish wool.—*Leaves* from three to four inches long, nearly an inch broad, slightly oblique, terminating in a short cylindric, somewhat thickened point, with a few unequally remote obtuse sub-glandular toothlets, quite smooth, decurrent on the short flattish striated petioles, the base of which widens a good deal and is woolly within.—*Scapes* several, erect, slender, several times longer than the leaves, from one to three feet high, five-cornered, striated, while young thickly beset with greyish adpressed hairs, becoming much smoother when old.—*Spike* from one to three inches long, cylindric, densely covered with ash-coloured flowers.—*Bractes* membranaceous, ovate, acuminate, smooth, keeled.—*Calyx* unequally four-leaved, nearly transparent, slightly ciliated and tipped with a few white hairs. *Corolla* tubular, with reflected ovate acute laciniæ and a prominent mouth. —*Style* very long and villous.

Obs. This tall and slender species comes very near to P. *eriostachya* and *altissima, Jacqu.*; from the former it differs in the smoothness of its leaves and form of the spike, and smooth bractes; from the latter, in its hairy scape, acuminate bractes, and ciliate calyx. It is also akin to P. *lanceolata* and *eriophora*.—N. W.

3. P. *erosa, Wall.*

Leaves ovate-lanceolar, acute, grossly and unequally repand-dentate, smooth, sub-ciliate, five-nerved; *petioles* short with woolly base. *Scape* adscending, compressed, slightly angular and hairy, nearly equalling the leaves. *Spike* slender, equalling the scape in length, with distinct flowers, which are somewhat remote towards the base.

A native of Nepala, growing with the preceding species.

Root consisting of numerous sub-simple fibres, from the head of which issue several tufts of scapes and leaves.—*Leaves* almost smooth, about three inches long, tapering towards the base; their margins remotely and unequally sinuate, with large, unequal, triangular, acute teeth.—*Petiols* shorter than the leaves; their base widening, membrane-margined and woolly within.—*Scape* slender, adscending, somewhat compressed, when young, villous.—*Spike* much lengthened, round, erect.—*Flowers* distinct, becoming somewhat remote downwards, mostly geminate.—*Bractes* ovate, membrane-margined, with a green thick rib, shorter than the calyx.—*Style* villous.—*Capsule* ovate-oblong, with numerous flattish brown seeds.—N. W.

CISSUS. Schreb. Gen. N. 192.

Calyx four-toothed. *Corol* four-petalled. *Germ* two-celled; *ovula* two, erect. *Berry* superior, from one to four-seeded. *Embryo* erect, in the base of an ample perisperm.

1. C. *adnata, R.*

Leaves round-cordate, bristle-serrate, acuminate. *Stipules* orbicular, adnate, with gibbous centre, and scariose margins. *Cymes* flat, flowers drooping.

A native of the country about Dacca, from thence the late Colonel Peter Murray sent it to the Botanic Garden, where it thrives well, blossoms in May and June, and ripens its fruit in September.

Stem and large *branches* ligneous, climbing over trees to a considerable extent, young shoots round, a little downy and slightly furrowed.—*Leaves* alternate, round-cordate, and often obliquely so, acuminate, bristle-serrate, the largest sometimes slightly repand; pretty smooth above, but downy underneath, particularly the veins, from two to six inches long, one-fourth less in breadth.—*Peduncles* half the length of the leaves, downy.—*Stipules* nearly oval, adnate, elevated, and of a deeper green in the middle; *margins* scariose, and brown.—*Tendrils* opposite to the leaves where there is no cyme.—*Cymes* opposite to the leaves, on villous peduncles, as long as the petioles, very ramous, flat.—*Flowers* numerous, small, drooping. The germ has two cells, in each of which are two seeds attached from their lower end to the partition at the bottom of the cell.—*Berry* succulent, size of a pea, when ripe black, and smooth, generally one-celled.—*Seed* for the most part solitary, obovate, lower end pointed.—*Integuments* two, exterior nuciform, and rugose; *inner* a very thin, white membrane, which adheres firmly to the perisperm.—*Perisperm* conform to the seed, dull white, cartilaginous, divided by two fissures into three, nearly equal parts.—*Embryo* scarcely half the length of the perisperm. *Cotyledons* reniform. *Radicle* oblong, inferior.

2. C. *vitiginea*, *Linn. Sp. Pl. ed. Willd.* i. 655.

Leaves cordate-serrate. *Cymes* decompound. *Berries* one or two-seeded.

Hind. and *Beng.* Guali-lata.

Teling. Maratta-tiga.

This species is a native of forests, mountains, and way sides in uncultivated places.

Trunk woolly, branchy, climbing over trees and shrubs to a great extent. *Tendril* opposite to the leaves, two-cleft.—*Leaves* alternate, petioled, cordate, serrate, or dentate, pretty smooth on both sides, from four to six inches long, and from three to four broad.—*Stipules* cordate.—*Cymes* peduncled, leaf-opposed, more than compound.—

Flowers small, red.—*Germ* superior, half immersed in the four-sided, fleshy nectary.—*Berry* small, black, one or two seeded.

3. C. *glauca*, *R.*

Leaves cordate, bristle-serrate, smooth. *Stipules* broad-sulcate. *Cymes* decompound. *Berries* one-seeded.

Schunambu Valli, *Rheed. Mal.* vii. *p.* 21. *t.* 11.

This species is still larger than the last, and a native of the same or similar places; the shoots, the tendrils, petioles and peduncles are all quite smooth and somewhat glaucous, which does not admit of my considering it to be either *repanda*, or *latifolia*.

Tendrils leaf-opposed, undivided.—*Leaves* alternate, petioled, broad-cordate, more or less acuminate, sometimes when large slightly lobed, bristle-serrate, smooth, from four to eight inches each way. *Stipules* broad-fulcate, apex rounded, caducous.—*Cymes* with from three to five primary branches; ultimate divisions umbelliferous.—*Berry* nearly round, size of a pea, smooth, succulent, purple, rarely more than one-seeded.—*Seed* obovate. *Integument* a double membrane, with a spungy silver-coloured substance between.—*Perisperm* and *embryo* as in the genus.

4. C. *cordata*, *R.**

Shrubby, scandent, smooth. *Leaves* cordate, slightly and remotely bristle-serrulate. *Cymes* compound, leaf-opposed.

Meriam-Pulli, *Rheed. Mal.* vii. *p.* 91. *t.* 48.

Funis crepitans major, *Rumph. Amb.* v. *p.* 446. *t.* 164. *f.* 1.

A native of Amboyna, from thence introduced into the Botanic Garden, in 1796—7. Flowering time in Bengal, the rainy season.

Stem and *branches* long, slender, climbing over bushes to a considerable extent. *Bark* smooth; on the young parts polished and purplish.—*Leaves* alternate, petioled, cordate, remotely and slightly serrulate, serratures acute, both sides smooth, length from four to six inches.—*Tendrils* leaf-opposed, simple, coloured.—*Stipules* adjoin-

* C. repens, Lam. Encycl. i. 31; et Poiret in ejusd. Suppl. i. 105.—N. W.

ed, oval, coloured.—*Cymes* leaf-opposed, first divisions trichotomous.—*Flowers* small, dull purple.—Taste of the leaves very acrid, and continuing long.

5. C. *quadrangularis, Willd. Spec.* i. *p.* 657.

Perennial, scandent, jointed, four-winged. *Leaves* reniform-cordate, serrate. *Berries* sperical, one-seeded.

Beng. Harjora.

Funis quadrangularis, *Rumph. Amb.* v. *p.* 83. *t.* 44. *f.* 2.

This plant is one of the most common over every part of India; blossoms during the rains; and ripens its seed in the cool season.

Root fibrous.—*Stem* and branches perennial, scandent, and often of very great length, smooth, succulent, jointed, contracted at the joints, four-winged, wings large, and sharp, young parts succulent, tender, and very smooth. These are very generally eaten by the natives in their curries.—*Leaves* alternate, one at each joint, short-petioled, reniform, cordate, serrulate-dentate, smooth on both sides, an inch or an inch and a half each way.—*Stipules* lunulate, entire.—*Tendrils* solitary, simple, leaf-opposed.—*Umbels* solitary, opposite to the leaves, occupying the place on which the tendril would otherwise grow, compound. *Umbellets* generally three or four, sub-globular.—*Involucre* and *involucels* scarcely any.—*Germ* ovate, immersed in the nectary, two-celled, with two ovula in each, attached to the bottom of the partition; style short. *Stigma* simple.—*Berry* round, size of a small pea, smooth, red, succulent, one-celled.—*Seed* solitary, obovate, covered with a double, dark brown, spongy integument. —*Perisperm* conform to the seed, a double, dark-coloured fissure runs down the middle.—*Embryo* small, erect, lodged on one side of the bottom of the double fissure just mentioned. *Cotyledons* with minute points. *Radicle* oblong, inferior.

6. C. *pentagona, R.*

Young shoots lucid, with five rounded angles and grooved sides. *Leaves* cordate, rarely sub-lobate, serrulate, acuminate, smooth.

An extensive perennial climber, a native of the forests of Chittagong.

7. C. *angustifolia, R.*

Dioicous. *Leaves* ternate; *leaflets* lanceolate, serrate, smooth. *Stipules* ovate, acute. *Cymes* shorter than the petioles. *Berries* spherical, one or two-seeded.

A delicate, scandent, perennial species, a native of Sumatra. Flowers in February; seeds ripen in November.

Stem scandent, slender, flat like a piece of tape.—*Tendrils* simple.—*Leaves* ternate. *Leaflets* petiolate, lanceolate, remotely serrate, acuminate, smooth on both sides, from three to six inches long, and from one to one and a half broad.—*Petioles* shorter than the leaflets, channelled, smooth.—*Stipules* triangularly ovate, acute, smooth.—*Cymes* axillary, solitary, nearly round, small, and shorter than the petioles.—*Flowers* small, greenish yellow, male on one plant and female on another.

MALE.—*Calyx*, *Corol* and *Stamina* as in the genus.—*Pistil* no other than a small abortive roundish body on the centre of a four-lobed gland.

FEMALE.—*Calyx* and *Corol* as in the male.—*Germ* round, two-celled, with two seeds in each. *Style* none.—*Stigma* large, obscurely four-lobed, and somewhat lacerated.—*Berries* spherical, smooth, size of a pea, white, one or two-seeded, when cut or bruised very fetid.—*Perisperm* conform to the seed.—*Embryo* lodged in the lower half of the perisperm. *Cotyledons* nearly round. *Radicle* cylindric, inferior.

8. C. *carnosa, Linn. Sp. Pl. ed. Willd.* i. 658.

Stem flattened, perennial. *Leaves* petioled, ternate; *leaflets* oval, serrate. *Berries* four-seeded.

Tsjori-Valli, *Rheed. Mal.* vii. *p.* 17. *t.* 9.

Folium Causonis, *Rumph. Amb.* v. *p.* 450. *t.* 166. *f.* 2.

Teling. Kanapa-tiga.

Is common in hedges and forests. Flowering time the rainy season.

Stem perennial, flat like a piece of tape, climbing to an extent of many fathoms, young shoots, villous. *Tendrils* leaf-opposed, from two to three cleft.—*Leaves* alternate, petioled, ternate. *Leaflets* oval, grossly serrate, downy, from two to three inches long.—*Petioles* round, downy.—*Stipules* oblong, caducous.—*Cymes* as in the former species.—*Flowers* small, white.—*Berry* black, four-seeded.

9. C. *setosa*, *R.*

Herbaceous, scandent, bristly. *Leaves* sessile, ternate; *leaflets* grossly bristle-serrate. *Berries* one-seeded.

Teling. Barubutsali.

A native of forests and hedges in the Rajamundree Circar. Flowering time the rainy season.

Root fusiform, perennial.—*Stems* herbaceous, succulent, round, striated, and covered with distinct, glandular, round-headed, white bristles, as is every part of the plant, even to the fruit.—*Tendrils* leaf-opposed, generally two cleft.—*Leaves* alternate, sessile, ternate, rarely quinate. *Leaflets,* the lateral two (or four, when there are five) are sessile, the middle one is short-petioled; all are oval, waved, fleshy, irregularly and grossly bristle serrate, both sides armed with the forementioned bristles, size various, from one to five inches long.—*Stipules* cordate.—*Cymes* as in the former species.—*Petals* revolute.

Obs. Every part of the plant is exceeding acrid. I unfortunately tasted both the roots and berries.

The leaves toasted and oiled, are applied to indolent tumors to bring them to suppuration.

10. C. *feminea*, *R.*

Leaves digitate; *leaflets* petiolate, broad-lanceolar, entire. *Style* none. *Stigma* from four to five-lobed.

A large scandent woody species, a native of woody mountains. Flowers during the wet and cold seasons.

Stem woody, climbing over trees. *Branches* flexuose from leaf to leaf.—*Tendrils* simple, opposite to the leaves.—*Leaves* alternate, petioled, digitate. *Leaflets* oblong, and lanceolar, a little crenulate, pointed, smooth, shining, from four to six inches long, and one or two inches broad.—*Petioles* round, smooth, from one to two inches long.—*Cymes* short-peduncled, axillary.—*Bractes* small, falling.—*Flowers* small, green.—*Calyx* small.—*Corol* four or five-petalled, generally four. *Nectary* scarcely any.—*Stamens* none, or at most a small seemingly sterile one.—*Germ* superior, globose. *Style* none. *Stigma* four, or five-lobed, according to the petals in the corol.—I have not seen the ripe fruit.

11. C. *elongata, R.*

Tender shoots succulent, polished. *Leaves* digitate; *leaflets* petiolate, oblong, acute, remotely serrate, smooth.

Beng. Dhemna, which is also the name of a long serpent.

Found amongst the mountains of the Coast of Coromandel, as well as in the forests of Bengal. Flowering time the hot season.

Stem very long, running up and over the highest trees, a little compressed, and until many years old smooth, polished, succulent, deep green. *Branches* few or none, until the plant arrives at a great height, and is several years of age.—*Leaves* alternate, petioled, digitate. *Leaflets* five, petioled, oblong, the lower pair obliquely so, serrate, smooth on both sides, from two to six inches long, and from one to three broad.—*Petioles* round, smooth, from four to five inches long. *Petiolets* channelled.—*Stipule* cordate, small, sub-adjoined.—*Tendrils* leaf-opposed, simple.—*Cymes* trichotomous, and dichotomous; the last divisions short, and ending in small umbellets.—*Bractes* cordate, permanent.—*Calyx* four-toothed.—*Corol,* petals four, inserted in a belt which surrounds the body of the germ. *Nectary* a four-lobed, coloured elevation between the stamina and the insertion of the style.—*Filaments* four, inserted opposite to, and just within the petals. *Anthers* incumbent.—*Germ* oval. *Style* short. *Stigma* simple.—*Berry* the size of a cherry, round-turbinate, smooth, when ripe black.—*Seed* single, large, somewhat rugose.

12. C. *auriculata, R.*

Scandent, perennial. *Leaves* quinate ; *leaflets* oblong, acute, obtusely serrate, smooth above, villous underneath. *Stipules* ear-shaped. *Berries* round, smooth, one-seeded.

A native of Mysore, from thence sent to the Botanic Garden in 1802, by Mr. B. Heyne, where it flowers and ripens its fruit through the whole year.

Stem, in five years old plants, as thick as a man's wrist, and of great length, even up to the tops of high trees. *Bark* spungy, and deeply cracked, young shoots succulent, round, and villous.—*Leaves* alternate, remote, long-petioled, quinate. *Leaflets* petioled, oblong-obovate, or rhombic, acute, serrate, smooth and shining above, villous underneath ; the *superior* from two to three inches long, and two-thirds of that in breadth, the *exterior* much larger.—*Petioles* nearly as long as the leaflets, round, and villous.—*Stipules* ear-shaped, recurved.—*Tendrils* leaf-opposed, two or three-cleft.—*Cymes* axillary, or terminal ; dichotomous or trichotomous, these again several times divided.—*Germ* conical, containing four seeds in one cell, attached to its bottom.—*Berry,* the size, shape, and appearance of a red cherry, one-seeded.—*Seed* obliquely obovate, a little compressed, somewhat pointed at the base, and there attached to the bottom of the cell. *Integument* single, nuciform, rugose, covered with the green pulp of the berry, which adheres to it, and may, when dry, be considered a second integument.—*Perisperm* conform to the seed, cartilaginous, grey; from the inner edge two-fissures penetrate deeply, and divide it nearly into three portions.—*Embryo* erect, white, lodged in the base of the perisperm.—*Cotyledons* two ; pedicelled, reniform. *Plumula* minute. *Radicle* oblong, inferior, a little curved, with the apex to the umbilicus.

13. C. *lanceolaria, R.*

Dioicous. *Leaves* pedate-quinate ; *leaflets* lanceolar, smooth, remotely serrulate. *Stipules* conically lanceolate. *Cymes* the length of the petioles.

Vallia-Tsjori-Valli, *Rheed. Mal.* vii. *p.* 15. *t.* 8.

A native of the mountains along the eastern frontier of Bengal. In the Botanic Garden it blossoms in February and March.

Trunk scandent, stout, ligneous, in six year-old plants as thick as a man's wrist, flattened, and grooved on the two flat sides. *Bark* dark brown, rough or deeply cracked. *Branches* and *Branchlets* climbing, or pendulous when not supported; young parts round, green and generally smooth; extent of the whole plant several fathoms.—*Tendrils* leaf-opposed, generally undivided.—*Leaves* alternate, petioled, pedate. *Leaflets* generally five, lanceolar, recurvate, most remotely and sharply serrulate, long, taper, acute-pointed, both sides polished; texture rather fleshy, length from three to six inches, and from one to two broad.—*Petioles* and *petiolets* clothed with short, reddish hairs.—*Stipules* large, of a narrow conic shape.—*Cymes* sub-axillary, or close by the side of the base of the petioles, shorter than the petioles, or when most luxuriant much shorter than the leaves, the ultimate divisions each supporting a small globular umbellet, of small pale yellow flowers.—*Bractes* round the base of the peduncles, cordate.

MALE.—*Calyx* scarcely any.—*Corol* and stamina as in the genus.—*Pistil* none, an umbilicate, yellow, round, convex gland occupying its place.

14. C. *pedata, Linn. Sp. Pl. ed. Willd.* i. 659.

Perennial, scandent. *Tendrils* simple, or two-cleft. *Leaves* pedate; *leaflets* broad-lanceolate, serrate. *Berries* four-seeded.

Belutta-Tsjori-Valli, *Rheed. Mal.* vii. *p.* 19. *t.* 10.

C. heptaphylla, *Retz. Obs.* v. 22.

Beng. Gwalia.

This is another very extensively climbing species, as indeed they all are; a native of forests, hedges, &c. Flowering time the rainy season.

Stems woody, climbing over trees, &c. young parts covered with soft down.—*Tendrils* leaf-opposed, simple or two-cleft.—*Leaves* alternate, petioled, pedate. *Leaflets* oblong irregularly serrate, downy, size various, the intermediate one is from four to six inches long, and the lowermost of the lateral ones from two to three inches long, the rest between these sizes.—*Petioles* common, round, downy, from three to five inches long.—*Stipules* cordate, pointed, falling.—*Cymes* leaf-opposed, or axillary. *Peduncles* half the length of the petioles, jointed, and bracted a little above the middle.—*Berries* white, four-lobed, depressed, four-seeded.

15. C. *serrulata, R.*

Perennial, scandent, smooth. *Leaves* pedate-quinate; *leaflets* oblong, serrulate.

A native of Chittagong, where it blossoms in April and May.

CORNUS.

Calyx superior, four-toothed. *Petals* four. *Drupe* with a two-celled nut.

1. C. *oblonga, Wall.*

Leaves oblong, acuminate, acute at the base, glaucous and scabrous beneath, with many excavated glands along the axils of the rib and nerves. *Cymes* spreading, panicled.

Native of Nepala where it is pretty common about Katmandu; blossoming during the first half part of the year.—Specimens and seeds communicated by the Hon. E. Gardner, in 1818.

A midling sized tree with copious spreading branches; the young shoots covered with short adpressed hair.—*Leaves* in approximate pairs, spreading, from four to six inches long, acute at the base, an inch or an inch and a half broad, finely acuminate, beset with short adpressed hairs, dark-green and shining above, glaucous and scabrous beneath, with elevated rib and sub-opposite arched nerves, the latter having in the axils at their base and along their outer side numerous ciliated glandular excavations which, when the leaves are dry, become slightly visible on their upper surface.—*Petiols* from half an inch to an inch long, slender, channelled, rough.—*Cyme* terminal, sessile, broad-ovate with opposite spreading ramifications, which terminate in fascicles of short-pedicelled, crowded, sub-racemed white or pale-purplish, fragrant flowers.—*Bractes* opposite, ovate, acute, deciduous.—*Calyx* small, four-toothed, as well as the pedicels and petals clothed with adpressed, short, silvery hairs.—*Petals* linear-oblong, even, four times longer than the calyx, at last revolute.—*Anthers* purple.—*Ovarium* three-celled.—*Style* deciduous.—*Drupe* ovate-oblong, with an elevated umbilicus, smooth. *Putamen* two, sometimes three-celled.

Obs. Dr. G. Govan has favored me with fruit-bearing specimens collected to the Northward of Saharunpore of a tree, which in general appearance and leaves comes very near to this species; the latter are however larger and less glandular. The chief difference consists in the peduncled cymes with umbelled ramifications and the unripe fruit being compressed, oval, and resembling that of a Viburnum. I have not seen its flowers.—N. W.

2. C. *macrophylla, Wall.*

Leaves broad-ovate, acuminate, rounded at the base, smooth, glaucous and soft beneath, with axils slightly glandular. *Cymes* sub-umbelled; *flowers* racemed.

Communicated by Dr. Govan who found the tree to the Northward of Saharunpore.

Branches smooth.—*Leaves* six inches long, by four broad, very pale and glaucous below, slightly pubescent, soft and dotted; nerves copious, sub-opposite, oblique.—*Petiols* an inch and a half long, with a few hairs in their axils.—*Cyme* ample, terminal, sessile, flattish; the divisions terminated by several racemes of unilateral flowers.—*Calyx* with acute teeth.—*Petiols* linear-oblong, hairy on their outside.—*Ovarium* sulcated, hairy.—*Style* shorter than the stamina. *Stigma* capitate.—*Drupe* round, smooth, as large as a black pepper corn, with a two-celled nut.

Obs. This tree differs from all the species of Cornus in the size of its broad-ovate leaves.—N. W.

3. C. *capitata, Wall.*

Arborescent. *Leaves* lanceolate, acuminate, short-petioled, scabrous with adpressed hair. *Flowers* sessile, densely aggregate, forming a round head supported by a very large, four-leaved, scabrous involucre.

Specimens communicated by the Hon. E. Gardner from Gossain-Than in Nepala, where the tree is called *Chungwa;* by Sir Robert Colquhoun from Sirinagur, where it is called *Bhumowra;* and by Dr. G. Govan who gathered them between the Sutlej and Jumna.

It grows to be a small tree, which according to Sir R. Colquhoun's observation has much of the habit of the Custard-apple.—*Branches* woody, spreading, opposite, like all the other parts covered with short stiff adpressed hairs; the younger ones slightly compressed.—*Leaves* opposite, coriaceous, tapering at both ends, entire, about two inches long, scabrous, covered on both sides with reflexed hairs, which on being rubbed off leave an impression on the surface; glaucous, very pale beneath with elevated, sometimes pink-coloured, sublongitudinal arched nerves, having each a minute gland in their axils.—*Petioles* very short, furrowed. *Head* terminal, solitary, round, as large as a moderate-sized cherry, thickly covered with flowers, supported by a sulcated, club-shaped peduncle, which widens at the upper end into a convex ligneous receptacle for the insertion of the flowers.—*Involucre* consisting of four spreading, somewhat unequal, cariaceous, yellow, obovate-cuneate, sometimes broad-ovate, veined, scabrous leaves, an inch long, rounded at their ends, with a short slightly grooved point.—*Calyx* cylindric, pubescent, with four oblong, erect, obtuse segments.—*Corol* twice as long as the calyx, consisting of four spreading cuneate leaflets; their upper surface slightly keeled.—*Stamens* equalling the petals, subulate, recurved. *Anthers* ovate. *Ovarium* immersed in the receptacle, oblong, two-celled, with a pendulous ovulum in each cell. *Style* thick, clavate, striated, shorter than the stamens, surrounded at its base by a short, cylindric fleshy, crenulate *nectary*. *Stigma* truncate.

Obs. This beautiful tree is very distinct from all the other species. It approaches a little to C. *florida* and *canadensis*, but is easily recognised by its copious, harsh, adpressed hairs, its stature, which is that of a large shrub or small tree, and the position of its leaves.—N. W.

SAMARA. *Schreb. Gen. N.* 195.

Calyx four-parted. *Corol* four-petalled. *Stamina* inserted opposite to the centre of the insertions of the petals. *Drupe* superior, one-seeded.

1. S. *paniculata, R.*

Arboreous. *Leaves* opposite, entire, broad-lanceolate. *Panicles* axillary. *Stigma* two-lobed.

A tree, a native of the Circar mountains.

Leaves opposite, petioled, broad-lanceolate, tapering to a sharp narrow point, entire, smooth, from three to five inches long, and one and a half or two broad.—*Corymbs* axillary, small, peduncled, shorter than the leaves, divided by three-forked divisions.—*Drupe* dry, oblong.—*Nut* one-celled.

2. S. *polygama, R.*

Leaves opposite, short-petioled, broad-lanceolate, firm and polished. *Panicles* axillary and terminal. *Filaments* short, and inserted in a notch near the edge of the base of the petals.

A native of the Moluccas.

BLACKBURNIA. *Schreb. Gen. N.* 199.

Calyx four-toothed. *Corol* four-petalled. *Nectary* a staminiferous ring surrounding the base of the germ. *Berry* superior, one-seeded. *Seed* arilled.

1. B. *monadelpha, R.*

Arboreous. *Leaves* alternate, lanceolate. *Filaments* united at the base.

A large erect timber tree, a native of the mountainous parts of the Circars. Flowering time the beginning of the hot season.

Trunk erect. *Bark* thin, firm, yellowish green on the outside on the inside deep red, tinging the spittle with that colour. It is a strong astringent. *Branches* numerous, forming a large shady head. *Leaves* alternate, short-petioled, linear-oblong, smooth, below a little whitish, entire, waved, pointed; five or six inches long, and one and a half or two broad.—*Stipules* none.—*Racemes* axillary, small, erect, frequently compound.—*Bractes* small, subulate.—*Flowers* small, yellow.—*Calyx* four-toothed.—*Petals* four, oblong, expanding, apices revolute.—*Nectary*, a ring surrounding the base of the germ, staminiferous—*Filaments* four, shorter than the petals, hairy, inserted into the margin of the nectary. *Anthers* oval.—*Germ* superior, globular. *Style* short, stigma headed.—*Berry* vertically compressed, succulent, the size of a large cherry, when ripe it has a deep purple bloom, which makes it very inviting to the eye. *Seed* one, arilled, seemingly two-lobed, of a lively purple throughout.

Obs. The wood of this tree is white, close-grained, and durable, the natives employ it for a variety of purposes.

FAGARA. Schreb. Gen. N. 196.

Calyx four-parted, or four-leaved. *Corol* four-petalled *Germ* superior. *Capsule* superior, two-valved, one seeded. *Embryo* inverse, with but little perisperm.

1. F. *triphylla, Linn. Sp. Pl. ed. Willd.* i. 666.

Leaves opposite, ternate; *leaflets* oblong, entire. *Panicles* axillary, brachiate.

Ampacus angustifolia, *Rumph. Amb.* ii. *p.* 188. *t.* 62.

Found indigenous, by Mr. W. Roxburgh, on the Island of Pulo-Penang, and from thence sent to the Botanic Garden, where in several years it has only reached the height of four-feet, and for these last three years has blossomed regularly during the dry season.

Stem erect, ligneous, as thick as a walking cane, smooth, ash-coloured. *Branches* opposite; the tender parts round and slighty villous.—*Leaves* opposite, long-petioled, ternate. *Leaflets* lanceolar or oblong, entire, smooth on both sides, length from two to ten inches, and from one to four-broad.—*Panicles* axillary, and in our single young individual, shorter than the petioles, brachiate, villous. —*Flowers* numerous, minute white, inodorous. *Bractes* triangular, solitary below the divisions of the panicle.—*Calyx* small, one-leaved, four-toothed.—*Petals* four, oblong, white, smooth, spreading, length of the pistil, and four times longer than the calyx.—*Filaments* four, length of the petals. *Anthers* oblong, incumbent.—*Germ* superior, roundish, woolly, four-grooved for the filaments to rest in, four-celled, with two *ovula* in each cell, attached to the middle of the thickened axis. *Style* as long as the germ. *Stigma* large, four-lobed. —*Capsule* from one to four, distinct, obovate, smooth, size of a field bean, one-celled, two-valved, opening along the upper and inner margin, and round the apex; a portion on the back does not separate spontaneously.—*Seeds* generally two, roundish, size of a grain of black pepper, shining, blackish-chesnut colour, as in the allied *Xanthoxylum* alatum.

2. F. *Budrunga, R.*

Arboreous, armed with small incurved aculei. *Leaves* pinnate; *leaflets* five or six pair, unequally ovate, lanceolate, entire, accuminate; panicles crowded.

Budrung the vernacular name in Silhet, where it grows to be a large tree. Blossoms in March and April, and the seeds ripen in July and August; they are used medicinally by the natives, being of a warm spicy nature with the fragrance of Lemon peel.

Trunk straight, while young armed, but by age the aculei fall off. *Branches* and *branchlets* armed with innumerable, small, white, sharp, incurved prickles.—*Leaves* crowded, from six to ten inches long. *Leaflets* five or six pair, opposite or nearly so, unequally ovate-lanceolate, entire, finely acuminate, smooth, from three to four inches long, by one and a half broad.—*Panicles* terminal, much crowded, long-peduncled, trichotomous, the whole forming one large, very ramous globular panicle, which is shorter than the leaves.—*Calyx* four-toothed.—*Petals* four, elliptic, spreading.—*Stamina* four, much longer than the petals.—*Germ* ovate. *Style* short.—*Capsule* the size of a pea, drupaceous, one-celled, two-valved; the exterior tunic is marked with cells filled with a fragrant balsam.—*Seed* solitary, attached to the apex of the suture of the valves, round, smooth, polished, shining black. *Integuments* three; the exterior one thin and brittle, between it and the second is spread a quantity of dark orange-coloured pulpy matter, the second thick and hard, a perfect nut, and of a dull black-colour, the third or innermost membranaceous.—*Perisperm* in small quantity.—*Embryo* inverse. *Cotyledons* obovate. *Radicle* ovate, pointing to the umbilicus.

3. F. *Rhetsa*, *R.*

Arboreous, round. *Leaflets* from eight to sixteen pair, lanceolate, falcate, smooth, entire. *Panicle* terminal.

Teling. Rhetsa-maun.

Rhetsa means a committee, or select number of men assembled to settle disputes, &c. and maun, means trees of the largest size. Under the shade of this tree the Hill people assemble to examine, agitate, and determine their matters of public concern, deliver discourses, &c.

It is a very large tree, a native of the mountainous parts of the Coast. *Leaves* deciduous during the cold season, they come out again, with the flowers about the beginning of the hot season.

Trunk straight. *Bark* corky, deeply cracked, light grey. *Branches* numerous, spreading to a great extent. *Prickles* very numerous, scattered over every part of the tree, small, very sharp, incurved; on the old wood they acquire a conical solid base. *Leaves* about the extremities of the branchlets, abruptly pinnate, from twelve to twenty inches long. *Leaflets* opposite, from eight to sixteen pairs, short-petioled, oblong or lanceolate, waved, attenuated to a narrow point, entire, smooth, about five or six inches long, and two broad, all nearly equal in size, with nerves which divide them unequally, the exterior division being twice as broad as the interior.—*Petiols* round, smooth, a little channelled.—*Stipules* none.—*Panicles* terminal, and from the exterior arils, frequently cross-armed, particularly the extreme ramifications.—*Peduncles* and *Pedicels* smooth, sometimes compressed.—*Bractes* minute, falling.—*Flowers* very minute, yellow. —*Calyx* four-leaved.—*Filaments* shorter than the petals.—*Style* thick, length of the filaments. *Stigma* tapering.—*Capsule* round, size of a pea, dry, one-celled, two-valved; seed one, round, glossy black.

Obs. The unripe capsules are like small berries, they are gratefully aromatic, and taste like the skin of a fresh orange. The ripe seeds taste exactly like black-pepper, but weaker; from this circumstance I conceive this may be F. *Piperita*, yet I have always found the leaflets entire. The bark, I mean the inner lamina, is also acrid, with a mixture of bitter.

4. F. *nitida, Roxb.*

Shrubby. *Branchlets, petiols* and *ribs* armed. *Leaflets* two or three pairs, with a terminal odd one, oblong, lucid, remotely gland-crenate, apex rather long and emarginate. *Racemes* axillary.

Tcheum-tsew of the Chinese.

Introduced from Canton into the Botanic Garden by Mr. W. Kerr in 1812, and in January of the following year one of the plants blossomed abundantly.

The flowers as in the other species, but the pistillum is three or four-lobed, each to be appearing to end in a recurved style with a two-lobed stigma. The shrub, therefore, approaches nearer than any of the preceding species to Zanthoxylon.

Additional Remarks.

The shrubs have continued to blossom yearly, without producing any fruit. *Branches* long and slender, rambling over the nearest bushes, covered with shining brown bark; the younger ones pubescent. *Aculei* numerous, compressed, purple, recurved. *Leaves* petioled, from six to twelve inches long, with pointed, prickly petiols. *Leaflets* almost sessile, lanceolate-oblong, rounded and nearly equal at the base, apex tapering into a narrow acumen, about three inches long, with undulated margins, entire or notched with remote glandular indentures, especially while young, shining above, with numerous sub-marginal, glandular, transparent dots; pale beneath, with a more or less prickly rib.—*Racemes* from one to three, fascicled in the exterior axils, a couple of inches long, erect, cylindric, on short villous round peduncles, covered with numerous white, fragrant, short-pedicelled flowers. *Calyx* flat, with very small ovate acute segments. *Petals* lanceolate, slightly concave, four times longer than the calyx, ending in a short point. *Stamina* twice the length of the petals, spreading.—*Pistilla?* four, ovate, small, ending in a short recurved style; stigma acute, marked above with a short tooth.

It appears to be the same as Loureiro's *Fagara piperita*, Flor. Cochinch. ed. Willd. i. 101, differing from Linneus's plant, as has been justly remarked in Roem. and Schult. Syst. Veg. iii. 296. It has also some resemblance to Tetradium, Lour l. cit. 115. which Sir J. E. Smith considers as related to Fagara.—I have not been able to discover any ovula in the four bodies in the centre of the flower, which therefore may be considered as nectaries or abortive pistils, a supposition which is much strengthened by the circumstance that none of the shrubs in the garden have produced any fruit. This dioicous species ought therefore to be referred to Zanthoxylon as suggested by Dr. Roxburgh.—All its younger parts and leaves have a slightly pungent aromatic taste and smell.—N. W.

LUDWIGIA. Schreb. Gen. N. 204.

Calyx four-parted. *Corol* four-petalled. *Capsule* inferior, four-sided, four-celled. *Seeds* numerous.

1. L. *parviflora, R.*

Annual, branches diffuse. *Leaves* alternate, lanceolate, short-petioled, smooth. *Flowers* axillary, sessile. *Capsule* linear, obsoletely four-sided.

A small, ramous plant, a native of Bengal; flowering time the rainy season; seed ripe about October.

Root ramous, annual.—*Stem* short, round, in young plants erect. *Branches* many, diffuse, slightly striated; the whole plant of a dark-ferruginous colour.—*Leaves* alternate, short-petioled, lanceolate, entire, smooth, length from one to three inches.—*Flowers* axillary, small, sessile. The leaves about the ends of the branches are so small, and crowded, as to make the extremities of the branchlets appear like leafy spikes.—*Bractes*, a most minute one on each side of the insertion of each flower.—*Calyx*, the divisions taper to most acute points, and equal the oblong petals.—*Capsule* beneath, linear, obsoletely four-sided, smooth, four-valved.—*Seeds* coloured, numerous, very small.

2. L. *prostrata*, *R.*

Lower branches creeping. *Leaves* alternate, petioled, lanceolar. *Flowers* axillary, sessile. *Capsules* filiform, with one row of seeds in each cell, attached immediately to the axis.

A native of Pegu, from thence introduced into this garden, where it blossoms and ripens its seed during the rainy season.

Root annual.—*Stem* leaning much to one side, from one to two feet long. *Branches* numerous, the lower ones longer than the stem, spreading flat on the ground and striking root, smooth, striated, generally coloured with red. *Leaves* alternate, short-petioled, lanceolar, entire, smooth, size various, the largest from three to five inches long.—*Flowers* axillary, sessile, from one to several on the same axil, very small much elevated on the four-sided peduncle-like germ.—*Petals* lanceolar, longer than the calyx, yellow.—*Filaments* four, length of the style. *Anthers* roundish, two-lobed, always after expansion found adhering to the large clammy stigma.—*Capsules* slender, sub-filiform, more or less curved, smooth, four-cornered; *angles* rounded, sides grooved; four-celled.—*Seeds* oval, minute, pink-coloured, disposed in one vertical row and attached immediately to the axis, without the intervention of any thing like a receptacle.

Obs. This species is so much like my L. *parviflora* as not to be distinguished, except by attending to the germ, or capsule. In that plant it is shorter, thicker, always straight, and with numerous crowded seeds in each cell, attached to a projecting receptacle, as in L. *alternifolia, Gært. Carp.* i. 158. *t* 31. In this there is only one vertical row in each cell attached to the axis.

OLDENLANDIA. Schreb. Gen. N. 205.

Calyx four-parted, permanent. *Corol* one-petalled, four-parted. *Capsule* inferior, two-celled, many-seeded. *Receptacle* free, fastened to the partition by the base only.

Obs. Would it not be better to omit this Gærtnerian part of the generic character, and unite our Indian species of *Hedyotis* to Oldenlandia? For except the attachment of their respective receptacles I can find no other mark by which the two genera are to be distinguished, and surely this is too minute and inconspicuous for the travelling Botanist to be able at all times to discover.

1. O. *umbellata, Linn Sp. Pl. ed. Willd.* i. 676. *Roxb. Corom. Pl.* i. *N.* 3.*

Erect. *Leaves* opposed, tern, and quatern, linear, scabrous. *Umbels* terminal. Inside of the corol hairy.

Teling. Cheri-Velloo.

Tam. Saya-wer, or Imburel.

Much cultivated on the light sandy lands near the shores of Coromandel; where the root is employed to dye the best, and most durable red on cotton cloth.

2. O. *alata, König's Mss.*

Erect, four-sided. *Leaves* opposite, sub-sessile, elliptic, smooth. *Panicles* terminal; *corol* campanulate; mouth shut with hairs. *Capsules* turbinate.

Beng. Gundha-baduli.

Hedyotis *racemosa, Linn. Sp. Pl. ed. Willd.* i. 565.

* Hedyotis, Lam.—N. W.

Annual, a native of the borders of rice lands and moist places; appears and flowers during the rainy season.

Stems several, erect, with few branches, four-sided, smooth, about a foot or eighteen inches high.—*Leaves* opposite, sub-sessile, oblong, smooth, entire, succulent, about two inches long.—*Stipules*, connecting membrane broad, toothed.—*Umbels* terminal, few-flowered.—*Flowers* white, small. *Pedicels* four-sided.—*Capsules* turbinate, smooth; receptacle clubbed, free, affixed by the small end into the partition near its base.

3. O. *paniculata*, *Burm. Fl. Ind.* 38. *t.* 15. *f.* 1.

Biennial, creeping. *Leaves* ovate-lanceolate. *Peduncles* three-flowered, or three times that number.

Introduced into the Botanic Garden from the Moluccas, in 1798, where it blossoms the greatest part of the year.

Stems none; but numerous diffuse, four-sided, smooth, ramous branches, spread close on the ground in every direction, and strike root from their joints, their general length about twelve inches.—*Leaves* opposite, sessile, ovate-lanceolate, smooth, and of a firm thick texture, the largest about an inch long. *Connecting* membrane with subulate divisions. *Peduncles* axillary, and terminal, longer than the leaves, smooth, four-sided, with generally three, small, white flowers, on pedicels of unequal length.—*Corol; Tube* gibbous, mouth shut with fine white hairs.—*Stamens*, in the belly of the tube below the hairs which shut its mouth.—*Stamina* entire.

4. O. *crystallina*, *R.*

Annual, very ramous. *Pedicels* solitary, shorter than the leaves, generally two-flowered. *Leaves* sessile, lanceolate, oblong, underneath marked with crystalline dotts.

Beng. Punki.

Hedyotis *pumila*, *Linn. Sp. Pl. ed. Willd.* i. 566.

A small, very ramous, diffuse, annual plant, a native of Bengal appearing in the rainy season.

Stem none ; *branches* numerous, diffuse, about six inches long, somewhat four-sided.—*Leaves* opposite, approximate, sessile, oblong, and lanceolate-oblong, smooth and marked with beautiful crystalline specks underneath, length about half or three quarters of an inch.—*Peduncles,* from within the stipulary sheaths, about half the length of the leaves, generally bearing two minute, white flowers on their proper pedicels.—*Corol,* mouth of the tube hairy.—*Capsules* much compressed laterally.

Obs. It is readily distinguished from O. *biflora* by the general habit of the plant, this being short, diffuse, very ramous, with oblong or lanceolate-oblong leaves, while the other is slender with long straggling branches, and linear-lanceolate leaves. I place less confidence in the crystalline appearance of the underside of theleaves, as soil, &c. may remove it.

5. O. *diffusa, R.*

Annual, round, flaccid. *Leaves* sessile, linear-lanceolate, acute. *Peduncles* axillary, solitary, one-flowered ; *capsules* sub-globular, smooth.

Hedyotis *diffusa, Linn. Sp. Pl. ed. Willd.* i. 566.

Found growing in boxes, which were sent with Nutmeg plants to the Botanic Garden at Calcutta from Banda.

Root ramous, annual.—*Stems* short, flaccid, spreading, round, most slightly scabrous, from six to twelve inches long.—*Leaves* opposite, sessile, linear-lanceolate, acute, smooth on both sides.—*Connecting* membrane, from entire to many-cleft.—*Peduncles* axillary, solitary, four times shorter than the leaves,one-flowered.—*Flowers* pretty large, pure white, smooth.—*Stigma* deeply two-cleft.—*Capsule* roundish, smooth.

Obs. The Linnean definition of O. *capensis* would lead one to imagine this to be that plant ; but the observation *simillima* O. *repenti, at capsula glabra,* in the *Suppl. Plant.* p. 127, convinces me it cannot be so ; as this plant has not the smallest resemblance to *Oldenlandia repens*, now *Dentella*.*

6. O. *biflora, Linn. Sp. Pl. ed. Willd.* i. 676.

Peduncles solitary, two-flowered, shorter than the narrow lanceolar sessile leaves.

Sans. क्षेत्रपर्पटी, Kshetra-purputee.

Beng. Khet-papura.

Antirrhinum humile, &c. *Burm. Zeyl.* 22. *t.* 11; good.

Annual; native of most parts of India. A small weak straggling delicate plant, appearing in the rainy season.

7. O. *herbacea, R.*

Erect, four-sided, dichotomous. *Leaves* linear-lanceolate. *Peduncles* in pairs from alternate axils, one-flowered.

Hedyotis *herbacea, Linn. Sp. Pl. ed. Willd.* i. 566.

Teling. Veri-nella-vemoo.

A native of road sides under thin bushes, &c. Flowering time the rainy season.

Stem erect, four-sided, smooth, short, ramous. *Branches* always two-cleft; from six to twenty-four inches high.—*Leaves* opposite, *sessile*, linear-lanceolate, spreading, smooth.—*Stipules*, connecting membrane, finely-toothed.—*Peduncles* axillary, always two from the small axils, with a branchlet, or the rudiments of one between them, round, nearly as long as the leaves, one-flowered.—*Capsule* smooth.

8. O. *ramosa, R.*

Diffuse, ramous. *Leaves* narrow-lanceolate. *Peduncles* axillary, with from one to several flowers.

A native of Pegu, from thence introduced into this Garden by Mr. F. Carey, where it blossoms and ripens its seed during the rains, and the early part of the cool season.

* Conf. H. capensis in Rees's Cyclopedia. The reader is also referred to the history of the genus *Hedyotis* in that work for an elucidation of several of the above species of *Oldenlandia*.—N. W.

Root annual.—The *Stems* and numerous branches diffuse, somewhat four-sided, pretty smooth, spreading to the extent of one or two feet.—*Leaves* opposite, narrow-lanceolar, smooth, entire, whiter underneath; length about an inch.—*Connecting* membrane many-toothed.—*Peduncles* axillary, from one to many-flowered.—*Flowers* small, white.—*Calyx*, the divisions conical.—*Corol* with gibbous tube, mouth bearded.—*Filaments* short. *Anthers* within the tube, blue.—*Germ*, &c. as in the genus.

AMMANNIA. *Schreb. Gen. N.* 206.

Corol four-petalled, or none; when present inserted in the plaited mouth of the calyx which is from four to eight-toothed. *Capsule* superior, from one to four-celled; *seeds* numerous.

1. A. *rotundifolia, Buch.*

Diffuse. *Leaves* opposite, round, entire. *Flowers* spiked, petals oval. *Capsule* four-lobed.

Plants sent from Lukshmee-*poora* and Malda, where they are indigenous, flowered in the Botanic Garden during the cold season of 1797.*

Stems several, creeping, with ascending apices, round, smooth, and succulent; whole plant spreading from six to eight inches.—*Leaves* opposite, sub-sessile, orbicular, entire, smooth, and succulent.—*Spikes* terminal and axillary, the terminal often compound.—*Bractes* opposite; *exterior* cordate; *interior* lanceolate, and often very small, one-flowered.—*Flowers* numerous, small, pink-coloured.—*Calyx* four-parted, divisions obtuse.—*Corol* of four, large, oval, sessile, pink-coloured withering petals.—*Capsule* hid in the bottom of the remaining calyx, and corol, four-lobed.—*Seeds* numerous.

2. A. *octandra, Linn. Sp. Pl. ed. Willd.* i. 679. *Roxb. Corom. Pl.* 2. *N.* 133.

* I have had this elegant plant from Mr. Wm. Jack, who gathered it at the Turraye, where it is called *Thuntuneea* and *Mirkooa*; also from Nepala in the vicinity of Katmandu, from the Hon. E. Gardner.—N. W.

Annual, erect, brachiate, four-sided. *Leaves* sessile, narrow-sagittate; *peduncles* from two to three-flowered; *flowers* octandrous. *Capsule* four-celled.

Ammannia *octandra, Suppl. Plant. p.* 127.

Annual, native of moist places; appearing in the rainy and cold seasons.

Root fibrous.—*Stem* straight, four-sided, smooth; from one to three feet high.—*Branches* numerous, cross-armed, and like the stem.—*Leaves* opposite, sessile, linearly-sagittate, smooth, margins a little revolute, from one to two inches long, and about a quarter of an inch broad.—*Peduncles* axillary, very short, from two to three-flowered.—*Flowers* small, colour a beautiful bright red.—*Bractes* minute.—*Calyx* and *corol* as in the genus.—*Filaments* as long as the corol. *Anthers* incumbent.

3. A. *multiflora.*

Erect, four-sided, ramous branches simple. *Leaves* sub-sagittate, stem-clasping. *Peduncles* from three to seven-flowered; *petals* large; *capsule* one-celled.

A native of open lands about Calcutta, annual, from six to twelve inches in height, appearing in the wet and cold seasons.

Stem straight, four-sided; angles sharp; sides convex. *Branches* cross-armed. *Leaves* opposite, decussate, horizontal, sessile, linear with enlarged cordate, stem-clasping base, size very various. *Peduncles* axillary, almost always solitary, about as long as the smaller leaves, generally three-flowered, though in the luxuriant plants there are often two or three, with one between, the lateral flowers have pretty large long pedicels.—*Calyx* as in the family.—*Petals* large, roundish, red.—*Capsule* one-celled, one-valved, receptacle globular, affixed at the base only.

4. A. *vescicatoria, R.*

Erect, ramous, branches undivided. *Leaves* opposite, sessile, lanceolar. *Flowers* axillary. *Corol* none. *Capsule* one-celled.

Beng. Daud-maree.

Teling. Aghundra-pakoo.

An annual species; found in wet cultivated ground during the rainy and cold seasons.

Stem erect, from six to thirty-six inches high, obsoletely four-sided, very branchy. *Branches,* below opposite, decussated; above frequently alternate.—*Leaves* opposite, sessile, lanced, smooth; *floral leaves* many times smaller than those below.—*Corol* none.—*Pericarp* one-celled, one-valved, half covered with the calyx.

Obs. It differs from A. *baccifera* in having a capsule of one cell, in being evalvular, and in the leaves being lanced, in that species they are linear-cordate at the base, and embrace the stalk.

This plant has a strong muriatic smell, but not disagreeable. Its leaves are exceedingly acrid; they are used universally by the natives to raise blisters in rheumatic pains, fevers, &c. The fresh leaves bruised and applied to the part intended to be blistered perform their office in the course of half an hour or a little more, and most effectually.

5. A. *pentandra, R.*

In the lower part of the plant creeping and simple, erect portion ramous; four-sided; *branches* simple. *Leaves* sessile, lanceolate. *Flowers* axillary, solitary, sessile, pentandrous. *Capsule* one-celled.

A native of moist places; flowering in the rainy and cold seasons.

Stems creeping near the base, above erect; the erect part is branchy, four-sided, smooth, and coloured; from six to eight inches high.—*Leaves* opposite, sessile, lanceolate, smooth, entire.—*Flowers* axillary, solitary, sessile.—*Bractes* two, lanceolate, pressing the calyx laterally.—*Calyx* ten-toothed, alternately smaller and inflected.—*Corol,* petals five, very minute.—*Stamens* five.

6. A. *nana, R.*

Base creeping, erect portion ramous; *branches* compound. *Leaves* wedge-shaped, sessile. *Flowers* solitary. *Capsule* one-celled.

A native of such places as the last described, it is considerably smaller than that species, but more branchy.

Stems as in A. *pentandra.*—*Leaves* opposite, sessile, wedge-formed, smooth, entire.—*Flowers* axillary, solitary, sessile.

TRAPA. Schreb. Gen. N. 208.

Calyx four-parted, or four-leaved. *Corol* four-petaled. *Germ* inferior, two-celled ; cells from one to two-seeded, attachment of the ovulum superior. *Nut* turbinate, armed with two or four horns, (which were the leaflets of the calyx.) *Embryo* simple, lodged in the apex of the ample perisperm.

1. T. *bicornis.*

Nuts with two, opposite, recurved, very obtuse horns.

Chin. Lin-ko.

A native of China, where it is carefully cultivated in lakes, ponds and other receptacles of water.

2 T. *bispinosa, R.*

Nut with two opposite, straight, barbed, spinous horns. *Nectary* cup-shaped, curled.

Shringata, Asiat. Res. vol. iii. p. 253.

Sans. शृङ्गाटः, Sringat*a*, जलकण्टकः, J*u*l*u*k*u*nt*u*k*a*.

Beng. Panee-phul. Singhara.

Panover-tsjerana, *Rheed. Hort. Mal.* vol. xi. p. 64. *t.* 33. (evidently intended for this plant, and not T. *natans.*)

Found floating on sweet, standing water in Bengal and many other parts of India. Flowering time the rainy season.

Root consisting of numerous, simple, capillary fibres which issue from the under margins of the cicatrices of the fallen leaves.—*Stems* stoloniferous, often several feet long, floating.—*Leaves* petioled, alternate, approximate round the apex of the shoots, of a rhomboidal-reniform shape; with the posterior margins entire, and the anterior ones serrate-dentate, smooth above, of a deep green, villous and purple underneath; from three to four inches broad, and nearly as long. —*Petioles* villous, lengthening with the age of the leaves, towards the apex much smaller, this part contains many inflated vescicles, which render the whole plant very buoyant.—*Stipules* two pair, the superior ones simple, semi-lanceolate, and caducous, the inferior pair at first simple and filiform; but becoming ramous by age, permanent.—*Peduncles* axillary, solitary, shorter than the petioles, one-flowered, incurved until the blossoms begin to decay, after which they bend down under the surface of the water, where the fruit grows, and comes to maturity.—*Flowers* pure white, expanding above the surface of the water late in the afternoon.—*Calyx* four-leaved, in opposite pairs, leaflets lanceolate, villous, one of the pairs permanent.—*Corol* contracted; *petals* four, contracted into a tube at the base by the firmness of the calyx, above much broader, and expanding; margins much curled.—*Nectary* cup-shaped, on the top of the germ, round the swelled base of the style, eight-grooved, with the margins minutely lacerate, and curled.—*Filaments* four, incurved. *Anthers* oval, incumbent.—*Germ* inferior, obconical, two-celled, with one ovulum in each attached to the top of the partition.—*Nut* inferior, of a turbinate triangular shape, a hard ligneous texture, and a blackish-brown colour, armed with two, strong, straight, conic, barbed spines, from the upper angles; elegantly embossed, one-celled, not opening, but having a perforation in the top for the *Embryo* to pass, which is guarded with a circle of stiff converging bristles from its margin, forming a conical roof over the aperture. *Seed* single, conform to the nut. *Integument* single, of a slightly brown colour, thin, but rather tough, and adhering firmly to the perisperm.—*Perisperm* conform to the seed, white, and of a firm cartilaginous texture. *Embryo* in the apex of the perisperm.

Obs. This is evidently a species perfectly distinct from *natans*, and *bicornis*. From the former it differs in having only two horns, and from the latter in the horns being straight, very sharp, and barbed.

The nuts are sold in the markets when in season, their kernels being much esteemed by the Hindoos.

3. T. *quadrispinosa, R.*

Nut with two opposite pairs of straight, acute, spinous horns.

A native of Silhet, where it is found floating on lakes, and pools of fresh water, its kernels are also esculent.

POTHOS. *Schreb. Gen. N.* 210.

Spathe sub-cylindric. *Spadix* covered with florets. *Calyx* none, or four-petalled; *corol* none. *Stamina* indefinite. *Germ* one-celled, from one to three-seeded; attachment of the ovula inferior (in *heterophylla* superior.) *Berry* one-celled, rarely many-seeded. *Embryo* uncinate, without perisperm.

1. P. *scandens, Linn. Sp. Pl. ed. Willd.* i. 684.

Parasitic, rooting. *Petioles* as broad as the lanceolar leaves. *Flowers* axillary; *spadix* gobular, reflex.

Ana-Parua, *Rheed. Hort. Mal.* vii. *p.* 75. *t.* 40.

Adpendix duplo folio, *Rumph. amb.* v. 490. *t.* 184. *f.* 2.

A native of Chittagong, Malabar, Amboyna, and other eastern Islands. Flowering time in the former country, the month of April.

Stems as thick as a slender ratan, often ramous, rooting upon trees in the most shady forests, round, smooth, and less succulent than any other species which I have yet met with, except P. *gracilis.*—*Leaves* alternate, petioled, lanceolate, entire, smooth; length from two to four inches, breadth about an inch.—*Petioles* most amply winged, so as to be nearly as broad as the leaves themselves, almost as long, equally entire, smooth and marked with similar veins. —*Peduncles* axillary, solitary, clothed with many small, ovate, bractiform scales up to the spathe, where they are completely recurved. —*Spathe* boat-shaped, erect.—*Spadix* globular, reflex.—*Calyx,* or *corol,* some roundish scales mixed amongst the germs, and of nearly the same length, the exact number to each germ not ascertained.—*Stamina,* about four obovate scales to each germ, with two polliniferous pits on the inside under the retuse apex.—*Germs* many, oblong, one-celled, with one, two, or three *ovula* attached to the bottom of the cell, and immersed in a clear gelatinous liquid.—*Style* none. *Stigma*, an umblicated elevation on the apex of each germ. —*Berries* oblong, size of a French bean, red, pulpy, one-seeded.—*Seed* solitary, or two, conform to the berry.

2. P. *officinalis*, *R.*

Perennial. Parasitic, *stems* rooting. *Leaves* oblong, cordate, entire. *Flowers* terminal; *florets* naked, octandrous. *Berries* one-seeded.

Sans. गजपिप्पली, *Guja-pippulee*, करिपिप्पली, *Kuri-pippulee*, कपिवल्ली, *Kupi-vullee*, कोलवल्ली, *Kolu-vullee*, श्रेयसी, *Shreyusee*, वशिर, *Vushira.*

Beng. Guj-pippul.

A native of Bengal, and common in some parts of the Midnapore district, where it is cultivated for its fruit, which, cut into transverse pieces and dried, is an article of some importance in the Hindoo *materia medica*, called *Guj-pippul*, and sold by the druggists under that name. Flowering time the rainy season; seed ripe in January.

Stem perennial, creeping up to the top of large trees; and like the *Ivy*, taking firm hold with its innumerable roots; about as thick as the little finger, smooth, except the cicatrices of the fallen leaves, and the roots from the side next the tree which supports it. —*Leaves* alternate, sub-bifarious, petioled, oblong-cordate, entire, filiform, pointed, smooth on both sides; from six to ten inches long, and from three to six broad.—*Petioles* sheathing, channelled, smooth, shorter than the leaves.—*Peduncles* terminal, solitary, round, smooth, scarcely half the length of the petioles; when in flower erect, but often drooping, from the increasing weight of the fruit. *Spathe* sub-cylindric, opening a little on one side; when in flower, smooth on both sides; greenish without, and of a pale yellow colour within, apex filiform, and a little twisted; length about six inches.—*Spadix* sub-cylindric, obtuse, equalling the spathe; colour a pale greenish yellow, dotted with the dark-coloured stigmas of the numerous fructifications which every where cover it.—*Perianth* none.—*Corol* none.—*Filaments;* I conclude there are about eight to each germ, two on each face; but the want of both perianth and corol, renders it difficult to assign the exact number of stamens to each floret, they are very short, and seem inserted on the base of the germ. *Anthers* sub-sagittate, four-lobed. *Germs* numerous, parallelopiped, truncate, vertically attached to, and completely covering every part of the common-cylindric receptacle, one-celled, with one seed attached in an oblique manner to the bottom of the cell. The substance of the germs is replete with rigid, sharp, vertical bristles, which are readily detached, and stick in the skin, causing pain and itching. The virtue of the drug may reside in these, as it does in the short stiff hairs of the legumes of *Carpopogon pruriens*. *Style* none. *Stigma*, a channel filled with gelatinous matter, running from the cell of the germ to the centre of the truncate apex, where it ends in a small slit, embraced by two darker-coloured, somewhat callous lips.—*Pericarps (capsulæ circumscissæ)*, as numerous as the germs, and of the same form, only larger; a few only are fertile, of a soft fleshy texture, one-celled, one-valved; when the fruit is ripe they detach themselves from the receptacle, and drop off, leaving the seed behind still attached to it.—*Seed* single, ovate-cordate, somewhat two-lobed, covered at the base with a thin, red, succulent aril. *Integuments* two, exclusive of the aril, *exterior* pale yellow, polished, thin, hard, and elastic; *inner* membranaceous. *Perisperm* none. *Embryo* uncinate, white, very succulent. *Radicle* inferior, thickened, and in other respects exactly as in Gærtner's figures of *Potamageton natans*, vol. ii. p. 23. *t.* 84.

Obs. The plant described by Sir William Jones by the name of Guj-peepul, (see *Asiatic Researches*, vol. iv. p. 303.) is my *Tetranthera apetala*. He, however observes, that it was from a native gardener he obtained the Sanscrit name of his plant, but as he mentions that Valli means a creeper, he suspects in some degree, the accuracy of his information.

Chavica, or Chavya mentioned under the same head is a new species of Piper.

3. P. *Peepla, R.*

Perennial, sub-parasitic, rooting on trees. *Leaves* long-petioled, oblong, acuminate. *Flowers* sub-terminal; *florets* apetalous, tetrandrous.

Peeplee of the Hindoos in the Silhet district, where the plant grows. Flowering time, April and May.

Stem rooting on trees, like the common *Ivy* of Europe, scarcely so thick as the little finger, round and smooth, somewhat contracted between the leaves.—*Leaves* petioled, alternate, oblong, entire, smooth on both sides, dagger-pointed; length from four to twelve inches.—*Petioles* as long as the leaves, channelled, smooth.—*Peduncles* sub-terminal, round, smooth; the base inserted in sheaths from three to four inches long.—*Spathe* sub-cylindric, opening on one side, smooth, acute; length of the spadix.—*Spadix* cylindric, obtuse, length of the inside of the spathe; colour pale yellow, every part completely covered with the fructification, without any proper perianth, or corolla.—*Filaments* about four to each germ, and about half its length. *Anthers* linear.—*Germs* numerous, of a square, wedge shape, truncate, the whole of their substance replete with small pellucid, sharp bristles, which readily enter the skin, and produce pain and itching.—*Style* none. *Stigma* sub-rotund.

4. P. *gracilis, R.**

Scandent. *Leaves* lanceolate, entire, on short, round petioles above their sheaths. *Spadix* peduncled, cylindric, scarcely equalling its linguiform spathe.

A native of Amboyna; it has a great resemblance to *Rumphius's Adpendix arborum, Hort. Amb.* vol. v. *t.*181. *f.* 2.

* This name being pre-occupied by a species described in Rudge's excellent work, *Plantæ Gujanæ* may be changed to *tenera*.—N. W.

5. P. *gigantea, R.*

Perennial. *Stems* rooting. *Leaves* oblong, cordate, entire. *Flowers* axillary; *florets* naked, heptandrous.

This is the stoutest of these parasitic *Aroideæ* that I have met with. It is a native of Prince of Wales Island; from thence introduced into the Botanic Garden by Mr. W. Roxburgh. Flowering time in Bengal the close of the rains in October. It has not yet produced ripe seed there.

Stems perennial, creeping on the ground, or upon trees and plants; in a very thriving state, of six years growth, not more than three and four feet long, and about four inches in circumference, round, and strongly marked with the circular scars of the fallen leaves.—*Leaves* alternate, sub-bifarious, petioled, approximate, oblong, cordate, entire, smooth; length from one to two feet high, and about two-thirds thereof in breadth.—*Petioles* as long as the leaves, channelled, smooth, much enlarged at the base, and stem-clasping.—*Peduncles* axillary, solitary, very short.—*Spathe* sub-cylindric, short-pointed, opening a little on one side about the time that the anthers discharge their pollen, smooth on both sides, of a dirty dull purple colour, length about six inches.—*Spadix* cylindric, length of the spathe, obtuse, every part most closely covered with the sessile fructifications; colour a pale lively yellow, beautifully dotted with the dark linear stigmas.—*Filaments* about seven to each germ, short, broad, and thin. *Anthers* two-lobed.—*Germs* very numerous, parallelopiped, truncate, much flattened, vertically one-celled, with generally two ovula, obliquely attached to the bottom of the cell. *Style* none. *Stigma* a vertical, linear, grooved, dark-coloured projection on the middle of the truncate germ, as in the last species; the substance of the germ and spathe also abound in short stiff bristles, which readily enter the skin, causing pain and itching.—*Calyx* and *Corol* none.

6. P. *pertusa, R.*

Scandent, sub-parasitic, rooting on trees. *Leaves* cordate, perforated on one side, and pinnatifid on the other. *Corollets* apetalous, tetrandrous. *Anthers* two-lobed.

Elettadi-Maravara, *Rheed. Hort. Mal.* xii. *p.* 41. *t.* 20 and 21.

A very large perennial, sub-parasitical species, so much like *Dracontium* pertusum, that they may be readily taken for the same plant. It is a native of the mountainous parts of the Coast of Coromandel, running up, and over the highest trees, to which it adheres like *Ivy* by roots issuing from the joints.

Leaves alternate, petioled, cordate, smooth, generally perforated with linear oblong holes on one side, and on the other passing through the margins as in the pinnatifid leaves; their length from twelve to eighteen inches long.—*Petioles* nearly as long as the leaves, deeply channelled, wanting the stipule-like processes at the apex, which Miller's figure represents, in which particular this plant differs from that in the first instance.—*Scape* short.—*Spathe* gibbous, pointed, a little longer than the spadix, cylindric, obtuse, every where covered with the fructifications.—*Corol* none.—*Filaments* intermixed amongst the germs, (the number to each germ cannot be determined, but I think about four,) compressed; apex bifid. *Anthers* oval, two to each filament.—*Pistils* mixed with the membranaceous filaments, four-sided, lobed. *Style* none. *Stigma,* a knob on the centre of the lobed germ.

7. P. *pinnata, Linn. Sp. Pl. ed. Willd.* i. 686.

Sub-parasitic, rooting on trees, the apices of the shoots clothed with fibres. *Leaves* pinnatifid. *Flowers* terminal.

Adpendix laciniata, *Rumph. Hort. Amb.* v. 489. *t.* 183. *f.* 2.

Found in forests in the Malay countries, rooting up and running over trees of great size. It is readily distinguished from my *decursiva,* by the abundance of brown fibres which are found on the young shoots, and the lower part of the petioles.

8. P. *decursiva, R.*

Perennial, rooting on trees, smooth. *Leaves* sub-decursively pinnate; *segments* falcate, cuspidate. *Flowers* lateral, or axillary, long-peduncled.

Guj-pippul is the vernacular name of this very large, powerful species, in the Silhet district, where it is indigenous, and blossoms during the hot season.

Stems as thick as a stout cane, rooting on, and ascending trees like the common Ivy; round, smooth, and rather contracted between the leaves.—*Leaves* approximate, alternate, petioled, pinnatifid, or completely pinnate; from two to six, or seven feet long, including the petioles; *segments*, or leaflets, linear-falcate, cuspidate, entire, smooth, sometimes from three to four-nerved or triple-nerved; length from six to eighteen inches, and two broad.—*Petioles* from six inches, to three feet long, round, smooth, stem-clasping, and the lower portion more deeply channelled.—*Peduncles* solitary, axillary or lateral, about six inches long, substantially thick, round, smooth, and replete with the same while bristly spicula that are found in the germ and seed vessels.—*Spathe* sub-cylindric, length of the peduncle.—*Spadix* cylindric, length of the spathe, completely covered with the innumerable fructifications.—*Calyx* none. *Corol* none, nor any thing like either.—*Filaments* about four to each germ, and nearly of their length. *Anthers* cordate, appearing beyond the germ.—*Germs* numerous, the most general form a four-sided wedge, apex truncate.—*Style*, scarcely any. *Stigma*, an elevated vertically oblong opening.

9. P. *heterophylla*, *R.*

Caulescent, creeping, armed. *Leaves* from cordate-sagittate to pinnatifid. *Spathe* erect, spiral, many times longer than the short cylindric spadix. *Florets* tetrapetalous, tetrandrous.

Can it be *Dracontium spinosum*, *Flor. Zeyl.* 328?

A native of Bengal, delighting in a rich, moist soil; it flowers, though rarely, during the cool season.

Root fibrous.—*Stems* and *branches* creeping, armed with short, sharp prickles.—*Leaves* until the plants are old, and begin to flower, simple-sagitate, and cordate-sagittate, afterwards pinnatifid, with long lanceolate, acuminate segments, smooth above, nerve and larger veins armed on the under side; length from six to fourteen or sixteen inches.—*Petioles* completely armed; *base* sheathing; above round, generally shorter than the leaves.—*Scapes* axillary, solitary, one-flowered, length of the petioles, and armed like them.—*Spathe* sub-cylindric, acute, spirally twisted above the spadix; colour a dark dull reddish purple; length from eight to twelve inches long, and as thick as the little finger, opening at the base only where the spadix is lodged while the pollen is discharging, after which it shuts, as before that period.—*Spadix* cylindric, obtuse, completely covered with hermaphrodite florets; length little more than one inch.—*Calyx* none.—*Petals* four, obovate, concave.—*Filaments* four, broad, length of the petals. *Anthers* two-lobed.—*Germ* oblong, one-celled, containing one *ovula*, attached to the top of the cell. *Style* none. *Stigma* glandular, flat.

10. P. *Lasia, R.*

Armed. *Leaves* radical, pinnatifid. *Spathe* very long, and twisted above the spadix. *Berries* one-seeded, muricate.

Lasia aculeata, Loureir. Cochinch. ed. Willd. i. 103.

Teling. Moola-sari.

Beng. Kant*a*-kutchoo.

A native of boggy, or wet places among the Circar mountain.

Leaves radical, petioled, pinnatifid; *segments* lanceolate, smooth shining, the nerves on the under side armed with prickles; length about two feet, breadth about a foot and half.—*Petioles* erect, from two to four feet long, round, every were thickly armed with sharp prickles.—*Scape* the length of the petioles, round, and equally covered with prickles.—*Spathe* very long, erect, below gaping; above the spadix twisted, leathery, dull red colour.—*Spadix* about one-sixth of the length of the spathe, cylindric, every where covered with the fructifications.—*Corol, stamens,* and *pistil* as in the genus.—*Berries,* (or rather capsules,) wedge-formed, somewhat four-sided, murexed, one-seeded.—*Seed* two-lobed.

CHLORANTHUS. *Schreb. Gen.* 1730.*

Calyx none. *Corol,* a three-lobed, unilateral antheriferous *petal* on the outside of the pistillum. *Germ* semi-superum one-seeded; attachment of the ovulum superior. *Drupe* one-seeded.

C. *inconspicuus, Linn. Spec. Plant. ed. Willd.* i. 688.

Tcheu-lan, of the Chinese.

From Canton it has been received into the Botanic Garden at Calcutta, where it grows freely to be a small upright shrub, with many straight sub-simple smooth branches, of about two feet in height, with oblong opposite serrulate smooth leaves. *Flowering* time in Bengal the rainy season. It has not yet ripened its seeds there.

ELÆAGNUS. *Schreb. Gen. N.* 213.

Corol none. *Calyx* gibbous, border four-parted. *Germ* one-celled, one-seeded; attachment of the ovulum inferior. *Drupe* inferior, one-seeded. *Embryo* erect, without perisperm.

1. E. *triflora, R.*

Shrubby, scandent, armed. *Leaves* oblong, obtuse, scaly. *Flowers* axillary, three-fold, peduncled.

Elæagnus foliis oblongis acuminatis maculatis, *Burm. Zeyl. p.* 92. *t.* 39 *f.* 1.

A native of Sumatra, from thence introduced into the Botanic Garden in 1804, where it blossoms in December. Seeds ripen three or four months afterwards.

Stems and *branches* of considerable size, and great extent, climbing; *bark* of the ligneous parts a beautiful dark purplish colour, and smooth. The short lateral branchlets which issue from the young climbing shoots are more or less recurvate and answer the end of tendrils; when old, they become stout thorns. *Bark* of the tender

* *Nigrina,* Thunb. *Creodus,* Loureir.—N. W.

part most beautifully clothed with numerous ferruginous, stellate scales.—*Leaves* alternate, short-petioled, oblong, entire, obtuse, above smooth; underneath ornamented with numerous stellate, ferruginous and silver-coloured scales; from two to three inches long, and from one to two broad.—*Stipules* none.—*Flowers* axillary, threefold, small, white.—*Peduncles* longer than the corols, one-flowered.—*Calyx* with a short, four-sided tube; and border of four cordate segments, on the outside a little scaly; on the inside smooth.—*Filaments* none. *Anthers* four, sessile in the mouth of the tube of the calyx.—*Germ* superior, oblong, embraced by the contracted bottom of the calyx, one-celled, one-seeded. *Style* as long as the calyx.—*Stigma* oblong, glandular, oblique.—*Berry* oblong, succulent, when ripe pale orange colour, pulp soft, of rather a pleasant astringent taste, and not acid like E. *conferta;* one-seeded.—*Seed* sub-clavate. *Integuments* two; *exterior* eight-ribbed, soft, and lined with much wool; *inner* membranaceous.—*Perisperm* none.—*Embryo* erect. *Cotyledons* conform to the seed, amygdaline. *Plumula* two-lobed. *Radicle* oval, inferior.

2. E. *conferta, R.*

Shrubby, scandent. *Leaves* alternate, oblong-pointed, covered with silver-coloured scales underneath. *Flowers* axillary, crowded, short-peduncled.

Beng. Guara.

This is a very large and ramous, scandent shrub; a native of the mountains near Silhet, in Bengal.* Flowering time, the cold season. Seed ripe in March.

Young *shoots* covered with innumerable rust-coloured scales.—*Leaves* alternate, short-petioled, generally erect, oblong-pointed, entire, with innumerable small silver coloured scales, particularly underneath, which gives them a bright colour; from three to four inches long, and from one and half to two broad.—*Petioles* short, scaly.—

* Also Nepala, Siriuagur, Sirmore.—N. W.

Peduncles axillary, crowded, short.—*Flowers* small, straw-coloured. *Calyx; tube* somewhat gibbous, four-sided; outside scaly; border of four cordate segments; its mouth marked on the inside with four curved lines, on their pointed junctures the filaments sit.—*Style* hairy.—*Drupes* oblong, succulent, when ripe red.

Obs. The calyx of my plant cannot be said to be placed on the germ; it has, as it were a lower tube which envelopes the germ only, this part grows with the fruit, and finally becomes the pulpy part thereof as in the *Nyctagineae* of Jussieu.

3. E. *arborea, R.*

Arboreous. *Branchlets* often spinous, pointed. *Leaves* oblong and broad-lanceolate, entire, silvery underneath. *Flowers* in axillary fascicles. *Fruit* cylindrically oblong.

A large tree, native of the Garrow Hills and there called *Sheu-Shong*. Flowering time December. The fruit shaped like an olive, but much larger; it is eaten by the natives when ripe, which it is generally in March and April.

CANSJERA. Juss. Gen. ed. Ust. 85.

Calyx large, gibbous, four-toothed. *Corol* none. *Nectary* four-petalled, embracing the germ, and of the same length. *Berry* superior, one-seeded.

1. C. *scandens, Roxb. Corom.* ii. *N.* 103.*

Shrubby, scandent. *Leaves* alternate. *Spikes* axillary.

Scheru-valli caniram, *Rheed. Mal.* vii. *p.* 7. *t.* 4.

Tsjeru caniram, ibid. *p.* 3. *t.* 2. appears to be the same plant.

A large woody climbing shrub; a native of the Circar mountains.

Leaves alternate, two-faced, short-petioled, broad-lanceolate, smooth, entire, about three inches long, and one broad.—*Spikes* axil-

* C. malabarica, Lam. Encycl. iii. 433.—Daphne polystachya, Willd. spec. ii. 420.—N. W.

lary, generally single, short, every were covered with flowers.—*Bractes* solitary, small, oval, one-flowered.—*Flowers* small, sessile, yellow. —*Calyx* inferior, one-leaved, gibbous; *mouth* four-toothed.—*Corol* none, except the calyx, or nectary be accounted one.—*Nectary* four-leaved; *leaflets* cordate, acute, half the length of the calyx, inserted round the base of the germ and embracing it.—*Filaments* four, inserted into the base of the calyx, and rather shorter than it. *Anthers*, ovula within the calyx.—*Berry* oblong, smooth, red, juicy, size of a pea, one-celled.—*Seed* single, shaped like the berry.

SANTALUM. *Schreb. Gen. N.* 215.

Calyx semi-supera, campanulate, four-cleft. *Corol* none. *Nectarial* scales four, alternate with the stamina. *Germ* one-celled, one-seeded, at achment inferior. *Berry* one-seeded. *Embryo* inverse, and furnished with a perisperm.

1. S. *album, Linn. Sp. Pl. ed. Willd.* i. 691.

Leaves oblong.

Sandalum album, *Rumph. Amb.* ii. 42. *t.* 11.

Chandana, *Asiatick Researches,* vol. iv. p. 253.

Santalum verum, *Linn. Mat. Med.* 102.

Sans. गन्धसारः, G*u*ndh*a*sar*u*, मलयजः, M*u*l*u*y*u*j*u*, भद्रश्री, Bh*u*dr*u*shree, चन्दन, Ch*u*nd*u*n*a*.

Beng. Chundun.

Pers. Sundul-s*u*fed.

Sandal-wood, Buchanan's Journey i. 186, and 202. ii. 117, 132, and 536. iii. 192.

A native of the mountainous parts of the Coast of Malabar. That of Timor, and the neighbouring Islands is most probably the same species. In the Botanic Garden near Calcutta, the trees reared from seeds brought from the former country, thrive luxuriantly, and produce flowers and ripe seed nearly the whole year round.

Trunk in trees of twelve years growth in the Botanic Garden, only a few feet high, when they divide into branches; and there from two and a half to three feet in circumference. *Bark* brownish, and scabrous, with longitudinal fissures. *Branches* numerous, dividing much, spreading, and rising in every direction, forming nearly a spherical head. Young *shoots* round and smooth ; extreme height of our largest twelve or thirteen year-old trees, about twenty-five feet.—*Leaves* opposite, petioled, oblong, generally obtuse, smooth, entire, glaucous underneath : *length* from one and a half to three inches.—*Petioles* scarcely one fourth the length of the leaves, smooth.—*Stipules* none. *Thyrsi* axillary and terminal, much shorter than the leaves. *Pedicels* opposite, the lower pair of each thyrsus generally three-flowered.—*Flowers* numerous, small, when they first expand straw-coloured, changing to a deep ferrugenous purple, inodorous, as are all the exterior parts of the growing plant, even when bruised.—*Calyx* one-leaved, campanulate; border four-parted ; *segments* ovate, smooth, base of the bell only permanent.—*Corol* none, except the calyx, or nectarial glands are so callèd.—*Nectary* of four, roundish, sessile, thick, fleshy glands, on the mouth of the bell of the calyx, alternate with the segments of its border, and when the flower first expands, the most coloured part, entire. In *Sirium (*now *Santalum) myrtifòlium,* they are *lobate.*—*Filaments* four, alternate with the nectarial scales, and of the same length, smooth, but behind each there rises a tuft of white hairs from the disk of the segments of the calyx, perfectly distinct from the stamina, but of the same length, the ends of these long, white, woolly hairs adhere to the back of the small, two-lobed anthers.—*Germ* semi-superum, one-celled, containing one conical seed, attached to the bottom of the cell. *Style,* length of the tube of the calyx.—*Stigma* three-nerved, four-lobed. *Berry* pyrenous, round, smooth, when ripe black, and succulent, size of a cherry, and crowned with the remaining calyx, one-celled.—*Nut* solitary, spherical ; three small, equisdistant elevations from the apex run a little way down the sides.—*Seed* conform to the nut. *Integument* single, soft, white, spungy, lining the nut, and adhering to it and the seed also.— *Perisperm* conform to the seed, white, amygdaline.— *Embryo* inverse, subulate, nearly as long as the vertical diameter of the perisperm. *Cotyledons* two, unequal, linear-lanceolate. *Plumula* minute, semilunate. *Radicle* tapering, superior.

Obs. The descriptions and original figures of the following species, *Sirium myrtifolium,* Corom. Plants, vol. i. t. 2. were taken from the plants which grow wild, on the mountains north of the Rajamundri Circar, on the coast of Coromandel, where the trees are but of a small size scarcely more than a large shrub, or bush; with lanceolate leaves; and the wood of little value. It is now nearly twenty years since that description, and the drawings were made, and I have reasons to think the tree which produces the sandal wood of Malabar,* and Timor, to be different, and having had an opportunity of cultivating the former, viz. that of the Malabar mountains in the Botanic Garden for these thirteen years past, I do not hesitate to say, that the Circar tree it is at least a strongly marked variety of the Malabar sort. For that reason I have thought it adviseable to give the foregoing full description, of the true *sandal wood,* which I hope will tend to clear up the botanical history of this celebrated tree, and its associate *Sirium myrtifolium.* They evidently belong to one genus, to which I have thought it necessary to assign a new essential character, and whether I consider them to be distinct species of that genus, or varieties of one species, must be of very little consequence. For my own part I think it nearer the truth to consider them distinct species. I therefore call the narrow-leaved one *Santalum myrtifolium.*

The nature and history of the wood, white and yellow *Saunders,* or *Sandal* are now too well known to require any thing in addition from me.

2. S. *myrtifolium, R.*

Leaves opposite, lanceolate.

Sirium *myrtifolium, Roxb. Corom. Pl.* i. *N.* 2. *Willd.*

A native of the Circar mountains, where it is but of a small size, and the wood of little or no value.

* I include however the other Islands thereabout, because the leaves of our tree agree particularly well with Rumphius's figures above quoted.

Leaves opposite, short-petioled, spreading, lanceolate, entire, waved, smooth, shinning; about two inches long, and three-fourths of an inch broad.—*Stipules* none.—*Racemes* thyrsiform, terminal, compound, small, erect.—*Flowers* small, reddish.—*Calyx* superior. *Tube* short, a little gibbous. *Border* four-cleft ; *divisions* ovate, expanding, coloured.—*Corol* no other than the above, except the nectary be such.—*Nectary* four-leaved, obcordate, notched, inserted into the mouth of the calyx.—*Filaments* four, short, hairy, inserted into the calyx, alternate with the leaflets of the nectary.—*Germ* globular. *Style* length of the tube. *Stigma* four-lobed.—*Berry* globular, size of a large pea, smooth, juicy, black, when ripe one-seeded.

Obs. Birds greedily eat the berries, by which means it is propagated extensively. I do not know that the wood of this small tree is ever used as a perfume.

GYROCARPUS. *Jacq. Gært.*

Calyx superior, four-leaved, unequal. *Corol* none. *Nectary* of four clavate glands, alternate with the stamina. *Berry* dry, one-seeded, ending in two long wings. *Embryo* erect, spirally rolled up; no perisperm.

1. G. *Jacquini, Roxb. Corom. Pl.* i. *N.* 1.

Polygamous. *Panicles* dichotomous.

Gyrocarpus *asiaticus, Linn. Sp. Pl. ed. Willd.* iv. *p.* 982.

Teling. Tanukoo.

Grows to be a very large tree, is chiefly a native of the mountainous parts of the coast of Coromandel. Leaves deciduous about the end of the wet season. Flowers during the cold season when the trees are naked ; the leaves come out soon after.

Trunk generally erect. *Bark* smooth, greenish, ash-coloured.—*Branches* thin, irregularly spreading in every direction.—*Leaves* approximated, about the extremities of the branchlets petioled, broad-cordate, three-nerved, often slightly lobed, above smooth, below downy, with two pits on the upper side of the base; length and breadth various, but in general about five or six inches each way.—*Petioles* round, downy, three inches long.—*Panicles* terminal, umbel-like; divisions two-forked.—*Flowers* very small, yellow.

HERMAPHRODITE FLOWERS solitary, sessile in the divisions of the panicle.—*Calyx* superior, four-leaved; *leaflets* unequal; *exterior* pair small, oval, bractiform, caducous; *interior* pair large, wedge-shaped, three-toothed, permanent, increasing in size with the pericarp into two long, membranaceous wings.—*Corol* none.—*Nectary* consisting of four clavate, yellow glands.—*Filaments* four, longer than the calyx, inserted alternately with the nectarial glands into a perforated receptacle. *Anthers* quadrangular, opening on each side with an oval lid.—*Germ* inferior, ovate. *Style* none. *Stigma* small, immersed in the perforation of the receptacle of the filament and nectarial glands.—*Capsule* globular, wrinkled, one-celled, one-valved, not opening spontaneously, size of a cherry, ending in two long, obtuse, lanceolate, membranaceous wings.—*Seed* one.

Obs. The wood of this tree which is white and very light, is employed to make *Catamarans* (floats), when to be had, in preference to any other.

DIGYNIA.

CUSCUTA. Schreb. Gen. N. 227.

Calyx four or five-leaved or parted. *Corol* one-petalled. *Germ* two-celled, one or two-seeded, attachment inferior. *Capsules* superior, two-celled, circumcised. *Embryo* spiral, rolled round the perisperm, with an inferior radicle.

1. C. *reflexa, Roxb. Corom. Pl.* ii. *N.* 104.

Parasitical; *flowers* racemed, pentandrous. *Scales* of the nectary bearded. *Capsules* four-seeded.

Beng. H*u*ldi alg*usi-l*uta.

Telng. Sitama-poorgon*ooloo*.

This plant is parasitical, appearing on any bush in the cold season.

—*Stems* or branches filiform, succulent, twining, very ramous, leafless, smooth, yellow.—*Racemes* numerous.—*Flowers* white, short pedicelled.—*Calyx* five-leaved; *leaflets* cordate, permanent.—*Corol; tube* cylindric; mouth five-parted; *division* reflexed.—*Nectary* of five scales, which surround the germ, and are inserted into the bottom of the tube.—*Filaments* five, very short. *Anthers* oval, appearing half above the mouth of the tube.—*Style* scarcely any. *Stigma* two, large, fleshy, expanding, pointed.

2. C. *aggregata, R.*

Head round, sessile, composed of ten or twelve perfectly sessile, aggregate, pentandrous flowers.

Accidentally introduced into the Botanic Garden amongst Flax seed from Bagdad. When the *Flax* plants were a few inches high, this little delicate, leafless, filiform, round, yellow parasite began to make its appearance, adhering in such quantities to their stems and leaves, as to destroy whole beds in a very short space of time.

3. C. *sulcata, R.*

Peduncles in fascicles, each bearing a few pentandrous, sessile flowers, with a fleshy furrowed calyx. *Capsule* two-seeded, surrounded by the deeply grooved calyx.

A native of Bengal, appears adhering to *Amaranthus oleraceus,* and some other plants, during the cool season. Is a plant of small size, of a pale yellow colour, throughout perfectly smooth.

Peduncles generally collected in small, remote fascicles of two three, four, or more, each supporting from two to four, small, sessile, white flowers.—*Bractes* ovate.—*Calyx* five-toothed, permanent, becoming fleshy, and deeply ribbed, with from six to ten ribs, length of the tube of the corol.—*Corol.* *Tube* short. *Border* five-cleft; divisions oval, obtuse, reflexed.—*Nectarial* scales five, below the insertion of the filaments, divided nearly to the base, into many, fine, filiform segments.—*Filaments* five, inserted a little below the fissures of the corol.—*Germ* round. *Styles* two. *Stigma* headed.—*Capsules* two-celled.—*Seeds* solitary.

4. C. *capitata, R.*

Heads round, sessile, bracted, with from ten to twenty pentandrous flowers. *Calyx* and *corol* papillose, pellucid and white.

Hind. and *Beng.* Al*gusi.*

A most beautiful, small parasite, found in great abundance growing on the lower part of the stems of *Crotolaria juncea,* during the month of February and March.

Stems numerous, filiform, smooth, except at the parts where it grows to the stalks of its supporter, there tuberculate; extent from one to two feet high; colour pale yellow.—*Leaves* none.—*Flowers* from ten to twenty, forming most beautiful, round, sessile, white, pellucid heads.—*Bractes* a single one under each head, which is white, membranaceous, and oblong.—*Calyx* and *corol* five-cleft, swelled with pellucid cells, like the *Ice-plant.*—*Nectarial* scales reniform, ciliate.—*Stamens* five.—*Germ* two-celled, with two seeds in each, attached to the bottom of the cells. *Styles* two. *Stigma* simple.

COLDENIA. Schreb. Gen. N. 233.

Calyx four-leaved. *Corol* infundibuliform. *Germ* superior, four-celled, one-seeded, attachment superior. *Nuts* four. *Embryo* inverse, and furnished with a perisperm.

1. C. *procumbens, Linn. Sp. Pl. ed. Willd.* i. 712.

A small annual, spreading flat on the ground and in general very common on dry rice-fields during the cold season; but never to be seen from the beginning of the periodical rains till they are completely over.

TETRAGYNIA.

BRUCEA. Schreb. Gen. N. 1508.

Calyx four-parted. *Petals* four. *Germs* four, one-seeded, attachment inferior. *Berries* four, superior, one-seeded. *Embryo* inverse, with scarcely any perisperm. Sometimes a fifth part is added.

1. B. *sumatrana, R.**

Shrubby. *Leaflets* serrate. *Racemes* axillary, generally compound.

Lussa-Radja, *Rumph. Amb.* Auctuar. 27. *t.* 15.

Mal. Ampadoo-Barrowing.

A native of Sumatra. From thence Mr. Ewer sent the seeds to the Botanic Garden, where in one year the plants were about four feet high with a straight ligneous stem; and few branches. They then began to flower in May, and continued to blossom and ripen their seed all the year.

Leaves scattered, unequally-pinnate, from twelve to eighteen inches long. *Leaflets* from four to six pair, opposite, short-petioled, obliquely ovate-lanceolate, grossly and obtusely serrate, pointed, villous underneath, from three to six inches long, very bitter, and somewhat fœtid.—*Petioles* common, round, and villous.—*Stipules* none. —*Racemes* axillary, solitary, from one to six, or eight inches long, somewhat compound, dark purple, and clothed with a few white hairs.—*Flowers* numerous, very minute, dark purple, and in all the Sumatra plant completely hermaphrodite.—*Bractes* subulate, hairy, very small, and caducous.—*Calyx* four-leaved, or deeply four-parted. *Segments* small, ovate-lanceolate, deep purple, hairy on the out side and with the corols becoming reflexed.—*Petals* four, longer than the calyx, lanceolate, dark purple, and hairy on the outside.—*Nectary* a dark purple, four-lobed, fleshy cup surrounding the insertion of the germs.—*Filaments* four, short, purple, inserted under the margin of the nectary. *Anthers* oval, large, and deep purple till they discharge their pollen.—*Germs* four, superior, one-celled, with one *ovulum* in each attached to the upper and inner side of the cell.—*Styles* a single, short, recurved, one to each germ. *Stigmas* simple.—*Drupes* four, when all come to maturity, size of a small grain of black pepper smooth, dark purple, one-celled.—*Nut* conform to the drupe, rugose, one-celled, inner edge thinner, and the upper half of it marked with a white line, (the umbilicus).—*Seed* conform to the nut. *Integuments* single, adhering to the perisperm, thin and white.—*Perisperm* in very small quantity, indeed it may be called a fleshy inner

* Gonus amarissimus Lour. Cochinch. ed. Will. ii. 809.—N. W.

integument.—*Embryo* as long as the perisperm, straight, inverse. *Cotyledons* oval, fleshy. *Plumula* two-lobed. *Radicle* roundish, superior.

Obs. Since I first described this tree, and only from plants that were completely hermaphrodite I have found some plants perfectly male, and others as completely female. In this plant, therefore, we have another instance of Trioicous Polygamy.

From the sensible qualities of the green parts of this plant, being somewhat fœtid, and simply, though intensely bitter it promises to be as good an antedysenterical medicine as *Bruce's Abyssinian Wooginos* itself. As soon as the plants arrive at greater maturity the bark shall be tried, and compared with Angustura bark which has by some been thought that of *Brucea antedysenterica* of *Bruce's Travels,* &c. vol. v. *p.* 69. also of *I. F. Miller, Tab.* 25. and still better described and figured by L. *Heritier* under the name *Brucea ferruginea, Stirp. Nov.* 19. *t.* 10. See also *Linn. Sp. Pl. ed. Will.* iv. 742.

MYRIOPHYLLUM. Schreb. Gen. N: 1440.

Calyx four-parted. *Corol* four-petalled. *Style* none; *stigmas* four. *Seeds* four, naked. *Embryo* inverse.

1. M. *tetrandrum, R.*

Leaves verticilled, all filiform-pinnatifid. *Flowers* axillary, verticilled, tetrandrous. *Calyx* four-leaved.

Compare with Hottonia sessiliflora, *Linn. Sp. Pl. ed. Willd.* i. 813.

Teling. Neerla-pamarum.

It grows in sweet water; appearing during the rains.

Stems round, articulate, spongy below the water, length various, above the water from four to six inches, about the thickness of a crow-quill.—*Leaves* verticilled, pinnatifid, below the water they are much larger, with capillary segments; above the water they are small.—*Flowers* very small, white, verticilled, i. e. one in the axil of each leaf.—*Calyx* above, four-leaved; *leaflets* permanent.—*Petals* four, oblong.—*Filaments* shorter than the petals.

2. M. *tuberculatum, R.*

Leaves verticilled, pinnatifid. *Flowers* axillary, verticilled, tetrandrous; fruit tetraceous, tuberculed.

A native of borders of lakes, and other moist places near Calcutta. Is in blossom most part of the year, and very much like M. *tetrandrum.*

Stems creeping, round, jointed, length various, the erect extremities about five or six inches.—*Leaves* verticilled, pinnatifid, or acutely serrate.—*Flowers* verticilled, sessile, one in the axil of each leaf, small, pink-coloured.—*Calyx* none.—*Corol;* petals oblong, expanding, pink-coloured.—*Filament* short. *Anthers* linear.—*Seeds* united in the form of a four-lobed, tubercled pericarp.

POTAMOGETON. Schreb. Gen. N. 234.

Calyx none. *Corol* four-petalled. *Germs* four. *Styles* none. *Seeds* four. *Embryo* hooked, without perisperm.

1. P. *indicum, R.*

Creeping. *Leaves* alternate, from narrow-lanceolate to elliptically oval, opaque, glossy, many-nerved. *Anthers* four pair, attached to the claws of the round concave petals.

A native of the borders of fresh water lakes and ditches in the vicinity of Calcutta. Flowers in February and March.

Stems creeping. The small simple roots which issue from the joints are all I can discover, for while the plant advances from the apex, it decays at the opposite end, so that I do not think any part of them exists for one year. *Branches* few and like what I call the stems; they are all round, smooth and jointed at various distances.—*Leaves* alternate, except next to the flowers, and there often opposite, petioled, from narrow lanceolate, to elliptically oval, opaque, even, and very smooth, or rather glossy, entire, obtuse, obscurely many-nerved; size very various. Those that remain immersed are narrower, longer, more strongly marked with nerves, and subtransparent.—*Petioles* single, not sheathing, length various, in plants left by the water much shorter than the leaves.—*Stipules* solitary, axillary, sheathing, length of the petioles, slit halfway or more on the inside. —*Peduncles* solitary, axillary, or opposite to a leaf, round, rather thicker, smooth, and nearly as long as the cylindric spike, which each supports.—*Flowers* small, numerous, green.—*Calyx* none.—*Petals* long-clawed, nearly round, incurved, fleshy; concave within, convex on the back, permanent.—*Filaments* none. *Anthers,* four pairs attached to the claws of the petals on the inside.—*Germs* four, ovate-oblong, one-celled, with one ovula in each. *Styles* short, and thick. *Stigmas* a dark speck on the apex of the styles.—*Drupes* obovate, smooth. *Nut* conform to the drupe, one-celled.—*Embryo* curved, as represented by *Gærtner* in P. *natans,* vol. ii. *t.* 84.

2. P. *tuberosum, R.*

Leaves alternate, and only opposite at the floriferous forks, sessile, linear, much waved, entire, clear, when dry membranaceous.

Found immersed in extensive masses, in Ponds, Lakes, and receptacles of fresh water in Bengal, during the dry season, chiefly when in flower in February, it rises so near to the surface as to allow the little spikes of flowers to emerge completely.

Root of long simple fibres, brown, &c. according to the colour of the mud they enter; on the creeping joints are generally found small roundish-lobate, viviparous tubers.—*Stems* and branches of various lengths, according to the depth of the water, &c. a groove on each side gives them the appearance of being compressed, thick as a pack-thread, smooth, at the bottom of the water creeping, towards the extremities when in flower dichotomous; previously to that period the branches are alternate and axillary.—*Leaves* sessile, alternate, except at the floriferous fork of the branches, where they are opposite, linear, rather obtuse, much waved, finely and acutely serrulate, with two slender nerves near the margin, pellucid, from two to three inches long, by little more than a quarter of an inch broad.—*Stipules* sheathing, short, smooth, greenish like the other parts of the plant, and scarcely to be distinguished from them.—

Spikes solitary in the fork of the branches, rather long-peduncled, from four to eight-flowered, when in blossom emerged, afterwards immerged.—*Corol, stamina,* and *germ* as in the genus, but here there certainly is a style with four large, obliquely truncated stigmas.—*Seed* as in P. *natans,* Gært. sem. ii. 23 tab. 84.

ILEX.

Calyx four-toothed. *Corolla* rotate, four-cleft. *Style* none. *Berry* from two-to four-seeded.

1. I. *dipyrena, Wall.*

Leaves oblong, accuminate, short-petioled, smooth, scarcely undulated, remotely serrate, serratures spinous ;*flowers* sessile in axillary fascicles ; *berries* two-seeded.

A very large tree, native of Ne*pala*, where it is called *Caulah.* Blossoms from April to July, and produces fruit during the rainy season. I have also received specimens gathered by Dr. Govan about Saharunpore, and by Capt. W. S. Webb, on the confines of Chinese Tartary, where he informs me the tree is called *Kundar.*

Branchlets angular, with cinereous bark ; young shoots pubescent.—*Leaves* alternate, copious, approximated, oblong, sometimes lanceolate-oblong, attenuated into an acumen, sharp at the base ; margins slightly undulated, while young repand-serrate with strong spines, when old more remotely and slightly serrated, sometimes almost entire ; of a firm and leathery texture, perfectly smooth and polished ; rib very prominent, furrowed above ; nerves short, nearly transversal, divided towards the margin into rediculated veins ; from three to nearly five inches long, one to one and a half broad.—*Petioles* not above two lines long, pubescent, deeply furrowed. —*Flowers* subsessile, collected into rounded axillary subsessile fascicles.—*Pedicels* very short, each supported by a couple of minute rounded concave ciliated bractes.—*Calyx* divided into four broad-ovate, acute, ciliate teeth.—*Corolla* deeply four-cleft, flat and spreading ; *segments* ovate, four times longer than the calyx, obtuse, smooth.—*Filaments* four, inserted on the mouth of the corolla, most spreading, alternating with the segments and somewhat exceeding them in length. *Anthers* roundish, two-celled.—*Ovarium* very small. *Stigma* obscurely two-lobed.—*Berry* round, smooth, about four lines in diameter, when dry wrinkled and of a dark brown colour, crowned with a vestige of the two-lobed stigma, two-seeded; *pulp* scanty.—*Seeds* semi-ovate, obtuse, rounded at the base; the inner side flat with a gibbous protuberance above the centre, the outer one convex with about six unequal, very prominent ribs, separated by as many deep furrows.—The rest as in Aquifolium, *Gært. Carp.* ii. 72, *tab.* 92.

Obs. This species varies considerably in the shape and serratures of the leaves. Generally they are oblong and remotely serrate; sometimes, however, they are of a more ovate form, rounded at the base, the margins sinuated and marked with pretty strong unequal spinous serratures. But in all the varieties the berries are two-seeded; a character which sufficiently distinguish them from the varieties of the common Holly. I am indebted to my esteemed friend Capt. Webb for a specimen without flowers which, if not the last mentioned varitey of my I. *dipyrena,* most probably is the I. *Aquifolium;* it was communicated to me with the following remark: "This Holly is found near stre ams in the Purgunna Choudans; it is called *Gumshing* in the Bhotea language, and forms a bush of ten to twelve feet in height; its berries ripen in October."—N. W.

END OF VOLUME FIRST.

CORRIGENDA ET ADDENDA.

Page 13, *line* 23, *for* loothlets, *read* toothlets.

—— 47, —— 23, *for* zerumbet, *read* Zerumbet.

—— 97, —— 17, *for* lanceolaria, *read* lanceolarium.

—— 98, —— 8, *for* Sirinagore, *read* Sirinagur.

—— 25, *for* nearly as the tube of the calyx, *read* nearly as long as the tube of the corol.

—— 100, —— 9, after 521, add N. W.

—— 101, —— 10, *for* *Japonicum*, *read* japonicum.

—— 107, —— 12, *for* dichotomus, *read* dichotomous.

—— 116, —— 6, *for* sub-bilabiate clavate, *read* sub-bilabiate. *Capsule* clavate.

—— 120, —— 19, *for* woody, *read* woolly

—— 28, *for* broader a little, *read* a little broader.

—— 124, dele the last five lines beginning with *Bractes*, and ending with succession, p. 125. l. 1. and instead thereof insert *Bractes* opposite, sessile, entire; those of the flowers (better perhaps called involucels) linear, oblong, cordate, unequal: the uppermost being larger and incumbent over the fascicle of flowers. *Calycine* bractes small.

—— 125, *line* 13, *for* quotted, *read* quoted.

—— 136, —— 15, *for* *rol.* *read vol.*

—— 163, —— 5, from the bottom, *for* three-coloured, *read* three coloured.

—— 167, —— 7, *for* leaflets, *read* leaflet.

—— 3, from the bottom, *for* permament, *read* permanent.

—— 177, —— 20, *for* Tali-pulli, *read* Tali-pullu;—*ibid.* after 9, add p. 123.

Page 197, dele the last three lines beginning with but Retzius.
—— 200, *line* 2, *for* 336, *read* 330.
—— 227, —— 6, from the bottom, *for* oblong scales; ovate, *read* oblong; scales ovate.
—— 230, —— 4, *for* astivalis, *read* aestivalis.
—— 248, —— 2, from the bottom, *for* three, *read* three-,
—— 249, —— 3, *for* sara, *read* Sara.
—— 250, —— 5, *for* vallies; compare, *read* vallies. Compare.
—— 251, —— 20 and 21, *for* seven-flowered, hermaphrodite, *read* seven flowered. *Hermaphrodite flower*.
—— 260, —— ult. *for* no, *read* or
—— 273, —— 8, from the bottom, *for* cernuum *read* cernuus.
—— 7, from the bottom, *for* verticels, *read* verticils.
—— 288, —— 21, dele P.
—— 22, *for* grass, can, *read* grass. Can.
—— 23, *for* the, *read* The.
—— 292, —— 11, *for* dactylon, *read* Dactylon.
—— 293, —— 9, *for* dactylon, *read* Dactylon.
—— 382, —— 7, *for* In the, *read* A native of Nepal*a*. In the.
—— 456, Immediately before P. *pinnata,* insert:

P. *caudata, R.*

Perennial rooting. Extreme shoots leafless, flagelliform, and pendulous. *Leaves* ovate-cordate, from entire to pinnatifid, *segments* sub-ensiform, one-nerved.

From Prince of Wales Island slips were brought to the Botanic Garden at Calcutta, in 1797; and now, 1812, it is one of the most common and most extensive climbers; but has not as yet blossomed in the garden. The long flagelliform pendulous extremities of its shoots readily distinguish it from every other species I have seen; the leaves also are oftener undivided than otherwise.

Page 457, Immediately before P. *heterophylla,* insert:

P. *pinnatifida, R.*

Perennial, rooting on trees. *Leaves* pinnatifid; *segments* falcate, one-nerved.

A native of Sumatra ; from thence brought into the Botanic Garden at Calcutta, where it grows freely on the trunks of trees, to which its lateral roots adhere most firmly, the original one being still in the ground, where first planted. Although some of the plants are six or seven years old and about twenty feet high, they have not yet blossomed.

INDEX.

Note. The words in italics are synonyms.